DIGITAL CONTROL OF HIGH-FREQUENCY SWITCHED-MODE POWER CONVERTERS

DIGITAL CONTROL OF HIGH-FREQUENCY SWITCHED-MODE POWER CONVERTERS

LUCA CORRADINI
DRAGAN MAKSIMOVIĆ
PAOLO MATTAVELLI
REGAN ZANE

IEEE PRESS SERIES ON POWER ENGINEERING

IEEE PRESS

WILEY

Library of Congress Cataloging-in-Publication Data:

Corradini, Luca.
  Digital control of high-frequency switched-mode power converters / Luca Corradini, Dragan
Maksimovic, Paolo Mattavelli, Regan Zane.
    pages cm
  Includes bibliographical references and index.
  ISBN 978-1-118-93510-1 (cloth)
  1. Electric current converters–Automatic control. 2. Switching power supplies–Automatic control.
3. Digital control systems. I. Maksimovic, Dragan, 1961- II. Mattavelli, Paolo. III. Zane, Regan. IV. Title.
  TK2796.C67 2015
  621.381′044–dc23

                                                                                      2014033408

10 9 8 7 6 5 4 3 2 1

CONTENTS

PREFACE

Power electronics fundamentals have been established within the framework of continuous-time analysis, averaged modeling of switched-mode power converters, and analog control theory [1–5]. Ever more often, control and management functions around power converters are implemented digitally, expanding the field of fundamentals to discrete-time modeling and digital control concepts specific to power electronics. Standard textbooks and courses dedicated to digital control of dynamic systems in general provide the necessary background but seldom, if ever, address the specifics necessary to fully understand and successfully practice the design of digitally controlled power converters. We attempt, in this book, to fill the gap by treating the fundamental aspects of digital control of high-frequency switched-mode power converters in a systematic and rigorous manner. Our objectives are to put the reader in the position to understand, analyze, model, design, and implement digital feedback loops around power converters, from system-level transfer function formulations to coding practical designs in one of the mainstream hardware description languages (HDLs) such as VHDL or Verilog.

The book is intended for graduate students of electrical engineering pursuing a curriculum in power electronics and as a reference for engineers and researchers who seek to expand on the expertise in design-oriented knowledge of digital control of power electronics. It is assumed that the reader is well acquainted with foundations of the power electronics discipline, along with associated continuous-time modeling and control techniques [1–6]. Familiarity with sampled-data and discrete-time system analysis topics is helpful but not absolutely essential. Key concepts are developed from the beginning, including a brief review of the necessary discrete-time system fundamentals in Appendix A. For a more comprehensive background, the reader is referred to one of the standard textbooks, such as [7, 8].

The book is composed of eight chapters, structured as follows. The introductory chapter provides an overview of digital control of high-frequency switched-mode power converters, the motivation behind the surge of interest in the area, a summary of analysis, modeling, control, and implementation issues, as well as a summary of recent advances demonstrating potential advantages of digital controllers, including system power management interfaces, programmability of control functions, dynamic response and efficiency improvements, and practical autotuning techniques.

Chapter 1 provides a review of the continuous-time averaged modeling approach for switched-mode power converters. Averaged small-signal modeling is extensively covered by a number of authoritative textbooks [1–3], and the intention is not to replicate this subject in its entirety. Rather, the purpose of Chapter 1 is to focus the reader's attention on the methodology and *assumptions* behind the averaging approach. Understanding of the philosophy and limitations of the averaging technique is essential to appreciate the need for a different approach when it comes to digitally controlled converters.

Chapter 2 introduces the main elements of a digitally controlled converter, with the purpose of providing the reader with a quick overview of the main differences between analog and digital control without immediately entering into detailed modeling issues. This chapter ends with a discussion about the use of continuous-time averaged modeling for designing digital loops, an approach often employed in practice but which can only account for sampling effects and digital control delays in an approximate manner.

The discussion motivates the formulation of a *discrete-time* modeling approach, developed in Chapter 3, which correctly treats the digitally controlled converter as a sampled-data system and formulates its small-signal dynamics in the z-domain rather than in the Laplace domain. In addition to providing the theoretical framework of discrete-time modeling, a number of modeling examples are discussed in Chapter 3. Furthermore, it is shown that a direct link can be established between continuous-time modeling and discrete-time modeling for the converters that are *topologically invariant*, such as the Buck converter. In such cases, a simple and straightforward discretization rule can be formulated, which translates the converter averaged small-signal model into the exact discrete-time model.

Chapter 4 is devoted to direct digital compensator design, based on the discrete-time models developed in Chapter 3. Among many techniques discussed in the literature, the emphasis is given here to the so-called bilinear transform method. The main advantage of the approach is that the entire design procedure is formulated in an equivalent continuous-time domain, in which both the digitally controlled converter and the compensator under design assume the form of continuous-time systems. As a result, the direct digital design can take advantage of the familiar analog control design techniques with the design specifications formulated in the frequency domain. Standard digital proportional-integral (PI) and proportional-integral-derivative (PID) compensator designs are addressed in a number of examples, including voltage-mode, current-mode, and multiloop control of dc–dc converters and power factor correction (PFC) rectifiers.

Amplitude quantization effects introduced by analog-to-digital (A/D) converters and digital pulse width modulators (DPWMs) are discussed in Chapter 5. This chapter first clarifies how limit cycle oscillations can arise in a digitally controlled dc–dc converter in relation to the existence of a dc operating point for the closed-loop system. Secondly, basic design guidelines—referred to as *no-limit-cycling conditions*—are presented, which aim at preventing such generally undesired phenomena to occur. This chapter ends with a brief overview of DPWM and A/D architectures and associated implementation trade-offs.

The issue of compensator implementation is covered in Chapter 6. Scaling and quantization of compensator coefficients are treated first, with the goal of quantifying the quantization-induced errors on the loop gain magnitude and phase at the desired crossover frequency. Secondly, this chapter addresses implementation of the control law in a fixed-point arithmetic environment, providing a methodology for word length determination of the various signals inside the control structure. Given the focus of the book on high-frequency switched-mode power converter applications, the emphasis is on *hardwired* implementations of the control law, together with VHDL and Verilog coding examples. Nevertheless, the principles that apply to software-based, microprogrammed realizations are highlighted as well.

Autotuning is an advanced application of digital control, which brings up intriguing potentials and additional challenges. Because of the importance of this emerging topic, Chapter 7 is devoted to an overview of digital autotuning techniques for high-frequency switched-mode power converters. After a brief discussion about digital autotuning basics, two autotuning techniques are presented in more detail: an injection-based approach and a relay-based approach.

An objective in writing the book has been to emphasize the distinction between the fundamental, theoretical aspects of digital control design on one side and the application of these techniques on the other, demystifying the perception about discrete-time models or digital control as being exceedingly complex and difficult to employ in practice. In line with such philosophy, Matlab® script examples are systematically developed alongside the theory. A few Matlab® commands allow, in most situations, to straightforwardly carry out system-level compensator designs and rapidly proceed to HDL coding and implementation steps. Furthermore, throughout the book, a number of design examples are fully worked out and verified by simulations in the Matlab® environment.

INTRODUCTION

Efficient processing and control of electric power is required in applications ranging from submilliwatt on-chip power management to hundreds of kilowatt and megawatt power levels in motor drives and utility applications. The objectives of high efficiency, as well as static and dynamic control of inputs or outputs under a range of operating conditions, are accomplished using power electronics, that is, switched-mode power converters consisting of passive (capacitive and inductive) components, and power semiconductor devices operated as switches. In high-power applications, control and monitoring tasks are often more complex, while the power semiconductor devices are operated at relatively low switching frequencies, for example, up to tens of kilohertz. The controller cost and power consumption are relatively low compared to the overall system cost and power rating. In these applications, digital control offers clear technical and economic advantages in addressing complex control, management, and monitoring tasks. As a result, for many years now, digital control methods and digital controllers based on general-purpose or dedicated microprocessors, digital signal processor (DSPs), or programmable logic devices have been widely adopted in power electronics applications at relatively high power levels.

In ubiquitous low-to-medium power switched-mode power supply (SMPS) applications, including point-of-load (POL) regulators, nonisolated and isolated dc–dc converters, single-phase power factor correction (PFC) rectifiers, single-phase inverters, and lighting applications, adoption of digital power management and digital control has been slower. In these applications, switching frequencies are often in the range from hundreds of kilohertz to multiple megahertz, and much faster dynamic responses are required. The controller cost and the controller power consumption can easily present significant portions of the system cost and power dissipation. Furthermore, in many applications, control challenges have been successfully met by continuous advances of readily available analog controllers, using well-established analog analysis, modeling, and design techniques [1–5]. Nevertheless, practical digital control of high-frequency switched-mode power converters has moved from proof-of-concept demonstrations to digital pulse width modulation (DPWM) controller chips commercially available from multiple vendors, with growing adoption rates in many applications. Several factors have

Digital Control of High-Frequency Switched-Mode Power Converters, First Edition.
Luca Corradini, Dragan Maksimović, Paolo Mattavelli, and Regan Zane.
© 2015 The Institute of Electrical and Electronics Engineers, Inc. Published 2015 by John Wiley & Sons, Inc.

contributed to the increasing penetration of the concept of "digital power" in high-frequency power electronics applications:

- Ongoing advances in digital integrated-circuit processes have continued to increase processing capabilities while bringing the cost down.

- The needs for improved system integration and increasingly complex power management and monitoring functions have translated into the needs for digital interfaces and programmability in switched-mode power conversion applications [9–11].

- Practical high-performance digital control techniques have been introduced and demonstrated, together with innovative approaches offering performance gains or entirely new capabilities that would be difficult or impractical to realize using traditional analog techniques [12–14].

The "digital power" concept encompasses several aspects:

1. *Digital power management,* which refers to system-level control and monitoring of power conversion and distribution, usually over a serial communication bus [9–11]. Power management functions include turning on and off or sequencing system power rails, adjusting setpoints for converter control loops, programming control loop parameters, monitoring and reporting of measured status or variables, and so on [15, 16]. These functions are typically performed at timescales that are relatively long compared to a switching period.

2. *Digital control,* which includes time-domain and frequency-domain converter modeling and control techniques, with control actions performed at timescales comparable to a switching period.

3. *Digital implementation techniques,* which can be classified into two main groups:

   o *Software-based controllers,* where control algorithms are designed and implemented in code executed on general-purpose or specialized microcontrollers or DSP chips. An early example of application of microprogrammed digital control to power factor preregulators is presented in [17].

   o *Hardware-based controllers,* based on custom-integrated circuits or programmable logic devices such as field-programmable gate arrays (FPGAs) [18, 19]. Early examples of such hardware-based digital controllers can be found in [20–22].

This book is focused on the fundamental aspects of analysis, modeling, and design of digital control loops around high-frequency switched-mode power converters in a systematic and rigorous manner. The objectives are to enable the reader to understand, analyze, model, design, and implement digital feedback loops around power converters, from system-level transfer function formulations to practical implementation details. The purpose of this chapter is to introduce the topics covered in the book and to motivate the reader to pursue the theoretical and practical concepts covered in the remaining chapters of this book. Furthermore, this introductory chapter points to some of the more advanced digital techniques reported in the literature, including approaches to dynamic response improvements, system identification, autotuning of digital control loops, and on-line efficiency optimization.

DIGITALLY CONTROLLED SWITCHED-MODE CONVERTERS

A number of DPWM controller architectures and implementation strategies have been investigated and realized in practice. Many standard microcontrollers and DSP chips are now available, featuring multiple high-resolution PWM and analog-to-digital (A/D) channels, which allow software-based implementation of control and management functions. While advances in this area have been rapid, the software-based approaches are still better suited for applications where switching converters operate at relatively low switching frequencies. On the other hand, at switching frequencies in the hundreds of kilohertz to megahertz range, specialized hardware-based control loops are often preferred. This approach is illustrated in the architecture shown in Fig. 1 [12, 13]. The control loop is digital, using specialized, programmable A/D, DPWM, and compensator blocks to achieve high-performance closed-loop dynamic responses, while programmability, power management, and system interface functions are delegated to a microcontroller core. Similar combinations of programmable hardware peripherals specialized for switched-mode power converter applications, with software-based realizations of higher-level management and communication functions are often found in commercially available DPWM controllers.

Controllers of the type shown in Fig. 1 can be developed, realized, and tested using standard digital VLSI design flow starting from logic functions described using

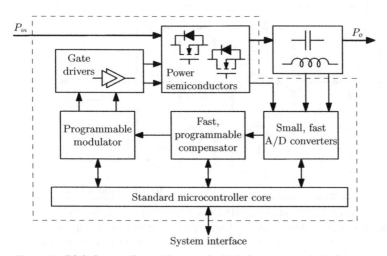

Figure 1   Digital controller architecture for high-frequency switched-mode power converters [13].

hardware description language (VHDL or Verilog), followed by prototyping and experimental verifications using FPGA development platforms, ultimately targeting relatively small, relatively low-gate-count integrated circuits capable of matching or surpassing the state-of-the-art analog solutions in terms of dynamic performance, power consumption, and cost. At the same time, digital PWM controllers offer digital system interface, programmability and flexibility, power management functions, reductions in the number of passive components, reduced sensitivity to process and temperature variations, and potentials for practical realizations of more advanced features.

Figure 2 shows a more detailed block diagram of a hardware-based digital controller around a POL synchronous Buck converter. Output voltage $v_o(t)$ is sampled by an A/D converter and compared to a setpoint reference V_{ref} to produce a digital voltage error signal $e[k]$. The error signal is processed by a discrete-time digital proportional-integral-derivative (PID) compensator to generate a duty cycle command $u_x[k]$. In the basic version of the controller, the compensator gains K_p, K_i, and K_d are found by design to meet control loop specifications, such as the crossover frequency and phase margin, as detailed in Chapters 1–6 of this book. Once a compensator is designed, the gains can be realized using digital multipliers, as shown in Fig. 2. As only a few bits are sufficient to represent the error signal $e[k]$, the entire compensator can also be implemented as a lookup table [21–24]. In a more advanced case, as illustrated by the digital autotuner block in Fig. 2, the compensator gains can

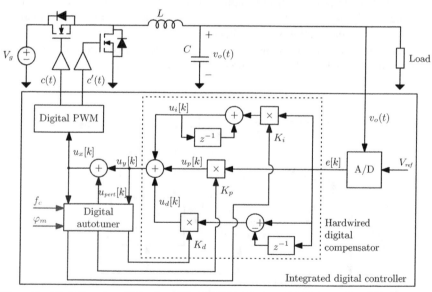

Figure 2   Digitally controlled point-of-load (POL) synchronous Buck dc–dc converter example. Analysis, modeling, design, and implementation of digital control loops are addressed in Chapters 1–6. An additional injection-based autotuning block is shown, which is further discussed in Chapter 7.

be tuned in response to the actual system dynamics to meet the desired specifications. Autotuning is addressed in Chapter 7. Finally, a DPWM block generates the complementary gate-drive control signals $c(t)$ and $c'(t)$ with duty cycle set by the digital command and with appropriate dead times. Together with various enhancements, such a controller can be realized in about 10,000 equivalent logic gates, which translates to about one-third of a square millimeter in a standard 0.35 μm CMOS process. Furthermore, higher-density CMOS processes with high-voltage extensions suitable for power electronics applications are now readily available, making power and cost-effective digital controllers for high-frequency switched-mode power converters a reality. Examples of integrated digital controllers can be found in [21, 22, 24–33].

It is of interest to examine a practical example. Following the block diagram shown in Fig. 2 (without the digital autotuner), a digitally controlled 5 to 1.6 V synchronous Buck converter prototype is described in [34]. The filter component values are $L = 1.1$ μH, $C = 250$ μF, and the switching frequency is 500 kHz. The A/D converter is a windowed converter [26], using threshold inverter quantization approach [35]. The A/D conversion range is approximately 200 mV, centered around the reference $V_{ref} = 1.6$ V, for an equivalent output voltage quantization step of 3 mV. A hybrid counter/ring oscillator DPWM is employed, with a time quantization of about 390 ps, that is, with 0.02% duty cycle resolution. A digital PID compensator designed for $f_c \approx f_s/10 = 100$ kHz crossover frequency is VHDL coded and implemented on an FPGA. Figure 3 illustrates an experimental 0 to 8 A load step response, with the voltage deviation and the response time comparable to responses expected from high-performance analog PWM controllers.

Analysis, Modeling, and Control Techniques

Referring to the example in Fig. 2, one may observe that the basic digital control loop is conceptually similar to the standard voltage-mode analog PWM control

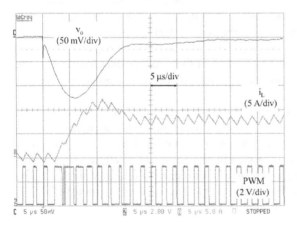

Figure 3   Experimental 0 to 8 A load step response in a digitally controlled POL Buck converter with conventional PID compensation [34] (v_o, 50 mV/div; i_L, 5 A/div; and timescale, 5 μs/div). © 2009 IEEE.

loop, with analog control techniques based on averaged converter models briefly reviewed in Chapter 1. As discussed further in Chapter 2, digital control differs from analog control in two key aspects: *time quantization* and *amplitude quantization*. Time quantization refers to the fact that the controller is a discrete-time system that processes *sampled* versions of sensed analog signals to be regulated and produces a discrete-time control output. In order to design high-performance control loops, it is necessary to understand and consider the resulting delays and aliasing effects. As discussed in Chapters 2 and 3, the use of continuous-time averaged modeling for designing digital loops, an approach often employed in practice, can only account for sampling effects and digital control delays in an approximate manner. A more rigorous approach is based on discrete-time modeling [36], which is described in detail in Chapter 3. This modeling approach enables direct-digital design of compensator transfer functions, which is presented in Chapter 4. The design specifications are presented in frequency domain, in terms of the quantities familiar to the analog designer: the loop-gain crossover frequency f_c and the phase margin φ_m.

Implementation Techniques

In digitally controlled converters, regulation precision and accuracy are determined by the resolutions of A/D and DPWM blocks, which introduce amplitude quantization effects discussed in Chapter 5. These nonlinear effects can lead to steady-state disturbances commonly referred to as limit cycling [37, 38]. Design guidelines to avoid limit cycling are also presented in Chapter 5, together with a brief summary of high-resolution DPWM and A/D implementation techniques.

Digital compensator implementation is addressed in Chapter 6. Scaling and quantization of compensator coefficients are treated first, with the goal of quantifying the errors introduced by coefficient quantization on the loop gain magnitude and phase at the desired crossover frequency. Chapter 6 then addresses PID compensator structures, such as the parallel PID realization shown in Fig. 2, and the implementation of the control law in a fixed-point arithmetic environment, providing a methodology for word length determination of the various signals inside the control structure. Given the focus of the book on high-frequency switched-mode power converter applications, the emphasis is on *hardwired* implementations of the control law, together with VHDL and Verilog coding examples, while the principles that apply to software-based, microprogrammed realizations are also highlighted.

The objectives of Chapters 1–6 are to enable the reader to successfully analyze, model, design, and implement voltage, current, or multiloop digital feedback loops around switched-mode power converters. Alongside, based on the theoretical concepts, Matlab® scripts are systematically developed, which allow one to rapidly perform discrete-time modeling and system-level compensator design steps, and proceed to implementation steps. Practical examples are used throughout the book to illustrate applications of the techniques developed.

SYSTEM AND PERFORMANCE GAINS VIA DIGITAL CONTROL

Increased flexibility, programmability, and integration of system interface and power management functions have been recognized as important advantages of digital controllers. Furthermore, as digital controller implementation opened opportunities for practical implementations of more sophisticated control approaches, considerable advances have been made in various directions. This section highlights some of the gains enabled by digital control in the areas of improved dynamic responses, integration of frequency-response measurements, autotuning of digital control loops, and on-line efficiency optimization.

Improved Dynamic Responses

Standard analog or digital converter controller design techniques are based on linear small-signal models and frequency-domain-based compensator designs. It has been recognized that considering the switching nature of the power stage directly, and operating with large-signal instantaneous state variables to provide on–off control actions accordingly, can result in improved dynamic responses. The switching surface control [39] and many other time-domain based approaches have been investigated both in analog and in digital domains. Digital implementation is particularly well suited for explorations of control techniques targeting improved dynamic responses. A case of special interest is a sequence of switching actions that result in minimum time, that is, time-optimal response to an external disturbance such as a step load transient. For example, for a Buck converter, a time-optimal response to a step load transient consists of a single precisely timed on/off sequence. Various digital control methods have been proposed to implement the time-optimal control [34, 35, 40–60].

These controllers have demonstrated step load transient responses that approach limits imposed by the converter passive LC filter components. For example, Fig. 4 illustrates the step load transient response of the parameter-independent time-optimal controller described in [34], for the same synchronous Buck prototype as in Fig. 3. The single switching action occurring immediately after the step load is visible, which quickly restores the output voltage to regulation. Compared to the response with the standard PID compensator in Fig. 3, both the voltage deviation and the response time are significantly reduced. As the control action is effectively saturated during time-optimal control events, overshoots may occur in converter internal states such as the inductor current. An extension of digital time-optimal control, including practical inductor current limitations, has been addressed in [57].

A number of other approaches to achieve improved dynamic responses have been explored. Multisampling techniques (where converter waveforms are sampled more than once per switching period [61–63]), asynchronous sampling techniques [35, 64, 65], and mixed-signal control techniques [66, 67] have been proposed to

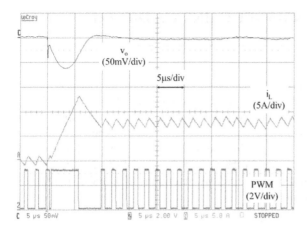

Figure 4   Experimental
0 A-8 A step load response
in a digitally controlled POL
Buck converter with the
time-optimal controller
described in [34] (v_o:
50 mV/div, i_L: 5 A/div,
timescale: 5 μs/div). © 2009
IEEE.

minimize the control loop delays. Multisampling techniques for constant on-time controllers can also be found in [68–70]. Furthermore, it has been shown that non-linear techniques can lead to dynamic performance improvements in dc–dc [71, 72] and PFC applications [73]. Possibilities for improved dynamic responses in multi-phase architectures have been investigated in [74, 75]. More complex controllers in conjunction with power stage modifications present another interesting direction in dynamic responses improvements. For example, it has been shown how digitally controlled power stages with additional auxiliary switches can offer significant dynamic response improvements [76], while [77] takes this approach a step further and proposes a much tighter load-controller interaction.

Integration of Frequency-Response Measurements

An important experimental verification step in traditional frequency-response-based controller designs includes measuring controller small-signal frequency responses using network analyzers [1, 78]. Feasibility of integrating such nonparametric frequency-domain system identification (system-ID) functionality into digital controllers has been demonstrated in [79, 80]. To briefly summarize the technique, Fig. 5 shows a block diagram of a digitally controlled converter with additional system-ID functions. The identification process consists of perturbing the duty cycle command with a pseudo-random binary sequence (PRBS), cross-correlating the perturbation with the measured output responses to obtain the system impulse response, and performing fast Fourier transform (FFT) to obtain frequency responses. Cross-correlation can be implemented efficiently using the fast Walsh–Hadamard transform (FWHT). Note that the approach reuses the DPWM and the A/D block already present in the digital control loop. To mitigate the effects of switching and quantization noise, the system-ID approach includes a sweep to optimize the perturbation magnitude, pre-emphasis and de-emphasis filters applied to the injected and the sensed signals, respectively, as well as smoothing in the frequency domain. This approach has been applied to a number of converter examples in [80],

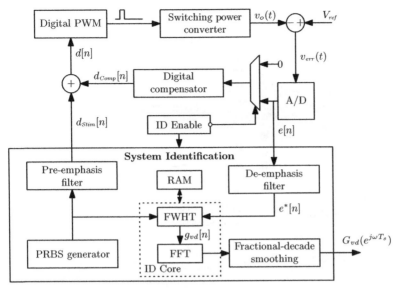

Figure 5 Digitally controlled PWM converter with integrated frequency response measurement capabilities [80]. © 2008 IEEE.

including a 15- to 30-V Boost dc–dc converter operating at $f_s = 195$ kHz. Figure 6 shows time-domain waveforms during the identification process, while Fig. 7 shows the identified frequency responses, which closely match discrete-time modeling predictions.

The resulting on-line identified frequency response can be used for design, diagnostic, or tuning purposes [81]. The success of these applications depends on the fidelity of the identified frequency responses and the degree to which the process is automated, as well as the costs, in terms of gate count or complexity, time duration of identification, and effects on output voltage, incurred to obtain the results. Reference [80] demonstrates the feasibility of incorporating fully automated frequency response measurement capabilities in digital PWM controllers at relatively low additional cost. The identification process can be typically accomplished in several hundred milliseconds, and the output voltage can be kept within narrow bounds during the entire process.

Autotuning

Taking advantage of the digital controller programmability, the overall objective of autotuning is to automatically tune the controller parameters in response to the actual system dynamics. An autotuning digital controller ideally becomes a "plug and play" unit capable of identifying the key characteristics of the power converter and the load and adjusting the controller parameters to achieve specified performance goals. This capability represents a significant departure from the conventional design flow.

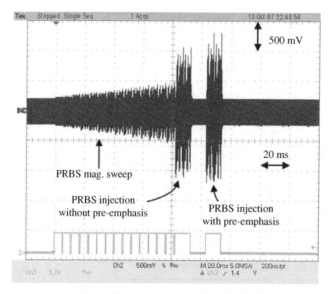

Figure 6 Output voltage during system identification process in a 15- to 30-V Boost dc–dc converter operating at $f_s = 195$ kHz [80]. © 2008 IEEE.

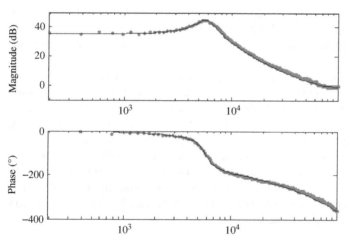

Figure 7 Control-to-output magnitude and phase responses determined by the system identification process illustrated in Figs. 5 and 6 [80]. © 2008 IEEE.

Numerous advances have been made in the area of practical autotuning digital control algorithms and implementation techniques [81–101], and the area is subject to ongoing research and development efforts. Chapter 7 presents an overview of digital autotuning techniques for high-frequency switched-mode power converters. Two autotuning techniques are presented in more detail: an injection-based approach and a relay-based approach.

The controller block diagram in Fig. 2 shows how a digital autotuner can be incorporated into the digital controller based on the injection approach [92, 93, 95, 97]. The autotuning system injects a digital perturbation signal $u_{pert}[k]$ into the feedback loop, superimposed to the PID compensator output $u_y[k]$. The overall control command $u_x[k] = u_y[k] + u_{pert}[k]$ modulates the converter. Simultaneously, signals $u_y[k]$ and $u_x[k]$ before and after the injection point are monitored and the tuning process adjusts the compensator gains until the correct amplitude and phase relationships are established between the ac components of $u_x[k]$ and $u_y[k]$ at the injection frequency, indicating that the loop gain meets the specifications given in terms of the desired crossover frequency f_c and phase margin φ_m. The hardware requirements for the entire adaptive tuning system are relatively modest, making it a practical solution.

Efficiency Optimization

Power conversion efficiency is a key performance metric in most applications. In the efficiency improvements area, potential advantages of digital controllers include abilities to precisely adjust switching frequency or other timing parameters of switch control waveforms [27, 102–107], abilities to reconfigure the power stage on-line either through power-switch segmentation [108, 109] or other gate-drive parameters [109, 110], phase shedding [58, 111], or control of current distribution in multiphase architectures [111–113], as well as abilities to implement algorithmic or prepro-grammed approaches to on-line efficiency optimization [102, 103, 114–117].

The shift from traditional analog techniques to digital control in high-frequency power electronics is making an impact on standard design practices, as well as in various applications. Programmability, monitoring, digital system interfaces, and system-level power management are becoming ubiquitous in power systems ranging from mobile electronics to desktop computing, data centers, and communication infrastructure. In these systems, digital control further brings opportunities for improved dynamic responses and correspondingly reduced size of passive filters, together with new approaches to converter-level and system-level efficiency optimizations. In response to increasing energy cost and environmental concerns, various energy efficiency initiatives and programs are addressing power conversion efficiency and power quality in data centers and computer power supplies. It is expected that future energy efficiency program specifications will be even more demanding in terms of efficiency, power factor, and harmonic distortion requirements for off-line power supplies over even wider load ranges. Further significant impact can also be expected in renewable energy applications. For example, distributed module integrated converters or micro-inverters in photovoltaic power systems can take advantage of digital control algorithms for improved maximum power point tracking, fault detection, and efficiency optimization. Similar impact can be foreseen in electric-drive vehicles, not just in inverter controls where digital control is already ubiquitous, but also in battery management and battery charger systems.

CHAPTER 1

CONTINUOUS-TIME AVERAGED MODELING OF DC–DC CONVERTERS

Converter systems rely on feedback loops to achieve the desired regulation performance. For example, in a typical dc–dc converter application, the objective is to maintain tight regulation of the output voltage in the presence of input voltage or load current variations. An accurate small-signal description of the converter control-to-output dynamics is the starting point for feedback loop design techniques based on frequency-domain concepts of loop gain, crossover frequency, phase margin, and gain margin.

The most successful and widespread modeling technique for switched-mode converters is based on *averaged small-signal modeling* [1, 118–120]. This technique is based on first averaging the converter behavior over a switching period with the purpose of smoothing the discontinuous, time-varying nature of the converter into a continuous, time-invariant nonlinear system model. A successive linearization step yields a linear, time-invariant model that can be treated using standard tools of linear system theory. The converter is described by a *continuous-time linear system*, often presented in the form of a linear equivalent circuit model, a natural representation in the context of analog control design.

The averaging approach is currently the most widely accepted way of *understanding* dynamics of switched-mode power converters. In addition to the relative simplicity and straightforwardness, popularity of the averaging approach has been reinforced by the success of innumerable practical designs supported by robust and easy-to-use integrated circuits for analog converter control.

The main purpose of this chapter is to revisit the main aspects of analysis and modeling techniques for switched-mode power converters. Averaged small-signal modeling, in particular, is reviewed in detail, highlighting the main assumptions behind the approach. This prepares the background necessary to understand the limitations of the averaged small-signal modeling in the context of digital control design and to allow subsequent developments of discrete-time models where these limitations are removed.

Digital Control of High-Frequency Switched-Mode Power Converters, First Edition.
Luca Corradini, Dragan Maksimović, Paolo Mattavelli, and Regan Zane.
© 2015 The Institute of Electrical and Electronics Engineers, Inc. Published 2015 by John Wiley & Sons, Inc.

A brief review of pulse width modulated (PWM) dc–dc converters is presented in Section 1.1, followed by a summary of steady-state analysis and modeling techniques in Section 1.2. Section 1.3 explains the need for dynamic modeling in the design of control loops around switched-mode power converters and introduces the small-signal averaged modeling approach. The method of state-space averaging [119, 120], a general approach to modeling switched-mode power converters, is summarized in Section 1.4. Analog control design examples are presented in Section 1.5. In the subsequent chapters, these examples are revisited to illustrate modeling and digital control design principles. To complete the background necessary to engage in developments of analysis, modeling and control techniques in the context of digitally controlled PWM converters, a discussion related to the nature of duty cycle, the control variable in PWM converters, is presented in Section 1.6. The key points are summarized in Section 1.7.

1.1   PULSE WIDTH MODULATED CONVERTERS

The focus of this book is on *PWM* converters, which are operated so as to alternate between two or more distinct subtopologies in a periodic fashion, with a fundamental *switching period* T_s. The Boost converter depicted in Fig. 1.1, for instance, operates with the switch in position 1 for a fraction DT_s of the switching period and with the switch in position 0 for the remaining fraction $D'T_s \triangleq (1 - D)T_s$. The quantity $0 \leq D \leq 1$ is the *duty cycle*, which determines the fraction of a switching period the switch is kept in position 1. In PWM converters, D is the *control input* for the system, which is adjusted by a controller in order to regulate a converter voltage or current.

Typical waveforms of a PWM converter are shown in Fig. 1.2, which exemplifies the gate driving signal $c(t)$ and one of the converter state variables, such as the output voltage $v_o(t)$. Assuming that the converter duty cycle is sinusoidally modulated at a frequency $f_m \ll f_s$, the output voltage consists of a low-frequency component $\bar{v}_o(t)$, plus a high-frequency *switching ripple*. The low-frequency component of $v_o(t)$ contains a dc term V_o and a spectral component at the modulation frequency f_m. Using the terminology of modulation theory, $\bar{v}_o(t)$ is the *baseband* component of $v_o(t)$. The high-frequency content, on the other hand, contains the switching frequency f_s and its harmonics, as well as all the modulation sidebands

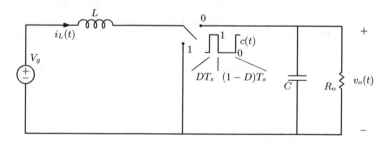

Figure 1.1   Pulse width modulated Boost converter.

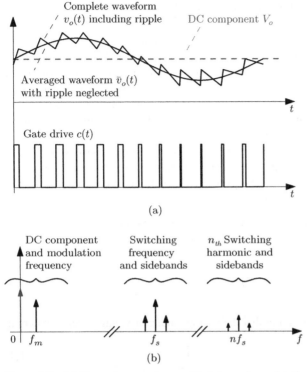

Figure 1.2   (a) Converter waveforms with duty cycle modulation and (b) qualitative spectrum of a pulse width modulated signal.

originating from nonlinear interactions between f_m and f_s components occurring as a result of the modulation process.

In the context of the averaged modeling approach, the separation between low-frequency and high-frequency portions of the converter signals is of central importance. To be more precise, the *moving average operator* $\langle \cdot \rangle_T$ is introduced,

$$\langle x(t) \rangle_T \triangleq \frac{1}{T} \int_{t-T/2}^{t+T/2} x(\tau) \, d\tau \, , \tag{1.1}$$

which averages signal $x(t)$ over a period T. With this definition, the low-frequency component $\bar{v}_o(t)$ of $v_o(t)$ illustrated in Fig. 1.2 is defined as its moving average over the switching period T_s,

$$\bar{v}_o(t) \triangleq \langle v_o(t) \rangle_{T_s} \, . \tag{1.2}$$

The fundamental simplification at the basis of the averaging method consists of describing the small-signal dynamics of $\bar{v}_o(t)$ rather than $v_o(t)$, therefore neglecting high-frequency components of the converter waveforms. Both the power and the limitations of the method reside in the averaging approximation.

1.2 CONVERTERS IN STEADY STATE

When a converter is operating in steady state, every converter state variable—and therefore every voltage and current—is *periodic* in time, with a period equal to the converter switching period T_s. Steady-state operation is reached when all the converter inputs are constant—including the duty cycle—and after all transients are extinguished. In the following text, the basic ideas behind steady-state analysis of PWM converters are summarized. More extensive and detailed treatments can be found in power electronics textbooks [1–5].

Steady-state analysis of switched-mode power converters consists of expressing the dc values of all the voltages and currents in terms of the converter inputs. The analysis is founded on two basic principles, which are direct consequences of the periodicity of the system waveforms:

- *Inductor volt-second balance.* As all the inductor currents are periodic, no net flux variation can occur in any inductor over a switching period,

$$L(i_L(T_s) - i_L(0)) = \int_0^{T_s} v_L(\tau) \, d\tau = 0. \tag{1.3}$$

This is equivalent to stating that the average inductor voltage over a switching interval is zero,

$$\boxed{\bar{v}_L(t) = 0}. \tag{1.4}$$

- *Capacitor charge (ampere-second) balance.* By a dual argument, as all the capacitor voltages are periodic, no net charge can be absorbed or delivered by any capacitor over a switching period,

$$C(v_C(T_s) - v_C(0)) = \int_0^{T_s} i_C(\tau) \, d\tau = 0. \tag{1.5}$$

This is equivalent to stating that the average capacitor current over a switching interval is zero,

$$\boxed{\bar{i}_C(t) = 0}. \tag{1.6}$$

The two above-mentioned conditions, combined with conventional circuit analysis, are sufficient to solve the steady-state problem. In practice, the calculations are greatly simplified by introducing the *small-ripple approximation*. By *switching ripple*, one refers to the ac component of a converter voltage or current. In steady state, the switching ripple is a periodic function with a fundamental frequency equal to the converter switching rate. The ripple peak-to-peak amplitudes of a capacitor voltage $v_C(t)$ and an inductor current $i_L(t)$ are denoted as Δv_C and Δi_L, respectively.

The small-ripple approximation states that the dc converter quantities can be approximately determined by neglecting both capacitors voltage ripples and inductors current ripples. This corresponds to considering every capacitor C as an ideal dc voltage source of unknown magnitude V_C and every inductor as a dc current source of unknown magnitude I_L,

$$\boxed{\frac{\Delta v_C}{\overline{v}_C} \ll 1 \Leftrightarrow v_C(t) = V_C = \text{constant}},$$

$$\boxed{\frac{\Delta i_L}{\overline{i}_L} \ll 1 \Leftrightarrow i_L(t) = I_L = \text{constant}}.$$

(1.7)

Contrary to the volt-second balance and the ampere-second balance, which follow directly from the characteristics of inductive and capacitive components and the periodicity of the steady-state operation, the small-ripple approximation is simply a convenient assumption that simplifies the steady-state solution and which is often satisfied in practical converter systems. A relaxed version of the small-ripple approximation, known as *linear-ripple approximation*, is also often employed. According to the linear-ripple approximation, ripple components of the $v_C(t)$'s and $i_L(t)$'s are allowed to be triangular waveshapes. It can be shown that the steady-state analysis proceeds as stated earlier for the small-ripple approximation. In practice, the linear-ripple approximation is easier to meet, especially when considering inductor current waveforms. As long as the small-ripple approximation is satisfied for capacitor voltages, in fact, inductor currents retain triangular waveforms even when the peak-to-peak ripple is not negligibly small compared with the dc component.

It is worth mentioning, at this point, that the above discussion is implicitly focused on the converters operating in *continuous conduction mode* (CCM), where the use of the small-ripple or linear-ripple approximation is well justified for *all* the converter state variables. As for converters operating in *discontinuous conduction mode* (DCM), on the other hand, the above-mentioned assumption does not hold and the analysis becomes somewhat more involved. Further details on DCM modeling can be found in [1, 121–123].

1.2.1 Boost Converter Example

As an example, consider the Boost converter depicted in Fig. 1.3. The physical inductor is represented by a series combination of an ideal inductor L and a resistive element r_L, modeling the inductor copper losses. Other converter components are assumed to be ideal.

With the switch in position 1 for an interval DT_s, the voltage across the ideal inductor L is

$$v_L(t) = V_g - r_L I_L,$$

(1.8)

where the small-ripple approximation $i_L(t) \approx I_L$ has been employed. In the same topological state and under the small-ripple approximation $v_C(t) \approx V_C$, the output capacitor current is

$$i_C(t) = -\frac{V_C}{R_o}.$$

(1.9)

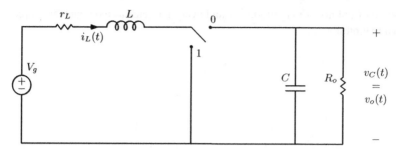

Figure 1.3 Boost converter example.

Similarly, with the switch in position 0 for an interval $D'T_s = (1 - D)T_s$, one has

$$v_L(t) = V_g - r_L I_L - V_C,$$
$$i_C(t) = I_L - \frac{V_C}{R_o}. \tag{1.10}$$

Waveforms $v_L(t)$ and $i_C(t)$, including the small-ripple approximation, are shown in Fig. 1.4. Imposing the volt-second balance (1.4) and the charge balance (1.6), one obtains

$$\bar{v}_L(t) = D(V_g - r_L I_L) + D'(V_g - r_L I_L - V_C) = 0,$$
$$\bar{i}_C(t) = D\left(-\frac{V_C}{R_o}\right) + D'\left(I_L - \frac{V_C}{R_o}\right) = 0, \tag{1.11}$$

the solution of which is

$$I_L = \frac{V_g}{D'^2 R_o} \frac{1}{1 + \frac{r_L}{D'^2 R_o}},$$
$$V_C = \frac{V_g}{D'} \frac{1}{1 + \frac{r_L}{D'^2 R_o}}. \tag{1.12}$$

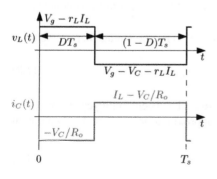

Figure 1.4 Boost converter example: waveforms based on the small-ripple approximation.

The converter *voltage conversion ratio* can be evaluated from the above-mentioned equations as

$$M(D) \triangleq \frac{V_o}{V_g} = \frac{V_C}{V_g} = \frac{1}{D'} \frac{1}{1 + \frac{r_L}{D'^2 R_o}}, \tag{1.13}$$

which reduces to the familiar Boost $M(D) = 1/D'$ for a lossless converter ($r_L = 0$).

1.2.2   Estimation of the Switching Ripple

Once the dc converter quantities are determined, one can go back to the converter topology and estimate both the waveshapes and the amplitudes of the steady-state inductor current and capacitor voltage ripples.

In the Boost converter example, as shown in Fig. 1.4, the inductor voltage waveform $v_L(t)$ is approximately a piecewise-constant signal. The inductor current ripple is therefore a triangular waveform with slopes determined by $v_L(t)$. Neglecting, for simplicity, the inductor series resistance r_L, the peak-to-peak current ripple Δi_L can be determined by integrating $v_L(t)/L$ over either one of the two switching subintervals,

$$\Delta i_L = \frac{1}{L} \int_0^{DT_s} v_L(\tau)\, d\tau = \frac{V_g}{L} DT_s = \frac{T_s}{L} V_g \left(1 - \frac{V_g}{V_o}\right). \tag{1.14}$$

Similarly, one can reconstruct the capacitor voltage ripple waveshape by integration of $i_C(t)$ shown in Fig. 1.4. More accurate results can be obtained by removing the small-ripple approximation and by deriving $i_C(t)$ using, this time, the triangular waveshape $i_L(t)$ determined earlier. The corresponding waveforms are depicted in Fig. 1.5.

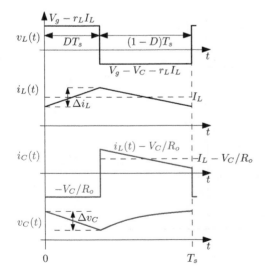

Figure 1.5   Boost converter example: estimation of the ripple waveshapes.

To determine the peak-to-peak output voltage ripple, similar to finding Δi_L, one can directly integrate $i_C(t)/C$ over either one of the two switching subintervals,

$$\Delta v_C = \frac{1}{C} \int_0^{DT_s} |i_C(\tau)|\, d\tau = \frac{V_o}{R_o C} DT_s = \frac{T_s}{R_o C} \left(V_o - V_g\right). \tag{1.15}$$

1.2.3   Voltage Conversion Ratios of Basic Converters

Systematic application of the volt-second and charge balance equations, along with the small-ripple approximation, allows straightforward steady-state analysis of any PWM converter. Table 1.1 reports the CCM conversion ratios of the three basic converter topologies in the ideal (lossless) case.

TABLE 1.1   Ideal Voltage Conversion Ratios of Basic Converters in CCM

Converter	Conversion Ratio

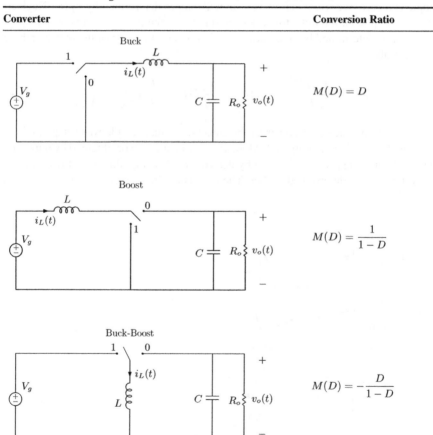

Buck: $M(D) = D$

Boost: $M(D) = \dfrac{1}{1-D}$

Buck-Boost: $M(D) = -\dfrac{D}{1-D}$

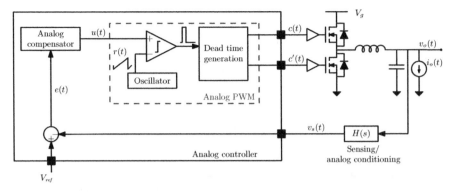

Figure 1.6   Analog voltage-mode control of a synchronous Buck converter.

1.3   CONVERTER DYNAMICS AND CONTROL

The main topic of this chapter—converter averaged small-signal modeling—is now discussed. Consider voltage-mode control of a synchronous Buck converter as a simple case study to review the basic concepts of the approach. A block diagram of the system is illustrated in Fig. 1.6. The term *synchronous* referred to the Buck converter is associated with the implementation of the rectifying, or secondary, switch: instead of the usual free-wheeling diode used as a passive rectifier, the Buck converter of Fig. 1.6 makes use of a *controlled* switch that is driven by the complementary version $c'(t)$ of the PWM signal,

$$c'(t) \triangleq 1 - c(t). \qquad (1.16)$$

A primary advantage of synchronous rectification is the smaller voltage drop across the rectifier switch during conduction, as opposed to the diode rectifier, an essential requirement when regulating low output voltages. Furthermore, the rectifier switch becomes *current bidirectional*, therefore guaranteeing CCM operation and converter controllability even at no load.

In Fig. 1.6, the load is represented by an independent current source rather than a resistance. This is an appropriate modeling choice for many digital loads in point-of-load applications, in which the converter output current depends on the load internal activity and is independent of the output voltage.

The converter is feedback-controlled in order to achieve regulation of the output voltage $v_o(t)$ at a constant reference value V_{ref}. To this end, a control error $e(t)$ is found as the difference between the analog setpoint V_{ref} and the sensed signal $v_s(t)$, where $v_s(t)$ is a scaled, filtered version of $v_o(t)$. In Fig. 1.6, sensing, scaling, and analog filtering of $v_o(t)$ are modeled by the transfer function $H(s)$.

The analog continuous-time compensator processes the error signal and outputs a control command $u(t)$. As exemplified in Fig. 1.7, $u(t)$ is then compared with the carrier $r(t)$ of a trailing-edge pulse width modulator, which in turn produces the modulated gate drive signal $c(t)$.

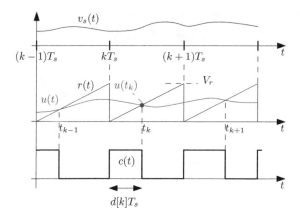

Figure 1.7   Typical analog control waveforms.

The end goal of the modeling step is the derivation of an equivalent small-signal model of the control loop. The process involves, as anticipated, averaging and linearizing the converter behavior around the above steady-state operating point. In the following, notation (1.2) is employed to denote converter quantities averaged over a switching period.

1.3.1   Converter Averaging and Linearization

Referring to the Buck converter shown in Fig. 1.8(a) and applying the moving average operator (1.1) to voltage v_x at the switching node, one has

$$\overline{v}_x(t) \approx d(t)\overline{v}_g(t), \tag{1.17}$$

while the averaged converter input current $\overline{i}_g$ is

$$\overline{i}_g(t) \approx d(t)\overline{i}_L(t). \tag{1.18}$$

These results[1] allow construction of an *averaged equivalent circuit*, as shown in Fig. 1.8(b) [1], which is now time-invariant but still nonlinear.

Perturbation of the circuit equations around the steady-state operating point and successive linearization yields

$$\hat{\overline{v}}_x(t) \approx D\hat{\overline{v}}_g(t) + V_g\hat{d}(t),$$
$$\hat{\overline{i}}_g(t) \approx D\hat{\overline{i}}_L(t) + I_L\hat{d}(t), \tag{1.19}$$

---

[1]Approximation $\langle c(t)x(t)\rangle_{T_s} \approx d(t)\langle x(t)\rangle_{T_s}$ is justified, in general, whenever $x(t)$ has negligible switching content, that is, when it can be regarded as an essentially baseband signal. One exception to this occurs when $x(t)$ has a triangular switching ripple, in which case the approximation is justified even in the presence of a large ripple component. In conclusion, one can safely assume $\langle c(t)x(t)\rangle_{T_s} \approx d(t)\langle x(t)\rangle_{T_s}$ under the small-ripple or linear-ripple approximations already discussed in Section 1.2.

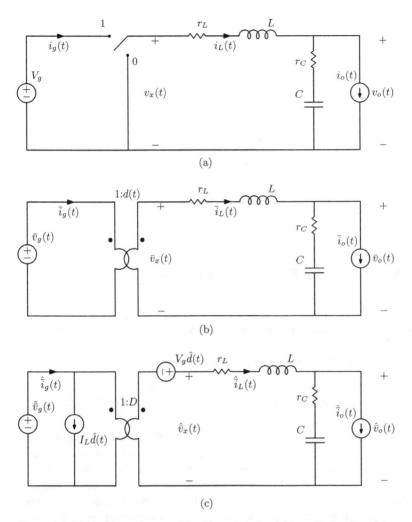

Figure 1.8   (a) Buck converter and its (b) averaged and (c) small-signal models.

where $\hat{\bar{x}}(t) = \bar{x}(t) - X$ denotes the small-signal component of $\bar{x}(t)$ with respect to the dc component X. Figure 1.8(c) illustrates the averaged, small-signal equivalent circuit of the Buck converter obtained after the linearization process. From the equivalent circuit model, evaluation of the control-to-output transfer function $G_{vd}(s)$ yields

$$G_{vd}(s) \triangleq \left. \frac{\hat{\bar{v}}_o(s)}{\hat{d}(s)} \right|_{\hat{\bar{v}}_g=0,\hat{\bar{i}}_o=0} = V_g \frac{1 + sr_C C}{1 + s(r_C + r_L)C + s^2 LC}$$

$$= G_{vd0} \frac{1 + \frac{s}{\omega_{ESR}}}{1 + \frac{s}{Q\omega_0} + \frac{s^2}{\omega_0^2}}, \qquad (1.20)$$

with

$$G_{vd0} \triangleq V_g,$$

$$\omega_{ESR} \triangleq \frac{1}{r_C C},$$

$$\omega_0 \triangleq \frac{1}{\sqrt{LC}}, \tag{1.21}$$

$$Q \triangleq \frac{1}{r_C + r_L} \sqrt{\frac{L}{C}}.$$

The converter small-signal behavior is therefore that of a second-order system with resonant frequency and Q-factor (ω_0, Q), and with a real left half-plane (LHP) zero located at $s = -\omega_{ESR}$. The zero originates from the equivalent series resistance (ESR) r_C of the output capacitor.

1.3.2   Modeling of the Pulse Width Modulator

A small-signal model of the pulse width modulator is necessary in order to develop a complete small-signal model of a converter system. This topic is particularly important as there are significant differences in the PWM small-signal dynamics between analog and digital control.

There are two main families of pulse width modulators:

- *Naturally sampled pulse width modulators (NSPWMs)* process a *continuous-time* modulating signal $u(t)$. They are commonly employed in analog controllers.

- *Uniformly sampled pulse width modulators (USPWMs)* are characterized by a *discrete-time modulating* signal $u[k]$, which is updated once every switching period and held constant throughout the entire switching interval during its comparison with the PWM carrier. USPWMs are most commonly employed in digital control loops, where the control signal is inherently discrete in time, as detailed further in the following chapters. It is worth mentioning, however, that it is possible to apply uniformly sampled modulation in analog control: the continuous-time control command $u(t)$ is in this case subject to a *sample & hold* operation, the output of which is then compared with the PWM carrier using an analog comparator.

Consider a naturally sampled PWM employed in analog, continuous-time control loop around a switched-mode converter. As illustrated in Fig. 1.7, the duty cycle $d[k]$ applied to the power converter during the kth switching cycle is equal to

$$d[k] = \frac{u(t_k)}{V_r}, \tag{1.22}$$

where t_k represents the instant at which $u(t)$ intersects the PWM carrier $r(t)$ during the kth switching cycle, while V_r is the PWM carrier amplitude. Duty cycle $d[k]$ during the kth switching cycle therefore corresponds to a *sampled* version of the modulating signal $u(t)$. Sampling occurs as a result of the intersection between u and r and is inherent to the PWM process–this is the main reason why these types of modulator are designated as *naturally sampled*. For small perturbations $\hat{u}$ around a steady-state value U, every sampling instant occurs at the same position in the switching interval, and the equivalent sampling performed by the PWM becomes uniform.

From (1.22), one also has that $d[k]$ is determined by the *instantaneous* value of $u(t)$ at its intersection with the PWM carrier. The absence of any delay between the natural sampling of $u(t)$ operated by the modulator and the generation of the PWM modulated edge justifies, at least on an intuitive level, the common practice in analog control modeling to treat the PWM as a simple *gain* block. Denoting with $\hat{u}$ and $\hat{d}$ the control command and duty cycle small-signal components with respect to their steady-state values, the PWM transfer function is therefore

$$G_{PWM}(s) \triangleq \frac{\hat{d}}{\hat{u}} = \frac{1}{V_r}. \tag{1.23}$$

It should be noted that (1.23) neglects propagation delays in the PWM comparator and in the gate driving circuitry between the pulse width modulator and the power switch. Such delays, however, are usually much shorter than the switching period T_s. It follows that:

Naturally sampled PWMs do not contribute to the small-signal dynamics of the control loop, except for a constant gain factor.

In contrast to the naturally sampled modulators, the uniformly sampled modulators do introduce dynamics in the loop in the form of an equivalent small-signal delay. This important distinction is further justified and explained in Chapter 2.

1.3.3 The System Loop Gain

Figure 1.9 shows a block diagram of the complete small-signal model of a closed-loop regulated converter. In the diagram, $G_c(s)$ represents the compensator transfer function to be designed.

From the block diagram, the system *loop gain* $T(s)$ can be defined by opening the feedback loop as suggested in Fig. 1.10 and by evaluating the resulting transfer

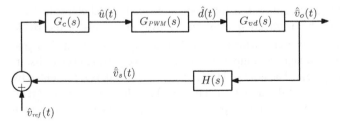

Figure 1.9   Small-signal block diagram of the analog voltage-mode control.

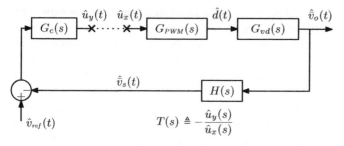

Figure 1.10 Definition of the system loop gain $T(s)$.

function between $\hat{u}_x$ and $\hat{u}_y$,

$$T(s) \triangleq -\left.\frac{\hat{u}_y(s)}{\hat{u}_x(s)}\right|_{\hat{v}_{ref}=0} = G_c(s)G_{PWM}(s)G_{vd}(s)H(s). \tag{1.24}$$

The *uncompensated loop gain* $T_u(s)$, on the other hand, is defined as the system loop gain when a unity compensation is employed, that is, when $G_c(s) = 1$,

$$\boxed{T_u(s) \triangleq G_{PWM}(s)G_{vd}(s)H(s)}. \tag{1.25}$$

From (1.20), (1.23), and (1.24), one has

$$T_u(s) = \frac{G_{vd0}}{V_r}\frac{1+\frac{s}{\omega_{ESR}}}{1+\frac{s}{Q\omega_0}+\frac{s^2}{\omega_0^2}}H(s). \tag{1.26}$$

Result (1.26) represents the starting point for commonly applied frequency-domain compensator design techniques. Analog compensator design proceeds with usual techniques of linear continuous-time control, with the main goals of ensuring sufficient stability margins for the closed-loop system and a control bandwidth adequate for the application.

1.3.4 Averaged Small-Signal Models of Basic Converters

The averaging and linearization steps carried out in Section 1.3.1 can be applied to any converter topology, resulting in a corresponding small-signal equivalent circuit. Figure 1.11 shows the averaged small-signal equivalent circuits of the Buck, Boost, and Buck–Boost converters. In the models, $\hat{v}_g(t)$ and $\hat{i}_o(t)$ are the small-signal components of the input voltage and output current, respectively, which act as disturbances for the control system. The control input, on the other hand, is represented by the small-signal component of the duty cycle command $\hat{d}$, which acts on the circuit via current and voltage generators having operating point dependent gains. Derivation

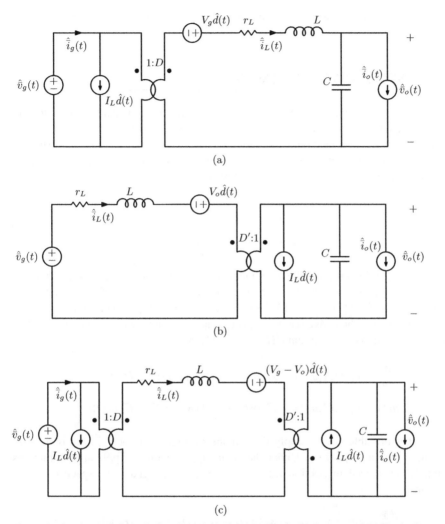

Figure 1.11   Averaged small-signal models of the (a) Buck, (b) Boost, and (c) Buck–Boost converters.

of the control-to-output transfer function, or any other input-output transfer function, can be accomplished via straightforward linear circuit analysis. The results and further details can be found in [1]. If needed, dynamic effects of the output capacitor ESR can be included as well, following [124].

Note that the above-mentioned small-signal models depend on the *average* converter operating point (V_g, I_o, D). This fact is compatible with the basic idea behind the averaged modeling approach that low-frequency dynamics are described accurately, while approximations inherent to the method are tolerated in the proximity and above the Nyquist rate. In contrast, as discussed further in Chapter 3, the discrete-time small-signal models depend on the converter waveforms *at a specific point in time.*

1.4 STATE-SPACE AVERAGING

State-space averaging [119, 120] presents a general mathematical formulation for the averaged small-signal modeling approach summarized in Section 1.3.1. In this formulation, the averaged model is derived in a state-space representation form.

Consider the converter operation as alternating between *two* topological states S_0 and S_1, each described by a *linear* set of state-space equations,

$$\frac{d\boldsymbol{x}}{dt} = \boldsymbol{A}_c \boldsymbol{x}(t) + \boldsymbol{B}_c \boldsymbol{v}(t),$$

$$\boldsymbol{y}(t) = \boldsymbol{C}_c \boldsymbol{x}(t) + \boldsymbol{E}_c \boldsymbol{v}(t),$$

(1.27)

where $\boldsymbol{x}$, $\boldsymbol{v}$, and $\boldsymbol{y}$ represent the state, input, and output vectors, respectively. Matrices $\boldsymbol{A}_c$, $\boldsymbol{B}_c$, $\boldsymbol{C}_c$, and $\boldsymbol{E}_c$ define the state-space model of the converter for each subtopology, with $c \in \{0, 1\}$ being the PWM signal denoting the topological state.

In general, the converter state-space equations can be written, using the PWM signal $c(t)$ and its complement $c'(t) = 1 - c(t)$, as

$$\frac{d\boldsymbol{x}}{dt} = c(t) \left[\boldsymbol{A}_1 \boldsymbol{x}(t) + \boldsymbol{B}_1 \boldsymbol{v}(t) \right] + c'(t) \left[\boldsymbol{A}_0 \boldsymbol{x}(t) + \boldsymbol{B}_0 \boldsymbol{v}(t) \right],$$

$$\boldsymbol{y}(t) = c(t) \left[\boldsymbol{C}_1 \boldsymbol{x}(t) + \boldsymbol{E}_1 \boldsymbol{v}(t) \right] + c'(t) \left[\boldsymbol{C}_0 \boldsymbol{x}(t) + \boldsymbol{E}_0 \boldsymbol{v}(t) \right].$$

(1.28)

It is possible now to apply the moving average operator $\langle \cdot \rangle_{T_s}$ to both sides of the foregoing equations. Under the small-ripple or linear-ripple approximations already introduced in Section 1.2, an averaged, large-signal state-space model is obtained,

$$\frac{d\overline{\boldsymbol{x}}}{dt} = \left[d(t) \boldsymbol{A}_1 + d'(t) \boldsymbol{A}_0 \right] \overline{\boldsymbol{x}}(t) + \left[d(t) \boldsymbol{B}_1 + d'(t) \boldsymbol{B}_0 \right] \overline{\boldsymbol{v}}(t),$$

$$\overline{\boldsymbol{y}}(t) = \left[d(t) \boldsymbol{C}_1 + d'(t) \boldsymbol{C}_0 \right] \overline{\boldsymbol{x}}(t) + \left[d(t) \boldsymbol{E}_1 + d'(t) \boldsymbol{E}_0 \right] \overline{\boldsymbol{v}}(t).$$

(1.29)

As expected, the moving average operator smooths out the time-varying nature of the system, and the system is modeled by a time-invariant, nonlinear set of state-space equations. From this point on, one proceeds with the evaluation of the converter steady-state operating point and with the perturbation/linearization step to obtain the small-signal model.

1.4.1 Converter Steady-State Operating Point

The average steady-state operating point is found from (1.29) by imposing constant inputs $d = D$ and $\overline{\boldsymbol{v}}(t) = \boldsymbol{V}$ and corresponding constant averaged state and output vectors $\overline{x}(t) = X$ and $\overline{y}(t) = Y$,

$$0 = [DA_1 + D'A_0]\,X + [DB_1 + D'B_0]\,V,$$
$$Y = [DC_1 + D'C_0]\,X + [DE_1 + D'E_0]\,V. \tag{1.30}$$

The first equation, in particular, expresses in a general form the inductor volt-second and capacitor charge balance principles. It corresponds to solving the converter network under the assumptions that $\overline{i}_L(t)$ and $\overline{v}_C(t)$ are *constants* of unknown magnitudes.

With the definitions

$$A \triangleq DA_1 + D'A_0,$$

$$B \triangleq DB_1 + D'B_0,$$

$$C \triangleq DC_1 + D'C_0, \tag{1.31}$$

$$E \triangleq DE_1 + D'E_0,$$

one finds the steady-state solution for the states and the outputs,

$$X = -A^{-1}BV,$$
$$Y = \left[-CA^{-1}B + E\right]V. \tag{1.32}$$

1.4.2   Averaged Small-Signal State-Space Model

One is now in the position to linearize (1.29) around the converter steady-state operating point (V, D). As usual, all the relevant quantities are expressed in terms of their steady-state value and small-signal ac component as

$$\hat{\overline{x}}(t) \triangleq \overline{x}(t) - X,$$

$$\hat{d}(t) \triangleq d(t) - D, \tag{1.33}$$

$$\hat{\overline{v}}(t) \triangleq \overline{v}(t) - V.$$

The state-space averaged, small-signal model of the converter is then

$$\boxed{\begin{aligned}\frac{d\hat{\overline{x}}}{dt} &= A\hat{\overline{x}}(t) + F\hat{d}(t) + B\hat{\overline{v}}(t),\\ \hat{\overline{y}}(t) &= C\hat{\overline{x}}(t) + G\hat{d}(t) + E\hat{\overline{v}}(t),\end{aligned}} \tag{1.34}$$

with

$$\boxed{\begin{aligned}F &\triangleq (A_1X + B_1V) - (A_0X + B_0V),\\ G &\triangleq (C_1X + E_1V) - (C_0X + E_0V).\end{aligned}} \tag{1.35}$$

Assume now an initial unperturbed condition $\hat{\bar{x}}(0) = 0$ and derive the system's forced response via Laplace transformation of (1.34),

$$s\hat{\bar{x}}(s) = A\hat{\bar{x}}(s) + F\hat{d}(s) + B\hat{\bar{v}}(s)$$

$$\hat{\bar{y}}(s) = C\hat{\bar{x}}(s) + G\hat{d}(s) + E\hat{\bar{v}}(s)$$

$$\Rightarrow \hat{\bar{y}}(s) = \left(C\left(sI - A\right)^{-1}F + G\right)\hat{d}(s) + \left(C\left(sI - A\right)^{-1}B + E\right)\hat{\bar{v}}(s). \tag{1.36}$$

From this result, one can derive transfer functions needed for control design purposes. For instance, the *control transfer matrix*, which relates the effect of the control command on the converter outputs, is

$$\boxed{W(s) \triangleq \left.\frac{\hat{\bar{y}}(s)}{\hat{d}(s)}\right|_{\hat{\bar{v}}=0} = C\left(sI - A\right)^{-1}F + G}, \tag{1.37}$$

whereas the *disturbance transfer matrix* is

$$\boxed{W_D(s) \triangleq \left.\frac{\hat{\bar{y}}(s)}{\hat{\bar{v}}(s)}\right|_{\hat{d}=0} = C\left(sI - A\right)^{-1}B + E}. \tag{1.38}$$

1.4.3   Boost Converter Example

As an example, the state-space averaged small-signal model of the nonideal Boost converter illustrated in Fig. 1.3 is derived in this section.

With the switch in position 1, one has

$$\frac{di_L}{dt} = \frac{v_g(t) - r_L i_L(t)}{L} \tag{1.39}$$

for the inductor loop equation and

$$\frac{dv_C}{dt} = \frac{dv_o}{dt} = -\frac{v_C(t)}{R_o C} \tag{1.40}$$

for the capacitor node equation. Observe that, in this example, $v_o(t) = v_C(t)$ as zero ESR is assumed for the output capacitor.

Having defined the state vector as $x \triangleq [i_L \quad v_C]^T = [i_L \quad v_o]^T$, the state equation of subtopology 1 is

$$\frac{dx}{dt} = \underbrace{\begin{bmatrix} -\dfrac{r_L}{L} & 0 \\ 0 & -\dfrac{1}{R_o C} \end{bmatrix}}_{A_1} x(t) + \underbrace{\begin{bmatrix} \dfrac{1}{L} \\ 0 \end{bmatrix}}_{B_1} v_g(t). \tag{1.41}$$

With the switch in position 0, on the other hand, one has

$$\frac{di_L}{dt} = \frac{v_g(t) - r_L i_L(t) - v_o(t)}{L} \cdot \tag{1.42}$$

and

$$\frac{dv_o}{dt} = \frac{i_L(t)}{C} - \frac{v_o(t)}{R_o C} \cdot \tag{1.43}$$

The state equation relative to subtopology 0 is then

$$\frac{d\boldsymbol{x}}{dt} = \underbrace{\begin{bmatrix} -\dfrac{r_L}{L} & -\dfrac{1}{L} \\[2mm] \dfrac{1}{C} & -\dfrac{1}{R_o C} \end{bmatrix}}_{\boldsymbol{A}_0} \boldsymbol{x}(t) + \underbrace{\begin{bmatrix} \dfrac{1}{L} \\[2mm] 0 \end{bmatrix}}_{\boldsymbol{B}_0} v_g(t). \tag{1.44}$$

Define now the system output to coincide with the state vector, that is, $\boldsymbol{y}(t) = \boldsymbol{x}(t)$, and therefore $\boldsymbol{C}_1 = \boldsymbol{C}_0 = \boldsymbol{I}$ and $\boldsymbol{E}_1 = \boldsymbol{E}_0 = 0$. Matrices $\boldsymbol{A}$, $\boldsymbol{B}$, and $\boldsymbol{C}$ of the averaged model can then be evaluated. The result is

$$\boldsymbol{A} \triangleq D\boldsymbol{A}_1 + D'\boldsymbol{A}_0 = \begin{bmatrix} -\dfrac{r_L}{L} & -\dfrac{D'}{L} \\[2mm] \dfrac{D'}{C} & -\dfrac{1}{R_o C} \end{bmatrix},$$

$$\boldsymbol{B} \triangleq D\boldsymbol{B}_1 + D'\boldsymbol{B}_0 = \begin{bmatrix} \dfrac{1}{L} \\[2mm] 0 \end{bmatrix}, \tag{1.45}$$

$$\boldsymbol{C} \triangleq D\boldsymbol{C}_1 + D'\boldsymbol{C}_0 = \begin{bmatrix} 1 & 0 \\ 0 & 1 \end{bmatrix}.$$

Solving for the converter operating point according to (1.32) yields

$$\boldsymbol{X} = \begin{bmatrix} I_L \\[2mm] V_o \end{bmatrix} = \begin{bmatrix} \dfrac{1}{r_L + D'^2 R_o} \\[4mm] \dfrac{1}{D'} \dfrac{1}{1 + \frac{r_L}{D'^2 R_o}} \end{bmatrix} V_g. \tag{1.46}$$

As expected, this is the same result as (1.12).

As for the small-signal model, as $B_1 = B_0$, matrix F evaluates as

$$F = (A_1 - A_0)\, X = \begin{bmatrix} \dfrac{V_o}{L} \\[2ex] -\dfrac{I_L}{C} \end{bmatrix}, \tag{1.47}$$

whereas $G = 0$ as $C_1 = C_0$. The control transfer matrix is

$$W(s) = C\,(sI - A)^{-1}\, F + G = \begin{bmatrix} G_{id}(s) \triangleq \dfrac{\hat{\bar{i}}_L(s)}{\hat{d}(s)} \\[3ex] G_{vd}(s) \triangleq \dfrac{\hat{\bar{v}}_o(s)}{\hat{d}(s)} \end{bmatrix}$$

$$= \begin{bmatrix} \dfrac{2V_o}{r_L + D'^2 R_o} \dfrac{1 + s\frac{R_o C}{2}}{\Delta(s)} \\[3ex] \dfrac{V_o}{D'} \dfrac{1 - \frac{r_L}{D'^2 R_o}}{1 + \frac{r_L}{D'^2 R_o}} \dfrac{1 - s\frac{L}{D'^2 R_o - r_L}}{\Delta(s)} \end{bmatrix}, \tag{1.48}$$

with

$$\Delta(s) \triangleq 1 + s\dfrac{r_L}{D'^2 R_o}\left(\dfrac{R_o C + \frac{L}{r_L}}{1 + \frac{r_L}{D'^2 R_o}}\right) + s^2 \dfrac{LC}{D'^2}\dfrac{1}{1 + \frac{r_L}{D'^2 R_o}}. \tag{1.49}$$

1.5   DESIGN EXAMPLES

This section presents some examples of analog control designs based on the converter small-signal models developed in Sections 1.3.1 and 1.4 and standard frequency-domain-based compensator design techniques.

1.5.1   Voltage-Mode Control of a Synchronous Buck Converter

Figure 1.12 illustrates an implementation example for an analog voltage-mode controller in the system of Fig. 1.6. The system makes use of an analog integrated circuit containing an error amplifier and an analog pulse width modulator. The control compensation is implemented via an external passive network, which shapes the response of the error amplifier.

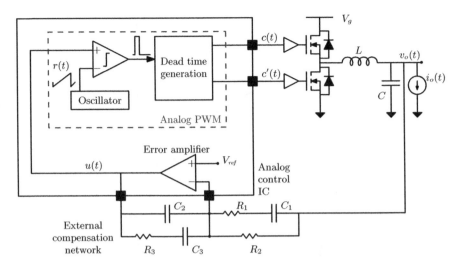

Figure 1.12   Synchronous Buck example: analog voltage-mode control scheme.

TABLE 1.2   Synchronous Buck Example Parameters

Parameter	Value
Input voltage V_g	5 V
Output voltage V_o	1.8 V
Load current $I_{o,max}$	5 A
Switching frequency f_s	1 MHz
Filter inductance L	1 μH
Inductor series resistance r_L	30 mΩ
Filter capacitance C	200 μF
Capacitor equivalent series resistance r_C	0.8 mΩ
PWM carrier amplitude V_r	1 V
Voltage sensing gain H	1 V/V

Design specifications and power stage parameters are summarized in Table 1.2. As reported in the table, the power stage nonidealities include a nonzero inductor series resistance r_L and a nonzero capacitor ESR r_C.

The small-signal model of the system is presented in Section 1.3, and the uncompensated loop gain expression is given in (1.26). The magnitude and phase Bode plots of $T_u(s)$ are shown in Fig. 1.13. The dc value of the uncompensated loop gain is

$$T_{u0} \triangleq T_u(s=0) = \frac{G_{vd0}}{V_r}H = \frac{V_g}{V_r} = 5 \Rightarrow 14 \text{ dB}. \qquad (1.50)$$

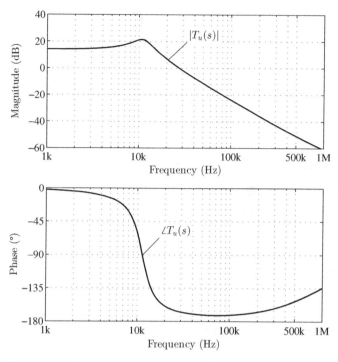

Figure 1.13   Synchronous Buck example: Bode plots of the uncompensated loop gain $T_u(s)$.

The system resonance occurs at

$$\omega_0 = \frac{1}{\sqrt{LC}} \approx 2\pi \cdot (11 \text{ kHz}),$$ (1.51)

while the r_C-related zero is located at

$$\omega_{ESR} = \frac{1}{r_C C} \approx 2\pi \cdot (1 \text{ MHz}).$$ (1.52)

As a design goal, the target crossover frequency is set at $f_c = 100$ kHz, that is, $1/10$ of the converter switching frequency, and a phase margin target is set at $\varphi_m = 55°$. At $f = f_c$, the uncompensated loop gain exhibits a phase of about $-171°$, implying that a *lead* type of compensation is required in the neighborhood of f_c to boost the phase margin by $\theta = 46°$. Such compensation is obtained by forming a pole-zero pair

$$G_{PD}(s) \triangleq G_{PD0} \frac{1 + \frac{s}{\omega_z}}{1 + \frac{s}{\omega_p}}.$$ (1.53)

Subscript *PD* stands for *proportional-derivative*, which is a term commonly used for the lead compensation.

The maximum phase boost generated by the *PD* pole-zero pair occurs at

$$\omega_{max} = \sqrt{\omega_z \omega_p} \tag{1.54}$$

and that it equals

$$\angle G_{PD}(j\omega_{max}) = \arctan\left(\sqrt{\frac{\omega_p}{\omega_z}}\right) - \arctan\left(\sqrt{\frac{\omega_z}{\omega_p}}\right) = \frac{\pi}{2} - 2\arctan\left(\sqrt{\frac{\omega_z}{\omega_p}}\right). \tag{1.55}$$

As the phase lead provided by the compensator should be equal to $\theta = 46°$, the required ω_z/ω_p ratio can be found from (1.55) which, using $\omega_c = \omega_{max} = \sqrt{\omega_z \omega_p}$, yields the values of both ω_z and ω_p,

$$\omega_z = \omega_c\sqrt{\frac{1-\sin\theta}{1+\sin\theta}} = 2\pi \cdot (40 \text{ kHz}),$$

$$\omega_p = \omega_c\sqrt{\frac{1+\sin\theta}{1-\sin\theta}} = 2\pi \cdot (250 \text{ kHz}). \tag{1.56}$$

The dc gain G_{PD0} of the lead action is determined by imposing unity loop gain at the desired crossover frequency f_c,

$$|T(j\omega_c)| = |T_u(j\omega_c)|G_{PD0}\sqrt{\frac{1+\left(\frac{\omega_c}{\omega_z}\right)^2}{1+\left(\frac{\omega_c}{\omega_p}\right)^2}} = 1, \tag{1.57}$$

which yields

$$G_{PD0} = \frac{1}{|T_u(j\omega_c)|}\sqrt{\frac{1+\left(\frac{\omega_c}{\omega_p}\right)^2}{1+\left(\frac{\omega_c}{\omega_z}\right)^2}} \approx 6.2 \Rightarrow 15.8 \text{ dB}. \tag{1.58}$$

Figure 1.14 illustrates the magnitude and phase Bode plots of the lead compensation.

As a last design step, adding an integral action, that is, a compensation pole at dc nulls the steady-state regulation error and, more generally, improves the regulation by increasing the low-frequency loop gain magnitude. This is accomplished by including a *lag* term of the type

$$G_{PI}(s) \triangleq G_{PI\infty}\left(1 + \frac{\omega_l}{s}\right). \tag{1.59}$$

Such term is also known as *proportional–integral* (PI) compensation.

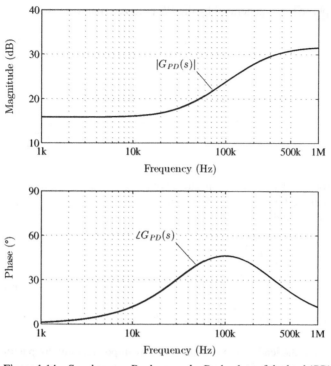

Figure 1.14   Synchronous Buck example: Bode plots of the lead (*PD*) compensation transfer function.

As a general rule, the *PI* term should not significantly affect either the system crossover frequency or its phase margin. Therefore, the high-frequency gain $G_{PI\infty}$ is set to one, and the zero corner frequency ω_l is selected such that $\omega_l \ll \omega_c$. A good choice is to let $\omega_l < \omega_c/10 = 2\pi \cdot (10 \text{ kHz})$. In this design example, choose

$$\omega_l = 2\pi \cdot (8 \text{ kHz}) . \tag{1.60}$$

The complete *proportional-integral-derivative* (PID) compensator transfer function is therefore

$$G_{PID}(s) = \underbrace{\left(1 + \frac{\omega_l}{s}\right)}_{PI} \cdot \underbrace{G_{PD0}\frac{1 + \frac{s}{\omega_z}}{1 + \frac{s}{\omega_p}}}_{PD} , \tag{1.61}$$

where all the corner frequencies and gains are now determined. Figure 1.15 shows Bode plots of the designed PID compensator transfer function.

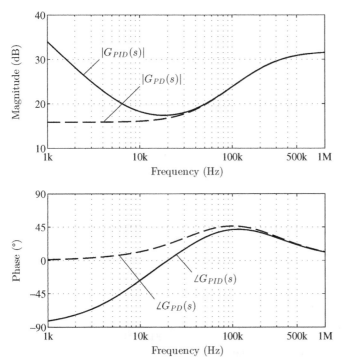

Figure 1.15   Synchronous Buck converter example: Bode plots of the lead (*PD*) and *PID* compensation transfer functions.

Going back to the external compensation network depicted in Figure 1.12, assuming $C_3 \gg C_2$ the corresponding transfer function is

$$G_c(s) \triangleq -\frac{\hat{u}(s)}{\hat{\tilde{v}}_o(s)} = \underbrace{\left(1 + \frac{1}{sR_3C_3}\right)}_{PI} \cdot \underbrace{\frac{R_3}{R_2}\frac{1 + s\left(R_1 + R_2\right)C_1}{1 + sR_1C_1}}_{PD} \cdot \underbrace{\frac{1}{1 + sR_3C_2}}_{HF\ Pole},$$

$$\tag{1.62}$$

where the different portions of the control action have been highlighted. Circuit-level design of the compensation network can now be performed by equating (1.61) and (1.62). Note that (1.62) allows for an additional high-frequency pole to be placed at

$$\omega_{p_2} \triangleq \frac{1}{R_3C_2}. \tag{1.63}$$

Such pole is commonly used to attenuate the gain of the compensator at high frequencies and prevent the propagation of switching harmonics produced by the converter through the feedback loop—a circumstance that can otherwise result in undesired effects. A good choice for ω_{p_2} is

$$\omega_{p_2} = 10\omega_c = 2\pi \cdot \left(1\ \text{MHz}\right), \tag{1.64}$$

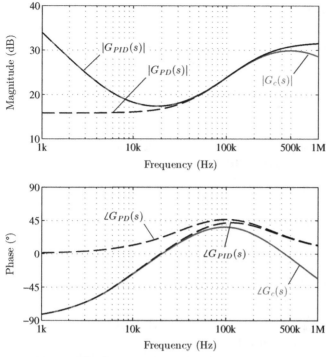

Figure 1.16 Synchronous Buck example: Bode plots of the lead (*PD*), *PID*, and overall compensation transfer functions.

which ensures that the added pole has limited impact on the designed phase margin. Figure 1.16 compares the Bode plots of $G_{PD}(s)$, $G_{PID}(s)$, and $G_c(s)$, while the magnitude and phase responses of the system loop gain are shown in Fig. 1.17. The combined effects of the low-frequency PI term and the high-frequency pole on the overall compensator transfer function lead to $\approx 10°$ phase margin loss with respect to the target $\varphi_m = 55°$. Such phase margin loss could be easily considered by imposing a correspondingly higher value on φ_m in the above-mentioned design procedure.

The above-mentioned compensator design can now be validated—and refined, if needed—via computer simulations. To this end, a Matlab® model of the voltage-controlled Buck converter pictured in Fig. 1.12 has been set up. The scheme depicted in Fig. 1.18 employs Middlebrook's approach [1, 78] to obtain $T(s)$ by simulation and to validate the averaged small-signal models employed in the design phase [1]. The closed-loop system is excited by a sinusoidal perturbation $u_{pert}(t)$ of small amplitude at frequency ω_{pert}. Signals $u_x(t)$ and $u_y(t)$ are acquired over a number of oscillation periods, and their Fourier components $u_x(\omega_{pert})$ and $u_y(\omega_{pert})$ at ω_{pert} are determined via an FFT-based postprocessing. The procedure is repeated for a number of perturbation frequencies, in order to extract the simulated loop gain

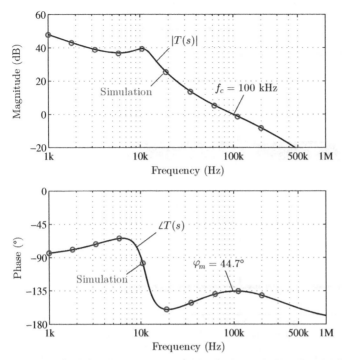

Figure 1.17   Synchronous Buck example: Bode plots of the theoretical and simulated system loop gain.

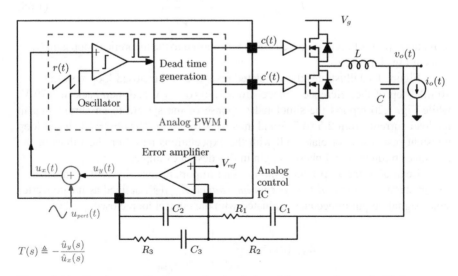

Figure 1.18   Synchronous Buck example: simulation of the system loop gain $T(s)$.

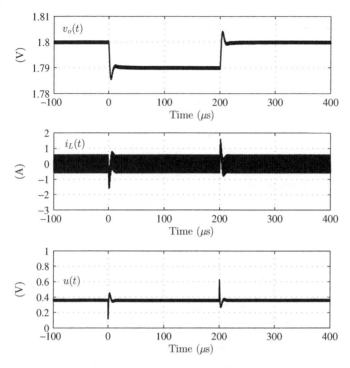

Figure 1.19 Synchronous Buck example: 1.79 V ↔ 1.8 V step reference responses.

$T_{sim}(j\omega_{pert})$ as

$$T_{sim}(j\omega_{pert}) = -\frac{u_y(\omega_{pert})}{u_x(\omega_{pert})}. \tag{1.65}$$

Simulation points thus determined are superimposed to the theoretical loop gain Bode plots in Fig. 1.17.

Figure 1.19 illustrates the simulated response of the closed-loop system to a 10 mV step of the reference voltage, from 1.79 to 1.8 V and then back to 1.79 V, while Fig. 1.20 reports the simulated response of the system to an abrupt step in the load current, from 2.5 to 5 A and then back to 2.5 A. The observed closed-loop transient responses correlate well with the expectations based on the values of the crossover frequency and phase margin in this design example.

Regarding the step load response, one important quantity to be evaluated at design stage is the *closed-loop output impedance* $Z_{o,cl}(s)$, defined as the converter small-signal output impedance evaluated with the control loop closed,

$$Z_{o,cl}(s) \triangleq -\frac{\hat{\bar{v}}_o(s)}{\hat{\bar{i}}_o(s)}\bigg|_{\hat{\bar{v}}_{ref}=0,\hat{\bar{v}}_g=0}. \tag{1.66}$$

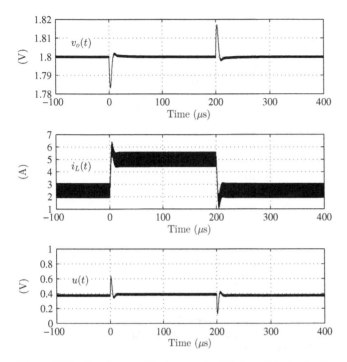

Figure 1.20   Synchronous Buck example: 2.5 A ↔ 5 A step load responses.

The closed-loop output impedance $Z_{o,cl}(s)$ can be expressed in terms of the converter *open-loop output impedance* $Z_o(s)$ and the system loop gain as [1]

$$Z_{o,cl}(s) = \frac{Z_o(s)}{1 + T(s)}, \tag{1.67}$$

with

$$Z_o(s) \triangleq -\left.\frac{\hat{\bar{v}}_o(s)}{\hat{\bar{i}}_o(s)}\right|_{\hat{u}=0,\hat{\bar{v}}_g=0}. \tag{1.68}$$

The open-loop output impedance is readily evaluated from the averaged small-signal equivalent circuit of the Buck converter (Fig. 1.11),

$$Z_o(s) = r_L \frac{(1 + sr_C C)\left(1 + s\frac{L}{r_L}\right)}{1 + s(r_C + r_L)C + s^2 LC}. \tag{1.69}$$

Bode plots of both $Z_{o,cl}(s)$ and $Z_o(s)$ for the voltage-mode control loop under consideration are shown in Fig. 1.21. Below the control bandwidth, the output impedance

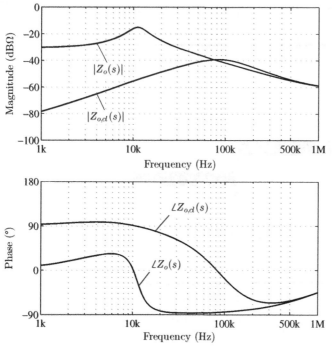

Figure 1.21 Synchronous Buck example: Bode plots of the open-loop and closed-loop output impedances.

is reduced by the feedback loop because of the large loop gain. At higher frequencies, on the other hand, $Z_{o,cl}(s)$ and $Z_o(s)$ practically coincide.

1.5.2 Average Current-Mode Control of a Boost Converter

As a second example, consider average current-mode control of a Boost converter depicted in Fig. 1.22. The converter parameters are listed in Table 1.3.

The Boost converter operates from a dc input voltage $V_g = 120$ V and delivers 500 W maximum power to a resistive load R_o. At the maximum power, the output voltage equals 380 V, so

$$P_o = \frac{V_o^2}{R_o} = \frac{(380\ \text{V})^2}{R_o} = 500\ \text{W} \Rightarrow R_o \approx 289\ \Omega. \tag{1.70}$$

The Boost inductor current $i_L(t)$ is converted into a voltage $v_s(t)$ by a 0.1 Ω shunt resistor R_{sense} and compared with the control setpoint V_{ref}. The regulation error is processed by an analog compensator implemented by an op-amp-based circuit. A symmetrical (triangle-wave) analog pulse width modulator converts the output $u(t)$ of the error amplifier into the logic gate-drive control $c(t)$.

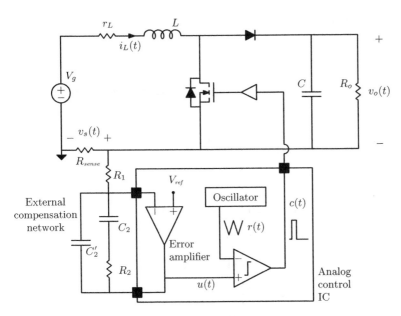

Figure 1.22   Average current-mode control of a Boost converter.

TABLE 1.3   Boost Converter Example Parameters

Parameter	Value
Input voltage V_g	120 V
Output voltage V_o	380 V
Power rating P_o	500 W
Switching frequency f_s	100 kHz
Filter inductance L	500 μH
Inductor series resistance r_L	20 mΩ
Filter capacitance C	220 μF
PWM carrier amplitude V_r	1 V
Current sensing gain R_{sense}	0.1 Ω

At maximum output power and neglecting parasitics, the steady-state duty cycle is determined from

$$M(D) = \frac{V_o}{V_g} = \frac{380 \text{ V}}{120 \text{ V}} = \frac{1}{1-D}$$

$$\Rightarrow D \approx 0.68. \tag{1.71}$$

The averaged small-signal model of the Boost converter has been derived in Section 1.4.3. Accounting for the additional sensing resistance R_{sense} is simply accomplished by substituting r_L with $r_L + R_{sense}$,

$$r_L \rightarrow r_L + R_{sense}. \tag{1.72}$$

From (1.48), the control-to-inductor current dynamics, described by the transfer function $G_{id}(s)$, has the form

$$G_{id}(s) = G_{id0} \frac{1 + \frac{s}{\omega_z}}{1 + \frac{s}{\omega_0 Q} + \frac{s^2}{\omega_0^2}}, \tag{1.73}$$

with

$$G_{id0} = 26.3 \text{ A} \Rightarrow 28.4 \text{ dB},$$

$$\omega_z = 2\pi \cdot (5 \text{ Hz}),$$

$$\omega_0 = 2\pi \cdot (152 \text{ Hz}), \tag{1.74}$$

$$Q = 3.7.$$

The uncompensated current loop gain is proportional to $G_{id}(s)$ and equals

$$T_u(s) = \frac{R_{sense}}{V_r} G_{id}(s). \tag{1.75}$$

Bode plots of $G_{id}(s)$ are illustrated in Fig. 1.23. Thanks to the LHP zero located at $s = -\omega_z$, the transfer function retains a -20 dB/decade slope above the system resonance, allowing for a high-bandwidth control to be designed using a simple PI compensation law,

$$G_{PI}(s) = G_{PI\infty} \left(1 + \frac{\omega_{PI}}{s} \right). \tag{1.76}$$

As discussed in the Buck voltage-mode control example, a high-frequency pole can be included in the compensator transfer function in order to provide some filtering action on the harmonic content of the sensed signal. For instance, set such high-frequency pole at half the switching rate, that is, at 50 kHz. With this choice, the compensator transfer function to be designed is

$$G_c(s) = \underbrace{G_{PI\infty} \left(1 + \frac{\omega_{PI}}{s} \right)}_{PI} \cdot \underbrace{\frac{1}{1 + \frac{s}{\omega_{HF}}}}_{HF\ Pole}, \tag{1.77}$$

$$\omega_{HF} = 2\pi \cdot (50 \text{ kHz}). \tag{1.78}$$

On the basis of the transfer function template, the objective is to design a $\omega_c = 2\pi \cdot (10 \text{ kHz})$ bandwidth compensation with a $\varphi_m = 50°$ phase margin. The

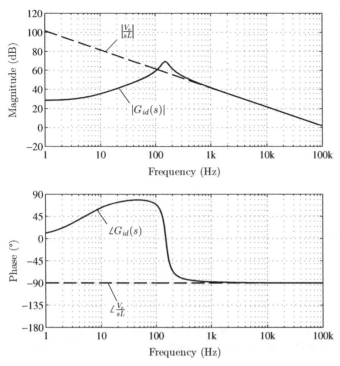

Figure 1.23   Boost converter example: Bode plots of the control-to-inductor current transfer function.

high-frequency pole located at 50 kHz introduces, at the desired control bandwidth, an additional phase *lag* equal to

$$\arctan\left(\frac{10\text{ kHz}}{50\text{ kHz}}\right) \approx 11°, \tag{1.79}$$

and therefore the PI portion of the compensation must be designed for a target phase margin of $\varphi'_m = 50° + 11°$. Having clarified this point, calculation of the unknown PI coefficients $G_{PI\infty}$ and ω_{PI} is straightforward once the magnitude and the phase of the system uncompensated loop gain are evaluated at the target control bandwidth, that is, at the target crossover frequency,

$$|T_u(j\omega_c)| \approx 1.2 \Rightarrow 1.6\text{ dB},$$
$$\angle T_u(j\omega_c) \approx -90°. \tag{1.80}$$

An alternative, quicker approach to estimate $T_u(s)$ is to employ a high-frequency approximation of $G_{id}(s)$. From (1.48), it is easy to see that, as long as $\omega \gg \omega_0$, one has

$$G_{id}(s) \approx \frac{V_o}{sL} \qquad (\omega \gg \omega_0), \tag{1.81}$$

and therefore the inductor current dynamics behaves almost ideally around the target control bandwidth. This approximation is illustrated in Fig. 1.23 as well. One can verify that (1.81) very accurately predicts the values reported in (1.80).

Derivation of ω_{PI} is based on the required phase margin,

$$-\frac{\pi}{2} + \arctan\left(\frac{\omega_c}{\omega_{PI}}\right) + \angle T_u(j\omega_c) = -\pi + \varphi'_m, \tag{1.82}$$

whereas the value of $G_{PI\infty}$ is imposed by the desired crossover frequency ω_c,

$$G_{PI\infty}\sqrt{1 + \left(\frac{\omega_c}{\omega_{PI}}\right)^2}|T_u(j\omega_c)| = 1. \tag{1.83}$$

Solving the above-mentioned equations yields

$$G_{PI\infty} = 0.73 \Rightarrow -2.7 \text{ dB},$$
$$\omega_{PI} = 2\pi \cdot (5.5 \text{ kHz}). \tag{1.84}$$

Bode plots of the compensator transfer function are shown in Fig. 1.24. Both the uncompensated and compensated current loop gains $T_u(s)$ and $T(s)$ are shown in

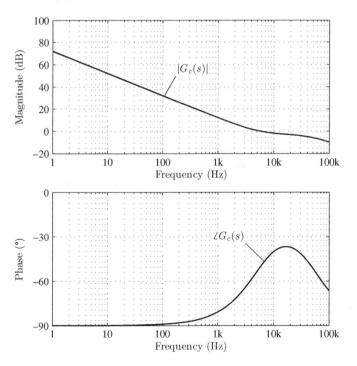

Figure 1.24 Boost converter example: Bode plots of the compensator transfer function.

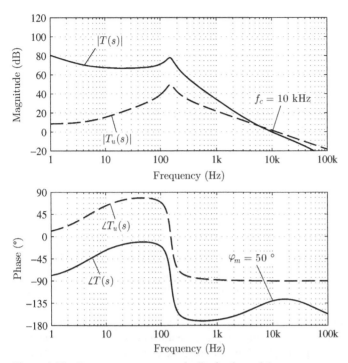

Figure 1.25   Boost converter example: Bode plots of the uncompensated and compensated current loop gains $T_u(s)$ and $T(s)$.

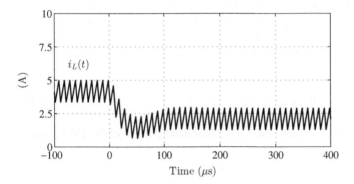

Figure 1.26   Boost converter example: 500 W→250 W step-reference response.

Fig. 1.25. Figure 1.26 illustrates the simulated closed-loop response to a step variation in the current setpoint corresponding to a 500 to 250 W reduction in the input power.

With the compensator transfer function so determined, one is in position to carry out the circuit-level design of the external compensation network (R_1, R_2, C_2, C_2') illustrated in Fig. 1.22. Within the finite gain-bandwidth product limitation of the error amplifier, the compensator transfer function is

$$G_c(s) \triangleq -\frac{\hat{u}(s)}{\hat{v}_o(s)} = \underbrace{\frac{R_2}{R_1}\left(1 + \frac{1}{sR_2C_2}\right)}_{PI} \cdot \underbrace{\frac{1}{1 + sR_2C_2'}}_{HF\ Pole} \qquad (C_2 \gg C_2'), \qquad (1.85)$$

and determination of the compensation network parameters starts by equating (1.85) to (1.77).

1.6   DUTY RATIO $d[k]$ VERSUS $d(t)$

In the foregoing discussions, the *cycle-by-cycle* duty ratio $d[k]$, intended as a discrete-time signal, and $d(t)$, that is, a continuous-time signal that acts as the control input for the converter in the context of averaged models, have intentionally been conflated. Before closing the review of analog (continuous-time) modeling and control in this chapter, it is useful to highlight a few aspects regarding the physical meaning of the duty cycle d as the control input, as well as relationships between $d[k]$ and $d(t)$.

Given the nature of the switched-mode power converter controlled by a PWM waveform, it is clear that a physical meaning can be attributed only to $d[k]$: the converter responds to $d[k]$, not to $d(t)$. Conceptually, the *duty cycle* is a property of a switching *interval* and not of a specific instant in time. Consequently:

The duty ratio is an inherently discrete-time signal, even in analog control.

Nonetheless, $d(t)$ as a continuous-time control signal has been employed in the context of averaged models and analog control of switched-mode converters. One interpretation of $d(t)$ is provided by [125]: $d(t)$ can be described as a baseband continuous-time signal interpolating $d[k]$ at the pulse width modulated switching events $T_k = DT_s + kT_s$,

$$d(t = T_k = kT_s + DT_s) = d[k]. \qquad (1.86)$$

More formally, one could intend $d(t)$ as the baseband portion of the PWM output $c(t)$,

$$d(\omega) \triangleq \mathcal{R}(\omega)c(\omega), \qquad (1.87)$$

where $\mathcal{R}(\omega)$ is the frequency response of the ideal brick-wall filter,

$$\mathcal{R}(\omega) = \begin{cases} 1, & -\frac{\omega_s}{2} < \omega < \frac{\omega_s}{2}, \\ 0 & \text{otherwise.} \end{cases} \qquad (1.88)$$

This interpretation of $d(t)$ highlights an advantage as well as a limitation of the averaged modeling approach: the converter behavior is studied at frequencies well below the Nyquist rate $f_s/2$, where it behaves as if *continuously* responding to the low-frequency portion of the PWM spectrum. The difference between $d(t)$ and the true converter control input $d[k]$ only becomes important in the proximity and above the Nyquist rate $f_s/2$.

For analog pulse width modulators, it can be shown that $d(t)$ coincides with $u(t)$ itself apart from a scaling factor equal to $1/V_r$ [126]. The control signal then appears unaltered in the baseband of $c(t)$, and the fact that $d(t) = u(t)/V_r$ is in agreement with (1.23). This justifies the common practice in analog modeling to treat $d(t)$ as a scaled version of the analog control command $u(t)$.

The situation is depicted by the qualitative spectra reported in Fig. 1.27. The comparison between analog control signal $u(t)$ and the carrier operated by the comparator produces the sequence $d[k]$ driving the converter. A certain PWM spectrum $c(t)$ and a certain converter response $v_s(t)$ can be associated with $d[k]$.

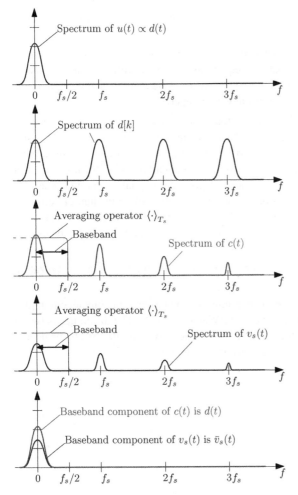

Figure 1.27   Qualitative signal spectra in analog control.

Averaged small-signal modeling, on the other hand, focuses on the baseband portion of $c(t)$ – that is, $d(t)$ – and puts it in relation with the baseband portion of $v_s(t)$, that is, with $\bar{v}_s(t)$.

The assumption that $u(t)$—and therefore $d(t)$—is a baseband signal is never strictly satisfied in practice, as $u(t)$ always includes some amount of switching content as a result of the switching harmonics not entirely filtered by the sensing path or the compensator. Differences between $u(t)$ and $d(t)$ arise when such switching frequency content is significant and make the PWM small-signal gain differ from $1/V_r$. One may note that the effect, usually undesired in PWM converters, is intentionally employed in analog peak current-mode controllers, where adding the compensation ramp to the modulating signal alters the small-signal gain of the modulator [1].

1.7 SUMMARY OF KEY POINTS

- PWM switched-mode power converters alternate between two or more subtopologies in a regular manner. In general, a switched-mode power converter is a time-varying nonlinear system.

- Steady-state analysis of switched-mode power converters is founded upon the volt-second and charge balance equations. Coupled with the small-ripple or linear-ripple approximations, averaged steady-state currents and voltages of the converter can be determined for any given operating point.

- Analysis of converter dynamics is founded upon applying an averaging operator to all converter waveforms. The resulting model is time-invariant but still nonlinear. Perturbation and linearization of the averaged model yields a linear small-signal model, which can be used to obtain all transfer functions relevant for the control design process. The averaged small-signal modeling framework formulates the above-mentioned concepts in terms of equivalent circuits that can be analyzed using conventional circuit analysis techniques. In general, the averaged models are intended to predict converter dynamics at frequencies well below the switching frequency.

- State-space averaging is a general formulation for the averaging/linearization process.

- The duty ratio $d(t)$ used in the continuous-time modeling is intended to represent the baseband component of the PWM signal $c(t)$. The difference between such control input and the true cycle-by-cycle duty ratio $d[k]$ can be neglected in the context of the *averaged* converter dynamics.

CHAPTER 2

THE DIGITAL CONTROL LOOP

The objectives of this chapter are to introduce the main blocks in the digital control loop around a switched-mode power converter and provide an overview of the design issues the digital scenario brings in from a system-level standpoint. Digital control of a switched-mode power converter differs from analog control in two aspects:

1. *Time Quantization.* The controller is a *discrete-time system* that processes a sampled version of the analog signal(s) to be controlled or regulated and produces a discrete-time control output.

2. *Amplitude Quantization.* Data processing inside the controller occurs *digitally*, that is, the signal amplitudes are quantized.

Time quantization affects the small-signal dynamics of the system, introducing *control delays* within the feedback loop, which are not found in analog control loops. Spectral aliasing due to the sampling process also comes into play in determining the frequency response of the digitally controlled converter. In general, the averaged small-signal modeling summarized in the previous chapter does not capture these effects. Given the popularity of the averaged modeling approach, the second part of this chapter presents several examples of how this approach can be used in digital design, but it also highlights some of its limitations. These limitations are removed in the exact modeling approach for digitally controlled converters, referred to as *discrete-time small-signal modeling*, which is discussed in Chapter 3.

Amplitude quantization is responsible for nonlinear effects, which may lead to degradation of static and dynamic regulation performance in digitally controlled converters. Such effects, together with related design guidelines, are discussed in Chapters 5 and 6.

The digital control loop is introduced in Section 2.1. Characteristics and modeling of the main blocks in the digital control loop, the analog-to-digital (A/D) conversion, the digital compensator, and the digital pulse width modulator (DPWM), are introduced in Sections 2.2, 2.3 and 2.4, respectively. Section 2.5 is devoted to a discussion of loop delays introduced by the time quantization and the sampling processes in the digital controller. Using averaged models in the digital design, together with limitations of this approach, are discussed in Section 2.6. The key points are summarized in Section 2.7.

Digital Control of High-Frequency Switched-Mode Power Converters, First Edition.
Luca Corradini, Dragan Maksimović, Paolo Mattavelli, and Regan Zane.
© 2015 The Institute of Electrical and Electronics Engineers, Inc. Published 2015 by John Wiley & Sons, Inc.

The reader is referred to Appendix A for a brief introduction to discrete-time systems and Z-Transform.

2.1 CASE STUDY: DIGITAL VOLTAGE-MODE CONTROL

As a case study, consider a digital version of the voltage-mode controlled Buck converter introduced in Section 1.5.1. The control system is illustrated in Fig. 2.1.

Similar to the analog case, the output voltage $v_o(t)$ undergoes a preliminary signal conditioning step, performed in the analog domain, which is modeled by the transfer function $H(s)$ of the sensing path. The sensed output voltage $v_s(t)$ is then A/D converted into a digital sequence $v_s^{\diamond}[k]$ having a sampling period T and a resolution determined by the A/D converter. In general, the sampled version of the sensed signal is denoted as $v_s[k]$,

$$v_s[k] \triangleq v_s(t_k), \tag{2.1}$$

where t_k are the sampling instants. Furthermore, whenever necessary, the digitized version of $v_s[k]$ is denoted as $v_s^{\diamond}[k]$.

The most common choice for the sampling period T is

$$\boxed{T = T_s}, \tag{2.2}$$

that is, the sampling process is *synchronized* with respect to the switching process of the power converter, and the sampling frequency coincides with the converter switching frequency. Consequently, the sampling instant always occurs at a fixed position within a switching period.

The control error $e[k]$ between the internal digital reference V_{ref} and the acquired signal v_s is then processed by a digital compensator, which calculates the digital control command $u[k]$ on a switching cycle basis. Following the calculation of $u[k]$, a DPWM produces a modulated output $c(t)$ by latching $u[k]$ every T_s

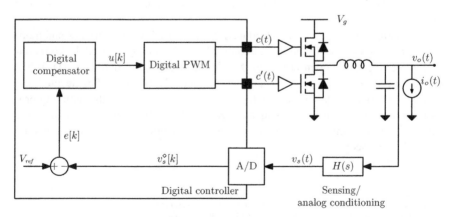

Figure 2.1　Digital voltage-mode control of a synchronous Buck converter.

seconds at the beginning of each modulation cycle and generating the output pulse $c(t)$ the duration of which is proportional to $u[k]$.

2.2 A/D CONVERSION

Quantization performed by the A/D converter on the analog sensed signal $v_s(t)$ is commonly referred to as *input quantization*. It is assumed that offset voltage and integral and differential nonlinearities can be neglected and that the effective number of bits of the converter coincides with its hardware resolution $n_{A/D}$.

Figure 2.2 shows a block diagram of the A/D converter and the waveforms illustrating its operation. The process of A/D conversion can be modeled as *sampling* of the analog input, followed by *amplitude quantization* of the acquired sample. Furthermore, regardless of the A/D converter architecture, a *conversion delay* $t_{A/D}$ is always present, representing the time between when the analog signal is sampled and the digital output signal is updated.

2.2.1 Sampling Rate

In a PWM switching converter, the state variables contain a baseband spectrum, consisting of dc and low-frequency components, as well as high-frequency spectral components centered around the switching frequency f_s and its harmonics, and associated with the switching activity of the converter. The prefiltering $H(s)$ by the analog sensing and conditioning circuitry usually attenuates the high-frequency content of the signal, but it does not qualitatively change the composition of the signal spectrum. Regardless of the analog prefiltering, the sensed signals *always* include switching content at high frequencies. As a result, *a certain degree of spectral aliasing always occurs, regardless of the sampling rate*. As anticipated by (2.2), the most common choice is to have the sampling rate equal to the switching frequency, a choice that is motivated by the following considerations.

Observe first that in steady state, the converter signals are strictly periodic with period T_s. In order to preserve the periodicity in the digital domain and not introduce

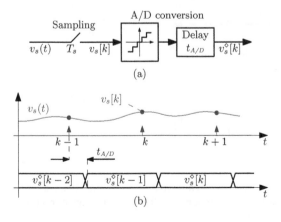

Figure 2.2   (a) Block diagram of the A/D conversion process and (b) associated waveforms.

sampling artifacts, the sampling rate is usually constrained to be a *multiple* of f_s. Such *synchronization* between pulse width modulation and sampling activity is of essential importance in digitally controlled power electronics.

Secondly, consider what happens when sampling a converter signal faster than the converter switching rate. Figure 2.3 illustrates the sampling of an analog signal $v_s(t)$ at $f_{sampling} = 3f_s$. Sampling causes a periodic superposition of the original spectrum, resulting in a postsampling spectrum which is periodic in frequency with period $3f_s$. The *Nyquist rate* f_N, which is equal to one-half of the sampling rate, represents the frequency above which spectral periodicity begins and is therefore the largest frequency "visible" to the digital system.

In the case shown, sampling involves superposition of the original spectrum around both dc and f_s. The dc components of $v_s(t)$ and $v_s[k]$ are different due to aliasing in the low-frequency range. Furthermore, spectral content around f_s in the sampled signal is the result of spectral aliasing of the switching ripple originally present in $v_s(t)$. Whenever $v_s(t)$ is sampled at a rate strictly higher than the

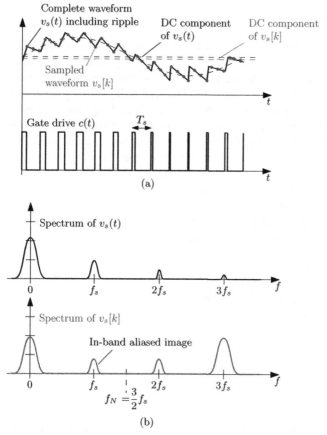

Figure 2.3   Sampling frequency higher than the switching frequency: (a) time-domain waveforms of the sensed signal before and after sampling and (b) qualitative signal spectra.

switching frequency, spectral aliasing of the switching ripple produces a frequency image below the Nyquist rate, which is therefore visible to the digital system. Such residual high-frequency component in the digital domain requires a filtering action to be performed in the digital controller. The situation is similar to the necessity of attenuating switching harmonics in analog compensators by purposely adding high-frequency poles, as discussed in Section 1.5.1. More details about sampling at a rate higher than the switching frequency, which is also referred to as *multisampling*, can be found in [61, 62].

Next, consider the case $f_{sampling} = f_s$, as shown in Fig. 2.4. Spectral aliasing now occurs only around dc, with no images of the original spectrum being created below the Nyquist rate as a result of sampling. It is instructive to consider a steady-state condition. In steady state, spectral aliasing manifests itself as $v_s(t)$ being converted into a *constant* signal, as illustrated in Fig. 2.5. In other words, switching harmonics present in the analog signal $v_s(t)$ only alias at dc. Dc spectral aliasing induced by

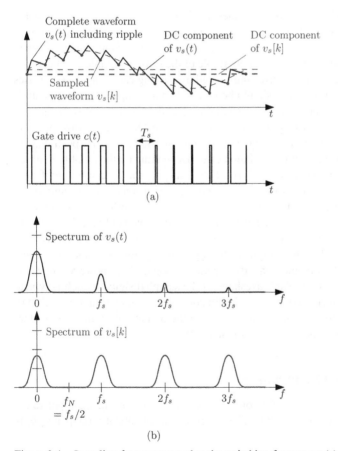

Figure 2.4   Sampling frequency equal to the switching frequency: (a) time-domain waveforms of the sensed signal before and after sampling and (b) qualitative signal spectra.

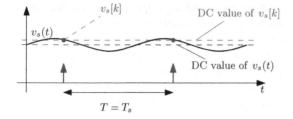

Figure 2.5   Example of dc aliasing in voltage sampling.

(2.2) has therefore the inherent advantage of removing switching harmonics from the feedback signal. It follows that high-frequency poles, which are purposely introduced in analog compensators in order to attenuate the switching noise, become unnecessary in digital control, as long as the sampling rate is equal to the switching rate. It is important to stress that synchronization between sampling and PWM operation is the common underlying assumption in all the above-mentioned considerations.

A drawback of (2.2) is, on the other hand, a corruption of the dc regulated value of the sampled variable $v_s[k]$, which in general differs, as anticipated, from the dc component of $v_s(t)$. Figure 2.5 shows an example of such aliasing effect as seen in the time domain. Whenever the switching ripple of the sensed variable $v_s(t)$ is much smaller than its dc value, the dc error introduced by aliasing can be neglected. Cases exist, however, where provisions should be undertaken to mitigate dc spectral aliasing. An example can be found in digital average current control in dc–dc or dc–ac converters, where it is desired to regulate the average value of the current waveform and where the current switching ripple may not be small. In such cases, dc spectral aliasing may induce an unacceptable regulation error. Depending on the amplitude of the switching ripple, the aliasing-induced error between $i_L[k]$ and I_L can be significant.

One commonly adopted solution to the problem of average current sampling is to employ a symmetrical PWM modulation and allocate the sampling instants either at the peak or at the valley point of the PWM carrier, as illustrated in Fig. 2.6. Thanks to the triangular shape of the current and the choice of the sampling instants, the average current can be sampled with negligible aliasing error regardless of the duty cycle and therefore regardless of the converter operating point. Note, however, that this approach relies on the triangular waveshape of the sampled signal, which is typical for inductor current in converters operating in continuous conduction mode (CCM). This is not the case, for instance, when sampling the inductor current in converters operating in discontinuous conduction mode (DCM).

2.2.2 Amplitude Quantization

The internal quantizer of the A/D converter operates over an analog input range $[0, V_{FS}]$ with a resolution of $n_{A/D}$ bits. The corresponding quantization step in $v_s[k]$ is

$$q_{v_s}^{(A/D)} \triangleq \frac{V_{FS}}{2^{n_{A/D}}}. \tag{2.3}$$

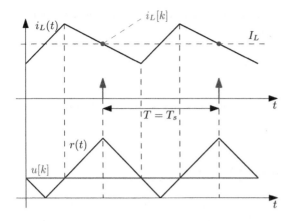

Figure 2.6   Commonly adopted PWM and sampling strategy for digital average current-mode control.

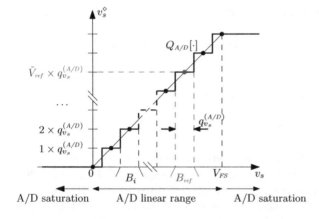

A/D saturation   A/D linear range   A/D saturation

Figure 2.7   A/D converter quantization characteristic.

The quantized range, often referred to as the *linear range* of the A/D converter, is thus subdivided into $2^{n_{A/D}}$ voltage intervals B_i, $i = 0 \dots 2^{n_{A/D}} - 1$, which are commonly called *bins*. Each bin of the A/D converter is $q_{v_s}^{(A/D)}$ volts wide. The quantization characteristic $Q_{A/D}[\cdot]$ of the A/D converter is illustrated in Fig. 2.7. In the linear range, the digital output $v_s^\circ[k]$ of the A/D converter is given by

$$v_s^\circ[k] \triangleq Q_{A/D}[v_s[k]] = q_{v_s}^{(A/D)} \tilde{v}_s[k], \qquad \tilde{v}_s[k] \in \mathbb{Z}, \tag{2.4}$$

where $\tilde{v}_s[k]$ is an integer that uniquely identifies the bin where $v_s[k]$ resides. The digital output of the A/D converter is a binary-coded version of $\tilde{v}_s[k]$. The specific code used may be 2's complement, offset binary, and so on. Outside the linear range, that is, when the analog input either exceeds V_{FS} or lies below 0, the A/D converter is *saturated* and its digital output typically remains at the highest or the lowest value.

The digital setpoint V_{ref} of the controller is expressed with the same resolution $n_{A/D}$ and the full-scale range V_{FS} of the A/D converter,

$$V_{ref} = q_{v_s}^{(A/D)} \tilde{V}_{ref}, \qquad \tilde{V}_{ref} \in \mathbb{Z}, \tag{2.5}$$

where the integer $\tilde{V}_{ref}$ identifies the *zero-error bin*, that is, the quantization interval B_{ref} to which $v_s[k]$ should be regulated.

2.3   THE DIGITAL COMPENSATOR

The compensator is a clocked digital logic that calculates the control command $u[k]$ based on the regulation error $e[k]$. As shown in Fig. 2.8, the compensator clocked on a sampling cycle basis produces the control command after a certain *computational delay* t_{calc}. For the moment, an idealized, instantaneous execution of the control calculations is assumed, while Section 2.5 presents a more general discussion about control delays.

A *linear* and *time-invariant* compensation law is described by a difference equation

$$u[k] = -a_1 u[k-1] - a_2 u[k-2] - \ldots - a_N u[k-N]$$

$$+ b_0 e[k] + b_1 e[k-1] + \ldots + b_M e[k-M]. \tag{2.6}$$

An important case of (2.6) is that of proportional-integral-derivative (PID) digital compensators, which represent the discrete-time counterpart of the popular PID analog regulators [127]. One can derive the general discrete-time PID equations by discretization of the differential equations of a continuous-time PID. Consider the block diagram in Fig. 2.9, which illustrates a continuous-time PID compensator in its so-called parallel realization. The parallel PID structure is often referred to as *noninteracting* because the proportional, integral, and derivative gains can be

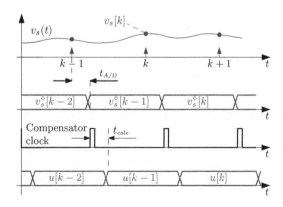

Figure 2.8   Timing diagram of the compensator operation.

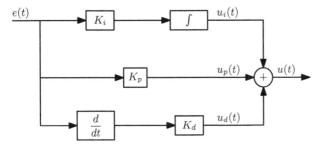

Figure 2.9   Block diagram of a continuous-time PID compensator in the parallel (noninteracting) form.

adjusted independently. Equations describing the analog PID compensator are

$$u_p(t) = K_p e(t),$$

$$u_i(t) = K_i \int e(\tau) \, d\tau,$$

$$u_d(t) = K_d \frac{de(t)}{dt},$$

$$u(t) = u_p(t) + u_i(t) + u_d(t),$$

(2.7)

or, in the Laplace domain,

$$G_{PID}(s) \triangleq \frac{\hat{u}(s)}{\hat{e}(s)} = \underbrace{K_p}_{\text{Proportional Term}} + \underbrace{\frac{K_i}{s}}_{\text{Integral Term}} + \underbrace{sK_d}_{\text{Derivative Term}} . \qquad (2.8)$$

For simplicity, (2.8) does not include any high-frequency poles or a filtering action usually embedded into the derivative term of an analog PID compensator.

Several discretization techniques are available to convert a continuous-time system into a discrete-time system having similar frequency response characteristics. One of the simplest is the *backward Euler* method, or *backward rectangular rule*, which is based on a rectangular approximation of the continuous-time integral operator. As shown in Fig. 2.10, the integral over one sampling step is approximated as

$$\int_{(k-1)T_s}^{kT_s} x(\tau) \, d\tau \approx T_s x(kT_s). \qquad (2.9)$$

In the Z-transform domain, the backward Euler discretization can be interpreted as an s-to-z mapping rule defined by the substitution

$$s \rightarrow \frac{1 - z^{-1}}{T_s}. \qquad (2.10)$$

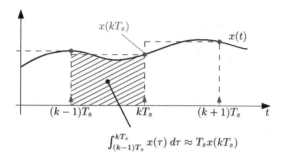

$$\int_{(k-1)T_s}^{kT_s} x(\tau)\, d\tau \approx T_s x(kT_s)$$

Figure 2.10   Backward Euler approximation.

Application of the backward Euler rule to (2.7) yields the equations of a discrete-time PID in additive form

$$u_p[k] = K_p e[k],$$

$$u_i[k] = u_i[k-1] + K_i T_s e[k],$$

$$u_d[k] = \frac{K_d}{T_s}(e[k] - e[k-1]),$$ (2.11)

$$u[k] = u_p[k] + u_i[k] + u_d[k].$$

In the z-domain,

$$G_{PID}(z) \triangleq \frac{\hat{u}(z)}{\hat{e}(z)} = \underbrace{K_p}_{Proportional\ Term} + \underbrace{\frac{T_s K_i}{1 - z^{-1}}}_{Integral\ Term} + \underbrace{\frac{K_d}{T_s}\left(1 - z^{-1}\right)}_{Derivative\ Term},$$ (2.12)

which is illustrated in Fig. 2.11 in the form of an equivalent block diagram.

Another frequently used discretization rule is known as *Tustin approach* or *trapezoidal rule*. This approach adopts a more accurate discrete representation of

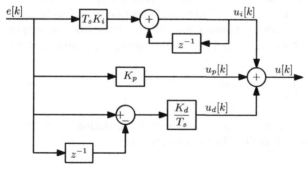

Figure 2.11   Block diagram of a discrete-time PID compensator obtained via backward Euler discretization of a continuous-time PID.

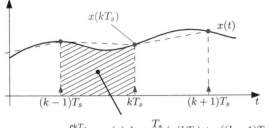

$$\int_{(k-1)T_s}^{kT_s} x(\tau)\, d\tau \approx \frac{T_s}{2}\left(x(kT_s) + x((k-1)T_s)\right)$$

Figure 2.12 Tustin approximation.

the integration operation by approximating the integral over the sampling step with the trapezoidal approximation of the integrand,

$$\int_{(k-1)T_s}^{kT_s} x(\tau)\, d\tau \approx \frac{T_s}{2}\left(x(kT_s) + x((k-1)T_s)\right), \qquad (2.13)$$

as illustrated in Fig. 2.12. With such approximation, the digital PID equations in the time domain become

$$u_p[k] = K_p e[k],$$

$$u_i[k] = \frac{K_i T_s}{2}\left(e[k] + e[k-1]\right) + u_i[k-1],$$

$$u_d[k] = \frac{2K_d}{T_s}\left(e[k] - e[k-1]\right) - u_d[k-1], \qquad (2.14)$$

$$u[k] = u_p[k] + u_i[k] + u_d[k].$$

The Tustin discretization can be interpreted as a s-to-z mapping rule defined by the substitution

$$s \to \frac{2}{T_s}\frac{1 - z^{-1}}{1 + z^{-1}}. \qquad (2.15)$$

Application of (2.13) to (2.7) yields

$$G_{PID}(z) \triangleq \frac{\hat{u}(z)}{\hat{e}(z)} = \underbrace{K_p}_{\text{Proportional Term}} + \underbrace{K_i \frac{T_s}{2}\frac{1 + z^{-1}}{1 - z^{-1}}}_{\text{Integral Term}} + \underbrace{K_d \frac{2}{T_s}\frac{1 - z^{-1}}{1 + z^{-1}}}_{\text{Derivative Term}}. \qquad (2.16)$$

A block diagram corresponding to this realization is illustrated in Fig. 2.13.

Figure 2.14 compares the Bode plots of the continuous-time PID compensator designed in Chapter 1 with the Bode plots of its Tustin discretization. The frequency responses of the two compensators essentially coincide throughout most of the frequency range and start departing from one another only in close vicinity of the Nyquist rate of the system, which is $f_s/2 = 500$ kHz in this example.

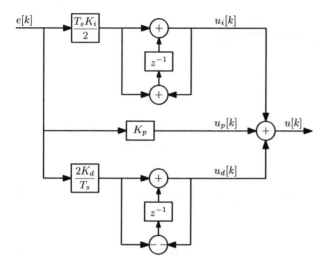

Figure 2.13   Block diagram of a discrete-time PID compensator obtained via Tustin discretization of a continuous-time PID.

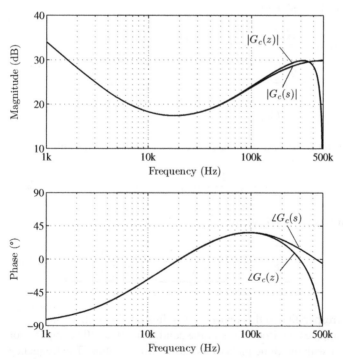

Figure 2.14   Comparison of Bode plots of the analog PID designed in Section 1.5.1 and its Tustin discretization.

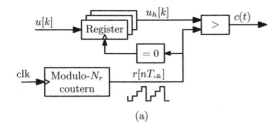

(a)

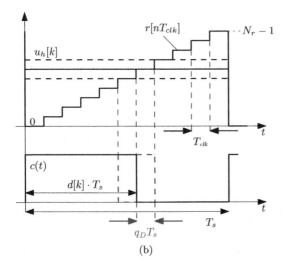

Figure 2.15   (a) Block diagram
of a counter-based digital pulse
width modulator and (b)
associated waveforms.

(b)

2.4   DIGITAL PULSE WIDTH MODULATION

The role of the DPWM is to acquire the control command $u[k]$ on a sampling cycle
basis and generate a train of pulses having duty cycle $d[k]$ proportional to $u[k]$.

Figure 2.15 shows a block diagram of the standard *counter-based* DPWM
architecture. This architecture is a digital realization of a conventional analog PWM.
A modulo-N_r counter, which is clocked at frequency $f_{clk} = 1/T_{clk}$, generates the
PWM carrier $r[nT_{clk}]$, which resembles a periodic staircase with period T_s. Every
switching period consists of N_r clock intervals,

$$T_s = N_r T_{clk}. \tag{2.17}$$

At the beginning of every switching cycle, for example, when the counter output
is zero, the input control command $u[k]$ is latched into the DPWM input register and
held constant throughout the rest of the interval. During this time, the content of the
register (signal $u_h[k]$ in Fig. 2.15) is compared with $r[nT_{clk}]$ on a clock cycle basis,

and a pulse is generated with duty cycle

$$d[k] = \frac{u[k]}{N_r}.$$ (2.18)

Regardless of the implementation, a DPWM can only produce a *finite* set of duty cycles. By *time resolution* of a DPWM, one commonly refers to the smallest variation of the pulse duration the modulator is capable of. This important quantity is denoted as

$$\Delta t_{DPWM}.$$ (2.19)

The modulator *duty cycle resolution* q_D is, accordingly, the smallest duty cycle variation

$$q_D \triangleq \frac{\Delta t_{DPWM}}{T_s}.$$ (2.20)

Both Δt_{DPWM} and q_D quantify the coarseness of the DPWM as an actuator.

When N_r is a power of two, the resolution of a modulator can be conveniently expressed by an *equivalent number of bits* n_{DPWM}, which is related to Δt_{DPWM} and q_D by

$$n_{DPWM} \triangleq \log_2\left(\frac{T_s}{\Delta t_{DPWM}}\right) = \log_2\left(\frac{1}{q_D}\right).$$ (2.21)

The value of and the expression for Δt_{DPWM} depends on the specific DPWM architecture. For the counter-based DPWM, the time resolution clearly coincides with the clock period,

$$\Delta t_{DPWM} = T_{clk},$$ (2.22)

and consequently

$$q_D = \frac{T_{clk}}{T_s} = \frac{1}{N_r}.$$ (2.23)

For instance, a 12-bit DPWM – $N_r = 4096$ – operating at $f_s = 200$ kHz has a time resolution $\Delta t_{DPWM} = 5\ \mu s/4096 \approx 1.2$ ns and a duty cycle resolution $q_D \approx 0.024\%$.

Quantization of the duty cycle due to the DPWM granularity can be represented by a suitable quantization characteristic $Q_D[\cdot]$,

$$D^\diamond = Q_D[D],$$ (2.24)

where the desired duty cycle $D = U/N_r$ depends on the control command U and the carrier amplitude N_r and where $D^\diamond$ is the duty cycle generated by the DPWM.

The duty cycle quantization characteristic $Q_D[\cdot]$ depends on the DPWM implementation. An example is illustrated in Fig. 2.16, which reports a possible quantization characteristic of a 3-bit DPWM—an intentionally low resolution

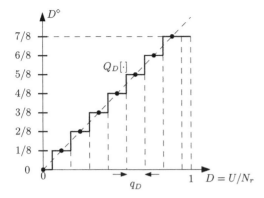

Figure 2.16 Duty cycle quantization operated by the DPWM.

chosen for ease of representation. This DPWM is capable of representing eight duty cycle levels ranging from 0 to 7/8 with a resolution $q_D = 1/8 = 12.5\%$.

2.5 LOOP DELAYS

The most important difference between the analog and digital control loops is the presence of delays of various nature, which affect frequency response and need to be modeled and accounted for in the design of digital control loops.

2.5.1 Control Delays

As seen in the previous sections, regardless of the specific A/D and compensator structures, the digital control command $u[k]$ is available at the compensator output after a certain delay with respect to the sampling instant. By *control delay* t_{cntrl}, one refers to the time interval separating the sampling event from the instant when the digital modulator latches the corresponding control command $u[k]$ calculated by the digital compensator.

The origin and the value of t_{cntrl} depend on the digital controller implementation. A distinction can be made between *hardware-based* digital controllers and *software-based* controllers:

- In *hardware-based controllers* implemented as custom-integrated circuits or in field programmable gate arrays (FPGAs), the digital compensator is custom designed into a hardwired combination of digital arithmetic blocks and registers. Fast, application-optimized A/D converters are employed.
  In such implementations, t_{cntrl} is determined by the A/D conversion time followed by the propagation time of the combinational portion of the compensator network, and it can easily be reduced to a small fraction of the switching period even when the switching frequency is in the order of hundreds of kHz or higher. A control timing diagram typical for hardware-based controllers is shown in Fig. 2.17. The fast calculation time offered by the hardware implementation allows the control command to be delivered to the digital pulse width modulator

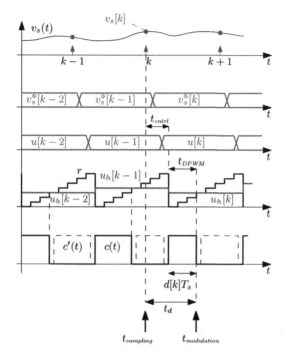

Figure 2.17 Typical timing diagram of a hardware-based controller.

within the same switching interval, so that

$$d[k] = \frac{u[k]}{N_r} \qquad (2.25)$$

as depicted in Fig. 2.17.

- In a *software-based controller*, the control law is software-programmed and executed by the central processing unit (CPU) of a microcontroller or a digital signal processor (DSP). In this case, the control delay t_{cntrl} corresponds to the A/D conversion time followed by the execution time of the control algorithm by the CPU, which can be comparable to the switching period. If the delay equals a full modulation cycle,

$$d[k] = \frac{u[k-1]}{N_r}, \qquad (2.26)$$

which corresponds to the timing diagram of Fig. 2.18.

2.5.2 Modulation Delay

In addition to the control delay, the *modulation delay* plays an important role in digital control modeling and design. By modulation delay, one refers to a *small-signal* delay introduced by the digital modulator as a consequence of the sampled nature of the modulating signal.

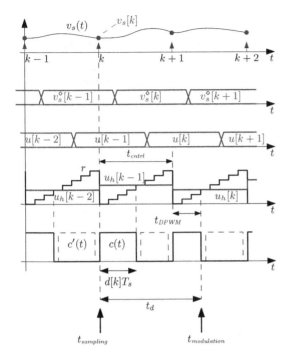

Figure 2.18 Typical timing diagram of a software-based controller.

To better understand the origin of the modulation delay, consider Fig. 2.17 and 2.18: both diagrams differ substantially from Fig. 1.7 in that a nonzero time delay exists between the latching of the control command $u[k]$ operated by the modulator and the generation of the corresponding modulated edge. Such delay, denoted as t_{DPWM} in both Figs. 2.17 and 2.18, is inherently associated with the sampled nature of the modulating signal. The modulators that sample the modulating signal before comparing it to the carrier are referred to as *uniformly sampled pulse width modulators* (USPWMs). Digital pulse width modulators discussed in Section 2.4 all belong to this class. On the other hand, naturally sampled pulse width modulators commonly employed in analog control and discussed in Chapter 1, which perform a *continuous* comparison between $u(t)$ and $r(t)$, do not exhibit modulation delay [126].

The small-signal frequency response of an USPWM can be written in the form

$$G_{PWM}(j\omega) = A_{PWM}(j\omega)e^{-j\omega t_{DPWM}}, \qquad (2.27)$$

where A_{PWM} is a real positive function of the angular frequency ω, whereas t_{DPWM} represents the equivalent small-signal delay introduced by the modulator. Table 2.1 reports the main waveforms and expressions of $G_{PWM}(j\omega)$ for the trailing-edge, leading-edge, and symmetrical (triangle-wave) modulations. In general, A_{PWM} is equal or close to $1/N_r$ over a wide range of frequencies, so that in most cases one can assume $A_{PWM} \approx 1/N_r$ with negligible loss in modeling accuracy. On the other hand, t_{DPWM} depends, in general, on the type of modulation and the steady-state

TABLE 2.1 Small-Signal Frequency Responses of Uniformly Sampled Pulse Width Modulators

Modulation Type	Frequency Response
Trailing-edge 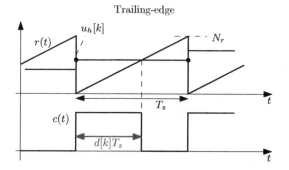	$G_{PWM,TE}(j\omega) = \dfrac{e^{-j\omega D T_s}}{N_r}$
Leading-edge 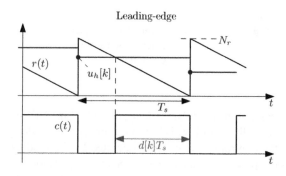	$G_{PWM,LE}(j\omega) = \dfrac{e^{-j\omega(1-D)T_s}}{N_r}$
Symmetrical 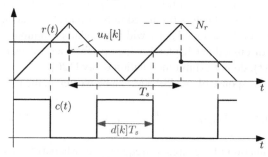	$G_{PWM,Sym}(j\omega) =$ $= \dfrac{\cos(\omega D T_s/2)}{N_r} e^{-j\omega \frac{T_s}{2}}$ $\approx \dfrac{e^{-j\omega \frac{T_s}{2}}}{N_r}$

duty cycle D, with the notable exception of the symmetrical modulation, in which $t_{DPWM} = T_s/2$ independent of the duty cycle.

With a detailed derivation given in Appendix C, the result (2.27) can be arrived at through the following argument. Referring to Fig. 2.19, consider a trailing-edge USPWM driven by a constant command U superimposed to a Kronecker pulse $\hat{u}[k] = \hat{u}[0]\delta[k]$ of amplitude $\hat{u}[0]$ applied at $k = 0$. The modulator output $c(t)$ consists of a steady-state PWM signal $c_s(t)$ having duty cycle $D = U/N_r$, plus a perturbation $\hat{c}(t)$ due to $\hat{u}[k]$. Such perturbation increases the duty cycle to $D + \hat{d}$ during the modulation period immediately following the application of the Kronecker pulse, with $\hat{d} = \hat{u}[0]/N_r$. Laplace transform of the output perturbation is

$$\hat{c}(s) \triangleq \int_0^{+\infty} \hat{c}(\tau)e^{-s\tau}d\tau$$

$$= \int_{DT_s}^{(D+\hat{d})T_s} e^{-s\tau}d\tau$$

$$= \frac{1 - e^{-s\hat{d}T_s}}{s}e^{-sDT_s}. \tag{2.28}$$

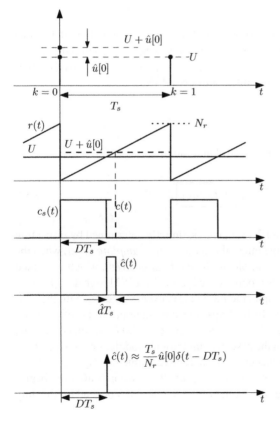

Figure 2.19   Dirac approximation of the uniformly sampled pulse width modulator output.

Given the small-signal assumption for $\hat{u}[0]$, a first-order approximation in $\hat{d}$ yields

$$\hat{c}(s) \approx e^{-sDT_s}\hat{d}T_s = e^{-sDT_s}\frac{T_s}{N_r}\hat{u}[0], \tag{2.29}$$

which in time domain corresponds to

$$\hat{c}(t) \approx \frac{T_s}{N_r}\hat{u}[0]\delta(t - DT_s). \tag{2.30}$$

In other words, the USPWM response to the input Kronecker pulse is a Dirac delta function delayed by DT_s and scaled by T_s/N_r. From (2.30), one recognizes that $t_{DPWM} = DT_s$.

For an arbitrary small-signal input perturbation $\hat{u}[k]$, the USPWM output is a modulated train of delayed Dirac delta pulses,

$$\hat{u}(t) \approx \frac{T_s}{N_r}\sum_{k=-\infty}^{+\infty}\hat{u}[k]\delta(t - kT_s - DT_s). \tag{2.31}$$

The modulation delay can be a significant fraction of the total loop delay. In a hardware-based controller implementation, for instance, where the control delay t_{cntrl} can be reduced to a small fraction of the switching interval, t_{DPWM} easily becomes the major limiting factor to the achievable control bandwidth.

2.5.3 Total Loop Delay

The total loop delay is the sum of the control and modulation delays,

$$\boxed{t_d \triangleq t_{cntrl} + t_{DPWM}}. \tag{2.32}$$

As suggested in both Figs. 2.17 and 2.18, t_d can be directly determined by inspection from the timing diagram as the time interval separating the *sampling event*, when the A/D converter acquires the analog sample, to the *modulation event*, which is defined as the steady-state position of the modulated edge of the PWM signal $c_s(t)$. In a trailing-edge modulation, for example, the modulation event coincides with the generation of the PWM falling edge. In leading-edge modulation, on the other hand, the modulated edge is the rising edge, which occurs $(1 - D)T_s$ seconds after the falling edge. Care must be taken in the case of symmetrical modulation, in which both edges of $c_s(t)$ are modulated. In this case, from Table 2.1, one has $t_{DPWM} = T_s/2$. In this case, the "equivalent" modulation event occurs $T_s/2$ seconds after the beginning of the switching interval, and the modulator delay does not depend on the duty cycle.

2.6 USE OF AVERAGED MODELS IN DIGITAL CONTROL DESIGN

As summarized in Chapter 1, averaged small-signal modeling has been historically developed in close relationship with analog control design. Given the body of knowledge available, along with successful experiences with the averaging modeling approach in power electronics practice, it is tempting to apply the same models in the design of digital control loops. The objectives of this section are to discuss approximations involved and highlight fundamental limitations of this approach, based on the following observations:

- As discussed in the previous section, unlike their naturally sampled counterparts commonly found in analog controllers, uniformly sampled pulse width modulators introduce additional small-signal dynamics in the control loop in the form of a modulation delay. Embedding the delay effects into the averaged modeling framework is possible but only in an approximate manner.

- Neglecting high-frequency converter dynamics, as is implied by averaging, does not account for aliasing effects which occur as a result of the sampling operation and which can manifest themselves even in the *low-frequency* range.

Using averaged continuous-time models, both of the above-mentioned shortcomings can be addressed in an approximate manner only in cases when it is possible to assume that the digital controller samples, at least approximately, the average value of the sensed signal, so that aliasing effects can be neglected. Otherwise, results can be unpredictable. The digital control design must then be based on the discrete-time modeling approach introduced in Chapter 3.

In this section, the limitations of the averaged small-signal modeling as related to digital control loop design are first exemplified by a design example. Possible modifications to the averaged small-signal models are then discussed, along with the conditions under which these approximate modeling and control design approaches can be applied.

2.6.1 Limitations of Averaged Modeling

To exemplify the modeling inaccuracies of the averaged small-signal modeling framework when it comes to digital control dynamics, consider digital voltage-mode control of the Buck converter depicted in Fig. 2.1, assuming the same parameters as in the analog control example introduced in Section 1.5.1 and summarized in Table 1.2. Assume that the digital controller employs a trailing-edge DPWM and that the control timing diagram resembles the one shown in Fig. 2.17. For better comparison with analog control, also assume that $N_r = 1$ and neglect amplitude quantization. The control delay t_{cntrl} is considered a parameter, and various scenarios are examined.

Suppose that modeling of the power converter is carried out according to the averaging method, as described in Section 1.3. On the basis of the continuous-time averaged model of the converter, a continuous-time compensator can be designed exactly as outlined in Section 1.5.1, with no modifications. The end result of this procedure is the s-domain expression for the compensator transfer function $G_c(s)$, reproduced here,

$$G_c(s) = \underbrace{\left(1 + \frac{\omega_l}{s}\right)}_{PI} \cdot \underbrace{G_{PD0} \frac{1 + \dfrac{s}{\omega_z}}{1 + \dfrac{s}{\omega_p}}}_{PD} \cdot \underbrace{\frac{1}{1 + \dfrac{s}{\omega_{p_2}}}}_{HF\ Pole}, \tag{2.33}$$

with

$$\omega_l = 2\pi \cdot (8\ \text{kHz}),$$

$$\omega_z = 2\pi \cdot (40\ \text{kHz}),$$

$$\omega_p = 2\pi \cdot (250\ \text{kHz}), \tag{2.34}$$

$$\omega_{p_2} = 2\pi \cdot (1\ \text{MHz}),$$

$$G_{PD0} = 6.2.$$

The transfer function of the digital compensator can be obtained, as discussed in Section 2.3, by discretization of $G_c(s)$. A comparison between the frequency responses of $G_c(s)$ and its Tustin discretization has already been shown in Fig. 2.14. The discretization step completes the system-level design of the compensation.

Summarizing the above-mentioned procedure:

- The converter is modeled according to the standard averaging modeling approach. The result of this step is a Laplace-domain expression for the uncompensated loop gain $T_u(s)$.

- A continuous-time compensator is designed based on conventional analog techniques, yielding an s-domain expression for the compensator transfer function $G_c(s)$.

- The transfer function of the digital compensator $G_c(z)$ is obtained by Tustin discretization of $G_c(s)$.

Figure 2.20 compares the steady-state behavior of the analog-controlled and the digital-controlled systems for $t_{cntrl} = 400$ ns. While the analog controller regulates the dc value V_o of $v_o(t)$ to $V_{ref} = 1.8$ V, it can be observed that the digital controller regulates the *sampled* waveform. This is an example of the dc aliasing effect induced by the sampling operation and anticipated in Section 2.2.1.

Consider next the closed-loop responses to various kinds of disturbances. Figure 2.21 reports responses for a 1.79 to 1.8 V step variation of V_{ref}. A clear difference between the two cases can be appreciated in terms of both overshoot and settling time. The same considerations hold for the 2.5 to 5 A step load response depicted in Fig. 2.22. As the discrete-time and continuous-time compensators have essentially the same frequency response—see Fig. 2.14—the observed differences in the closed-loop behavior of the system must reside in the *plant*, that is, in

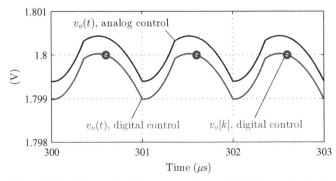

Figure 2.20   Comparison between analog and digital control: steady-state output voltage waveform.

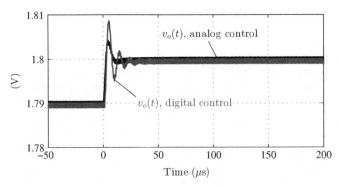

Figure 2.21   Comparison between analog and digital control: 1.79 V→1.8 V step-reference responses.

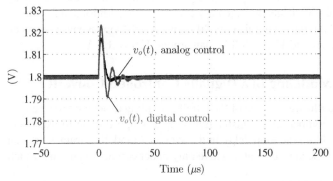

Figure 2.22   Comparison between analog and digital control: 2.5 A→5 A step-load responses.

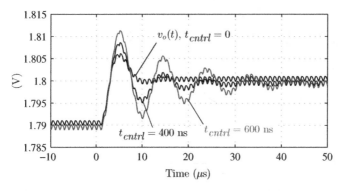

Figure 2.23   Effect of the control delay: 1.79 V→1.8 V step reference responses.

the modulator/power converter system. As anticipated, the unmodeled dynamics concerns the total loop delay t_d present in the digital control loop. Such delay contributes an additional phase lag, which ultimately results in a much reduced phase margin, which in turn corresponds to more significant overshoots and ringing in time-domain responses.

It is of interest to further examine the effects of $t_d = t_{cntrl} + t_{DPWM}$ on the designed control system. One expects both t_{cntrl} and t_{DPWM} to affect the system phase margin. Figure 2.23 compares three distinct step reference responses corresponding to $t_{cntrl} = 0$, 400, and 600 ns respectively. The effect of increasing t_{cntrl}, and therefore t_d, clearly goes in the direction of decreasing the system stability margin. Another comparison, even more effective in highlighting the differences between the analog and digital control dynamics, is illustrated in Fig. 2.24 and refers to a difference in t_{DPWM} rather than in t_{cntrl}. Suppose the control setpoint is $V_{ref} = 3.3$ V rather than 1.8 V. The steady-state duty cycle changes then from $D = 0.36$ to $D = 0.66$. In the framework of averaged modeling, such duty cycle change would *not* affect the small-signal dynamics of the Buck converter. However, in the digital controller, a change in the steady-state operating point produces a different modulation delay t_{DPWM}. From Table 2.1, one has $t_{DPWM} = DT_s$ and therefore the modulation delay increases proportionally with steady-state duty cycle D. The additional delay translates, as seen in Fig. 2.24, into a significantly different closed-loop response—a much less damped response around $V_{ref} = 3.3$ V than around $V_{ref} = 1.8$ V.

2.6.2   Averaged Modeling of a Digitally Controlled Converter

The previous example illustrates some of the pitfalls of failing to account for the loop delay in shaping the compensator transfer function. It is appropriate to ask, then, whether simply embedding the loop delay in the averaged modeling framework can mitigate the problem. The answer is affirmative only under certain conditions, which are now discussed.

Following the considerations developed in Chapter 1, the *averaged* small-signal dynamics of a converter is always determined by the *baseband* portion $\hat{d}(t)$ of the PWM signal $c(t)$, regardless of how the PWM signal is generated—by an analog or

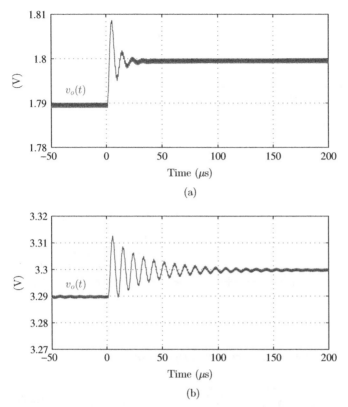

Figure 2.24   Effect of the operating point on the modulation delay: comparison between (a) the 1.79 V→1.8 V and (b) the 3.29 V→3.3 V step-reference responses.

a digital controller. As discussed in Section 2.5, the digital nature of the control is responsible for the total loop delay t_d, which separates the sampling event from the modulated edge of $c(t)$, that is, from the actual application of the modulated signal to the converter. This motivates the definition of an *effective uncompensated loop gain* where the averaged small-signal dynamics is combined with the response corresponding to the total loop delay,

$$\boxed{T_u^\dagger(s) \triangleq T_u(s)e^{-st_d}}. \tag{2.35}$$

To take the delay into account, one may then proceed to design a compensator based on $T_u^\dagger(s)$, instead of $T_u(s)$. The definition (2.35), however, is only useful for design purposes to the extent that it models the *actual* dynamics seen by the digital compensator. Consider Fig. 2.25. The digital compensator responds to the sampled signal $v_s[k]$. In general, the baseband content of $v_s[k]$ is *not* equal to the baseband component of $v_s(t)$ because of aliasing effects. In other words, the digital compensator responds to dynamics not given by $T_u^\dagger(s)$ but rather represented by the discrete-time model developed in Chapter 3. Only in cases when the sampled signal

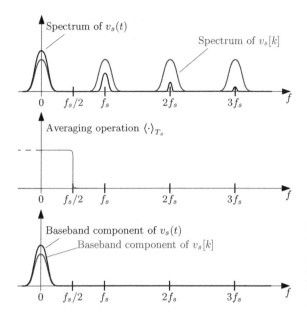

Figure 2.25 Qualitative spectra of (top) $v_s(t)$ and $v_s[k]$, (middle) the averaging operation, and (bottom) the baseband portions of $v_s(t)$ and $v_s[k]$.

can be assumed to closely follow the *averaged, sampled waveform*,

$$\boxed{v_s[k] \approx \overline{v}_s(t_k)}, \tag{2.36}$$

the baseband contents of $v_s(t)$ and $v_s[k]$ are approximately the same, and the digital compensator behaves *as if* it were sampling and controlling the system represented by the averaged converter model. Condition (2.36) is here referred to as *small-aliasing approximation*. In general, validity of the small-aliasing approximation should be examined case by case. Two scenarios of practical importance in which the small-aliasing approximation is applicable are as follows:

1. Control of a well-filtered converter state variable, such as the output voltage. In this case, the sampled dynamics is dominated by the dynamics of the averaged waveform, with negligible ripple contributions.

2. The controller is intentionally sampling a waveform at an instant *when the ripple component is zero*. As an example, this is the case when an inductor current is sampled using symmetrical modulation to achieve average current control, as shown in Fig. 2.6. In this case, the peak-to-peak amplitude of the ripple component can be large, but the small-aliasing approximation is still valid.

In those situations where (2.36) holds, the design procedure can follow the flow described in the previous subsection, with the only exception that $T_u^\dagger(s)$ is taken as the uncompensated loop gain rather than $T_u(s)$.

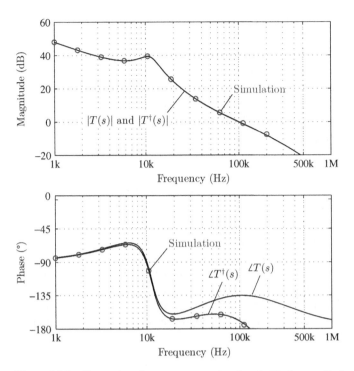

Figure 2.26   Comparison between conventional and effective small-signal loop dynamics.

In the previously discussed example, which is a case when the well-filtered voltage is sampled and regulated, the effective loop dynamics can be described by

$$T^\dagger(s) \triangleq G_c(s)T_u^\dagger(s) = G_c(s)T_u(s)e^{-st_d}. \tag{2.37}$$

Figure 2.26 compares the effective loop gain $T^\dagger(s)$ with the loop gain $T(s)$ predicted by conventional averaged small-signal modeling and already shown in Fig. 1.17. Furthermore, both $T^\dagger(s)$ and $T(s)$ are compared with the system loop gain obtained by numerical simulation. The effect of the additional phase lag due to the loop delay t_d ($t_d = 600$ ns in this example) is well described by the effective loop gain $T^\dagger(s)$.

A more in-depth discussion regarding the level of approximation introduced by $T_u^\dagger(s)$ will be undertaken in Chapter 3, with the aid of numerical examples.

The above-mentioned discussion suggests that the averaging approach is not always the most suitable tool for modeling and designing digitally controlled power converters. Chapter 3 presents a different modeling approach, known in the literature as *discrete-time modeling*, which correctly accounts for all of the effects highlighted in this section, and is well suited for the design of digital control loops in all cases.

2.7   SUMMARY OF KEY POINTS

- Main elements of a digital control loop of a power converter are the A/D converter, the digital compensator, and the digital pulse width modulator. The controller sampling frequency is synchronized to the converter switching frequency. Having the sampling frequency equal to the converter switching frequency results in less severe and more controllable aliasing effects.

- Each of the digital controller building blocks introduces a delay in the control loop, which must be accounted for at the design stage.

- Contrary to naturally sampled PWM typical for analog controllers, uniformly sampled PWM contributes to the overall loop delay. Such modulation-related delay depends on the PWM carrier and, in general, on the converter operating point.

- Conventional continuous-time averaged small-signal models can be augmented to account for the loop delay only in an approximated manner. Under the small-aliasing approximation, that is, when the sampled signal closely follows the averaged value of the sensed waveform, the total loop delay can be described by attributing a transport delay to the uncompensated s-domain loop gain. In general, however, this modeling approach is only an approximation.

CHAPTER 3

DISCRETE-TIME MODELING

Recent advances in digital control of high-frequency switched-mode power converters have renewed the interest in *discrete-time modeling techniques* as more natural and more accurate representations of the converter dynamics when controlled digitally [36, 125, 128–133]. Discrete-time modeling aims at describing the dynamics of the *sampled* converter waveforms, with no averaging step involved in the process.

It is interesting, from a historical standpoint, that sampled-data modeling had been recognized as an inherently accurate way of describing the linearized switched-mode power converter dynamics even before averaged small-signal modeling gained popularity [125, 128]. More recent works have formulated the discrete-time modeling in the context of digital control, embedding in the theory the important aspect of modulation delay introduced in Section 2.5.2 [36, 131], which assumes critical importance when designing for fast, wide-bandwidth control loops.

This chapter introduces the process of discrete-time modeling of dc–dc converters not only from a theoretical perspective but also from a practical standpoint. Section 3.1 lays the theoretical foundations of the discrete-time modeling technique. Section 3.2 provides a number of modeling examples and discusses the most interesting features of the discrete-time dynamics, some of which cannot be predicted well using averaged modeling techniques. As opposed to continuous-time averaged models, discrete-time analysis and the resulting mathematical models do not easily lead to design-oriented equivalent circuit models. In various examples, systematic Matlab® coding is instead presented as a fast and effective way to address derivations of discrete-time models and facilitate rapid evaluation of a converter's z-domain control-to-output transfer functions.

Section 3.3 deals with an important class of converters where the switching event does not alter their *topology*—a notable example being the Buck converter. For these converters, the exact discrete-time model can be obtained via *discretization* of the averaged small-signal model. Existence of such direct link between s-domain and z-domain models greatly simplifies the math and justifies a dedicated discussion.

Digital Control of High-Frequency Switched-Mode Power Converters, First Edition.
Luca Corradini, Dragan Maksimović, Paolo Mattavelli, and Regan Zane.

3.1   DISCRETE-TIME SMALL-SIGNAL MODELING

Consider the converter operation as alternating between *two* topological states S_0 and S_1, each described by a *linear* set of state-space equations

$$\frac{dx}{dt} = A_c x(t) + B_c v(t),$$
$$y(t) = C_c x(t) + E_c v(t), \tag{3.1}$$

where $c \in \{0, 1\}$ is the PWM signal that denotes the topological state, while x, v, and y represent the state, input, and output vectors respectively. With a focus on the control-to-output dynamics, the input vector is assumed constant from here on,

$$v(t) = V. \tag{3.2}$$

The basic idea behind discrete-time small-signal modeling is quite simple—perhaps simpler than the concept behind averaged modeling. The approach can be thought of as a three-step process:

1. One first expresses the *sampled* converter state vector x at instant $k + 1$ in terms of the state vector, input vector v, and control input u at instant k. In doing so, a *nonlinear* state equation is usually obtained,

$$x[k + 1] = f(x[k], V, u[k]), \tag{3.3}$$

where f is a nonlinear vector function.

2. Secondly, the converter operating point Q is determined by solving the above-mentioned equation for a constant sampled state vector $x[k + 1] = x[k] = X$ and a constant control input $u[k] = U$,

$$X = f(X, V, U). \tag{3.4}$$

The converter operating point Q is then defined as

$$Q \triangleq (X, V, U). \tag{3.5}$$

3. The nonlinear state equation is successively *perturbed and linearized* in the neighborhood of Q, yielding the small-signal state-space description of the sampled dynamics,

$$\boxed{\begin{aligned} \hat{x}[k + 1] &= \Phi \hat{x}[k] + \gamma \hat{u}[k], \\ \hat{y}[k] &= \delta \hat{x}[k], \end{aligned}} \tag{3.6}$$

where

$$\hat{x}[k] \triangleq x[k] - X,$$

$$\hat{u}[k] \triangleq u[k] - U, \qquad (3.7)$$

$$\hat{y}[k] \triangleq y[k] - Y$$

are the small-signal components of the sampled state vector, control command, and output vector, relative to their dc components X, U, and Y, respectively. Matrices Φ and γ represent the small-signal state matrix and the small-signal control-to-state matrix, respectively,[1]

$$\Phi \triangleq \left. \frac{\partial f}{\partial x} \right|_Q,$$
$$\qquad (3.8)$$
$$\gamma \triangleq \left. \frac{\partial f}{\partial u} \right|_Q.$$

Matrix δ, on the other hand, represents the converter output matrix *pertaining to the subtopology in which sampling occurs*,

$$\delta \triangleq \begin{cases} C_1 & \text{If sampling occurs during subtopology } S_1, \\ C_0 & \text{If sampling occurs during subtopology } S_0. \end{cases} \qquad (3.9)$$

Notice also that matrices E_c do not enter into the control-to-output small-signal model.

In the z-domain, (3.6) becomes

$$\boxed{\begin{aligned} \hat{x}(z) &= (z I - \Phi)^{-1} \gamma \hat{u}(z), \\ \hat{y}(z) &= \delta \hat{x}(z), \end{aligned}} \qquad (3.10)$$

---

[1]Recall that by derivative of a vector function f with respect to its arguments x, one refers to the matrix whose columns are the derivatives of f with respect to the successive components of x,

$$\frac{\partial f}{\partial x} \triangleq \begin{bmatrix} \dfrac{\partial f_1}{\partial x_1} & \dfrac{\partial f_1}{\partial x_2} & \cdots & \dfrac{\partial f_1}{\partial x_n} \\ \dfrac{\partial f_2}{\partial x_1} & \dfrac{\partial f_2}{\partial x_2} & \cdots & \dfrac{\partial f_2}{\partial x_n} \\ \cdots & \cdots & \cdots & \cdots \\ \dfrac{\partial f_n}{\partial x_1} & \dfrac{\partial f_n}{\partial x_2} & \cdots & \dfrac{\partial f_n}{\partial x_n} \end{bmatrix},$$

also referred to as the *Jacobian* of f with respect to x.

and in the z-domain the small-signal control-to-output transfer matrix $W(z)$ is

$$\boxed{W(z) \triangleq \frac{\hat{y}(z)}{\hat{u}(z)} = \delta \left(z I - \Phi\right)^{-1} \gamma}. \tag{3.11}$$

For example, defining the output vector in terms of the inductor current and the output voltage as $y = [i_L \ v_o]^T$, $W(z)$ becomes

$$W(z) = \begin{bmatrix} G_{iu}(z) \triangleq \dfrac{\hat{i}_L(z)}{\hat{u}(z)} \\[2mm] G_{vu}(z) \triangleq \dfrac{\hat{v}_o(z)}{\hat{u}(z)} \end{bmatrix}. \tag{3.12}$$

Transfer matrix $W(z)$ represents the end result of the discrete-time modeling approach and the starting point for the compensator design.

Before engaging in a study of the discrete-time modeling procedure in its general formulation, it is instructive to focus on a simple, yet meaningful example. The goal is to unravel the basic ideas of the approach without being distracted by mathematical complexity.

3.1.1 A Preliminary Example: A Switched Inductor

Consider the circuit illustrated in Fig. 3.1(a) with time-domain waveforms sketched in Fig. 3.1(b).

A resistor R and an inductor L connected in series are driven by two pulse width modulated voltage sources V_A and V_B. During the kth switching cycle, voltage across RL equals

$$v_{RL}(t) = c(t)V_A - c'(t)V_B = \begin{cases} V_A, & 0 < t < d[k]T_s, \\ -V_B, & d[k]T_s < t < T_s. \end{cases} \tag{3.13}$$

Assume that a trailing-edge modulator is employed, so that the falling edge is the only modulated edge of the PWM signal $c(t)$. Referring to Fig. 3.1(b), the converter periodic *steady-state* trajectory $i_{L,s}(t)$ is a result of the steady-state PWM command $c_s(t)$ with duty cycle $D = U/N_r$. The small-signal component $\hat{i}_L(t)$ results from the PWM perturbation $\hat{c}(t) = c(t) - c_s(t)$, where $\hat{c}(t)$ consists of narrow pulses the width of which is determined by the cycle-by-cycle control perturbation $\hat{u}[k] = u[k] - U = (d[k] - D)N_r$. Assume, as suggested in Fig. 3.1(b), that current $i_L(t)$ flowing through the inductor is sampled some time before the beginning of every switching interval. The objective is to describe the small-signal dynamics of the sampled current $i_L[k]$.

Start by writing the *continuous-time* differential equation governing the behavior of the circuit,

$$\frac{di_L}{dt} = -\frac{R}{L}i_L(t) + \frac{1}{L}v_{RL}(t). \tag{3.14}$$

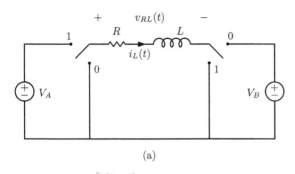

(a)

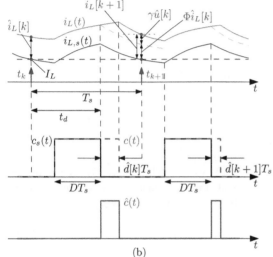

(b)

Figure 3.1   (a) The switched inductor and (b) waveforms associated with its discrete-time model derivation.

This simple first-order differential equation, in which $v_{RL}(t)$ acts as the input, can be directly solved for a given initial condition $i_L(t_0)$,

$$i_L(t) = e^{-\omega_p t} i_L(t_0) + \int_{t_0}^{t} v_{RL}(\tau) g(t - \tau) \, d\tau, \qquad (3.15)$$

where

$$\omega_p \triangleq \frac{R}{L} \qquad (3.16)$$

and where

$$g(t) = \frac{1}{L} e^{-\omega_p t}, \qquad t \geq 0 \qquad (3.17)$$

is the system impulse response.

From (3.15) and expressing current at instant t_{k+1} in terms of the current at instant t_k and the input acting between t_k and t_{k+1}, one has

$$i_L[k+1] = e^{-\omega_p T_s} i_L[k] + \int_{t_k}^{t_k + T_s} v_{RL}(\tau) g(T_s + t_k - \tau) \, d\tau = f(i_L[k], \boldsymbol{V}, u[k]).$$

$$(3.18)$$

This is the nonlinear discrete-time equation (3.3) for the RL network under consideration, with $\boldsymbol{V} = [V_A, V_B]^T$ defined as the input vector for the system. The equation describes the sampled inductor current dynamics in the most general case and without approximations.

To determine the steady-state operating point relative to the sampled current, let

$$i_L[k+1] = i_L[k] \triangleq I_L \tag{3.19}$$

and assume a constant control input U, that is, a constant duty cycle D,

$$v_{RL,s}(t) = c_s(t)V_A - c_s'(t)V_B = \begin{cases} V_A, & 0 \le t \le DT_s, \\ -V_B, & DT_s \le t \le T_s. \end{cases} \tag{3.20}$$

From (3.18), one has

$$I_L \left(1 - e^{-\omega_p T_s}\right) = \int_{t_k}^{t_k+T_s} v_{RL,s}(\tau)g(T_s + t_k - \tau)\, d\tau$$

$$\Rightarrow I_L = \frac{\displaystyle\int_{t_k}^{t_k+T_s} v_{RL,s}(\tau)g(T_s + t_k - \tau)\, d\tau}{1 - e^{-\omega_p T_s}}. \tag{3.21}$$

It is of no interest, in this context, to explicitly work out the integral. What is important to stress is that I_L represents the steady-state inductor current *at the sampling instant*. It is *not* the average value of the steady-state inductor current waveform $i_{L,s}(t)$,

$$I_L \triangleq i_{L,s}(t_k) \neq \bar{i}_{L,s}(t). \tag{3.22}$$

The last step in the derivation of the discrete-time model prescribes to perturb and linearize (3.18) around the steady-state operating point $\boldsymbol{Q} = (I_L, \boldsymbol{V}, U)$. To this end, make the substitutions

$$i_L[k] \to I_L + \hat{i}_L[k]$$

$$u[k] \to U + \hat{u}[k] \quad (\text{i.e., } d[k] \to D + \hat{d}[k]) \tag{3.23}$$

$$c(t) \to c_s(t) + \hat{c}(t).$$

The input perturbation $\hat{v}_{RL}(t)$ acting on the RL series is therefore

$$\hat{v}_{RL}(t) = \hat{c}(t)V_A - \hat{c}'(t)V_B = (V_A + V_B)\, \hat{c}(t), \tag{3.24}$$

and the state equation describing the perturbation dynamics becomes

$$\hat{i}_L[k+1] = e^{-\omega_p T_s}\hat{i}_L[k] + \int_0^{T_s} \hat{v}_{RL}(\tau)g(T_s - \tau)\, d\tau$$

$$= e^{-\omega_p T_s}\hat{i}_L[k] + (V_A + V_B)\int_{t_d}^{t_d+\hat{d}[k]T_s} g(T_s - \tau)\, d\tau. \tag{3.25}$$

This equation is still nonlinear in $\hat{d}$. The desired small-signal model of the sampled inductor current is derived via first-order Taylor expansion in $\hat{d}$,

$$
\hat{i}_L[k+1] \approx e^{-\omega_p T_s} \hat{i}_L[k] + (V_A + V_B)\, g(T_s - t_d)\hat{d}[k]T_s
$$

$$
= e^{-\omega_p T_s} \hat{i}_L[k] + \frac{V_A + V_B}{L} e^{-\omega_p(T_s - t_d)}\hat{d}[k]T_s, \tag{3.26}
$$

or

$$
\hat{i}_L[k+1] = \Phi \hat{i}_L[k] + \gamma \hat{u}[k], \tag{3.27}
$$

with

$$
\Phi \triangleq e^{-\omega_p T_s},
$$

$$
\gamma \triangleq \frac{T_s}{N_r} e^{-\omega_p(T_s - t_d)} \frac{V_A + V_B}{L}. \tag{3.28}
$$

In the z-domain, the small-signal discrete-time model of the switched RL series network is

$$
G_{iu}(z) \triangleq \frac{\hat{i}_L(z)}{\hat{u}(z)} = \frac{1}{z - \Phi}\gamma
$$

$$
= \frac{T_s}{N_r} \frac{V_A + V_B}{L} \frac{e^{-\omega_p(T_s - t_d)}z^{-1}}{1 - e^{-\omega_p T_s}z^{-1}}. \tag{3.29}
$$

A special case of interest arises when $R \to 0$. In this case,

$$
G_{iu}(z) \xrightarrow{R \to 0} \frac{T_s}{N_r} \frac{V_A + V_B}{L} \frac{z^{-1}}{1 - z^{-1}}. \tag{3.30}
$$

Not surprisingly, the discrete-time model of an ideal switched inductor is a discrete-time integrator. The presence of a *one-step delay* implies that a change in the control signal $u[k]$ at sampling instant k only manifests itself at $k + 1$.

3.1.2   The General Case

The above-mentioned derivation highlights the essential aspects of the discrete-time modeling process. One is now in a position to extend the procedure to the most general case.

Assume again that a trailing-edge modulator is employed. Extending the definitions given in the switched inductor examples, let $x_s(t)$ be the converter *steady-state* trajectory due to the steady-state PWM command $c_s(t)$ and let $\hat{x}(t)$ be the small-signal component that results from the PWM perturbation $\hat{c}(t) = c(t) - c_s(t)$. Further assume, without loss of generality, that sampling occurs during topological state S_0 and denote with t_d the time interval between the sampling event and the modulated edge of $c_s(t)$. It should be noted that t_d represents the total loop delay introduced in Section 2.5.

Within a given topological state S_c, the general solution of (3.1) from initial state $x(t_0)$ is

$$x(t) = e^{A_c(t-t_0)}x(t_0) + \int_{t_0}^{t} e^{A_c(t-\tau)}B_cV \, d\tau. \tag{3.31}$$

Step 1 of the discrete-time small-signal modeling derivation simply makes repeated use of (3.31) throughout the switching interval in order to express $x[k+1]$ in terms of $x[k]$, V, and $u[k]$. In doing so, one must remember that the state vector is a continuous function of time and therefore does not exhibit discontinuities at the switching instants.

Once an expression for f is available, steps 2 and 3 of the small-signal modeling procedure follow without conceptual difficulties. For the scenario depicted earlier and illustrated in Fig. 3.2, the result is

$$\boxed{\boldsymbol{\Phi} \triangleq \left.\frac{\partial f}{\partial x}\right|_Q = e^{A_0(T_s-t_d)}e^{A_1DT_s}e^{A_0(t_d-DT_s)}} \tag{3.32}$$

for the state matrix and is

$$\boxed{\gamma \triangleq \left.\frac{\partial f}{\partial u}\right|_Q = \frac{T_s}{N_r}e^{A_0(T_s-t_d)}\left[\left(A_1X_\downarrow + B_1V\right) - \left(A_0X_\downarrow + B_0V\right)\right]} \tag{3.33}$$

for the input matrix. Vector $X_\downarrow \triangleq x_s(t_\downarrow)$ is the value of $x_s(t)$ at the modulated edge of $c_s(t)$, as indicated in Fig. 3.2. $X_\downarrow$ is derived from the converter operating point

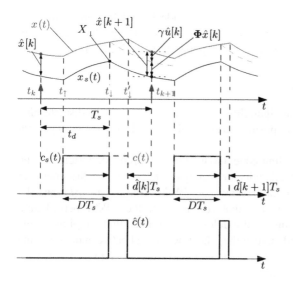

Figure 3.2   Waveforms illustrating the discrete-time model derivation.

and equals

$$X_{\downarrow} = \left(I - e^{A_1 DT_s} e^{A_0 D'T_s}\right)^{-1} \cdot$$

$$\left[-e^{A_1 DT_s} A_0^{-1} \left(I - e^{A_0 D'T_s}\right) B_0 - A_1^{-1} \left(I - e^{A_1 DT_s}\right) B_1\right] V. \quad (3.34)$$

In spite of the complex expressions, both (3.32) and (3.33) can be given simple physical interpretations. First, note that expression of Φ can be thought of as the repeated application of the exponential operator $e^{A_c t}$ to the initial state $\hat{x}[k]$,

$$\hat{x}[k+1] = \underbrace{e^{A_0(T_s - t_d)}}_{\text{From } t_\downarrow = t'_\downarrow \text{ to } t_{k+1}} \cdot \underbrace{e^{A_1 DT_s}}_{\text{From } t_\uparrow \text{ to } t_\downarrow = t'_\downarrow} \cdot \underbrace{e^{A_0(t_d - DT_s)}}_{\text{From } t_k \text{ to } t_\uparrow} \cdot \underbrace{\hat{x}[k]}_{\text{State at } t_k} ,$$

$$\underbrace{}_{\text{State at } t_{k+1}}$$

(3.35)

when $\hat{u} = 0$. Therefore, expression of Φ simply reflects how the state perturbation propagates through the various topological states of the converter, throughout the switching interval, in the absence of a control perturbation.

As for matrix γ, its expression is nothing but the linear propagation of the state small-signal component during the infinitesimally short interval $\hat{d}[k]T_s$, followed by the propagation of such perturbation from $t'_\downarrow$ to t_{k+1},

$$\hat{x}[k+1] = \underbrace{e^{A_0(T_s - t_d)}}_{\text{From } t'_\downarrow \text{ to } t_{k+1}} \underbrace{\left[\left(A_1 X_\downarrow + B_1 V\right) - \left(A_0 X_\downarrow + B_0 V\right)\right]}_{\text{Slope of } \hat{x}(t) \text{ at } t_\downarrow} \underbrace{\frac{\hat{u}[k]}{N_r} T_s}_{\text{On}-\text{time perturbation}} ,$$

$$\underbrace{}_{\text{State at } t_{k+1}} \qquad\qquad\qquad\qquad \underbrace{}_{\text{From } t_\downarrow \text{ to } t'_\downarrow}$$

(3.36)

when $\hat{x}[k] = 0$. Equation (3.6), finally, is the linear superposition of the two propagations, which holds in the small-signal limit.

3.1.3   Discrete-Time Models for Basic Types of PWM Modulation

Tables 3.1–3.3 summarize expressions for small-signal matrices Φ, γ, and δ for the three most common types of pulse width modulation, obtained by straightforward applications of the procedure outlined in Section 3.1. The expressions refer to the sampling scheme sketched in each corresponding figure. Notice that for the trailing-edge and symmetrical modulation, sampling during subtopology S_0 is assumed, whereas the leading-edge modulation case assumes sampling during subtopology S_1. Other cases, not considered here, can be nonetheless derived without difficulty.

Note that all the matrices of the small-signal model are directly computed from matrices A_c and B_c. The calculation can be entirely carried out numerically, as exemplified in Section 3.2.

3.2 DISCRETE-TIME MODELING EXAMPLES

The discrete-time modeling procedure discussed in the previous section is now illustrated by several practical examples. Before doing so, however, it is appropriate to briefly address the problem of computational complexity. As one can imagine by looking at equations reported in Tables 3.1–3.3, *analytical* derivation of the discrete-time model of even the simplest converter represents a considerable computational effort. One key step involved in the modeling step is the calculation of the matrix exponential

$$e^{\boldsymbol{A}_c T}.$$

Hand evaluation of $e^{\boldsymbol{A}_c T}$ is a rather lengthy process regardless of the approach one undertakes. One method for evaluating the matrix exponential is to first perform a basis transformation and express $\boldsymbol{A}_c$ in a form for which the exponential is easily computable. This can be either a diagonal form or, more generally, the *canonical*

TABLE 3.1 Discrete-Time Small-Signal Model — Trailing-Edge Modulation

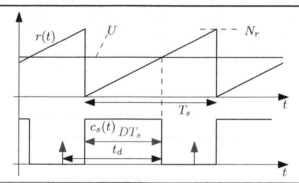

Discrete-Time Model

$$\boldsymbol{\Phi} = e^{\boldsymbol{A}_0(T_s - t_d)} e^{\boldsymbol{A}_1 DT_s} e^{\boldsymbol{A}_0(t_d - DT_s)},$$

$$\boldsymbol{\gamma} = \frac{T_s}{N_r} e^{\boldsymbol{A}_0(T_s - t_d)} \boldsymbol{F}_\downarrow,$$

$$\boldsymbol{\delta} = \boldsymbol{C}_0.$$

$$\boldsymbol{F}_\downarrow \triangleq \left(\boldsymbol{A}_1 \boldsymbol{X}_\downarrow + \boldsymbol{B}_1 \boldsymbol{V} \right) - \left(\boldsymbol{A}_0 \boldsymbol{X}_\downarrow + \boldsymbol{B}_0 \boldsymbol{V} \right),$$

$$\boldsymbol{X}_\downarrow = \left(\boldsymbol{I} - e^{\boldsymbol{A}_1 DT_s} e^{\boldsymbol{A}_0 D'T_s} \right)^{-1} \cdot$$

$$\left[-e^{\boldsymbol{A}_1 DT_s} \boldsymbol{A}_0^{-1} \left(\boldsymbol{I} - e^{\boldsymbol{A}_0 D'T_s} \right) \boldsymbol{B}_0 - \boldsymbol{A}_1^{-1} \left(\boldsymbol{I} - e^{\boldsymbol{A}_1 DT_s} \right) \boldsymbol{B}_1 \right] \boldsymbol{V}.$$

TABLE 3.2   Discrete-Time Small-Signal Model—Leading-Edge Modulation

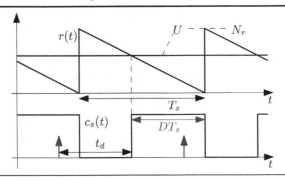

Discrete-Time Model

$$\Phi = e^{A_1(t_d - D'T_s)} e^{A_0 D'T_s} e^{A_1(T_s - t_d)},$$

$$\gamma = \frac{T_s}{N_r} e^{A_0(T_s - t_d)} F_\uparrow,$$

$$\delta = C_1.$$

$$F_\uparrow \triangleq \left(A_1 X_\uparrow + B_1 V\right) - \left(A_0 X_\uparrow + B_0 V\right),$$

$$X_\uparrow = e^{A_0 D'T_s} X_\downarrow - A_0^{-1}\left(I - e^{A_0 D'T_s}\right) B_0 V.$$

($X_\downarrow$ as defined for the trailing-edge modulation)

TABLE 3.3   Discrete-Time Small-Signal Model—Symmetrical Modulation

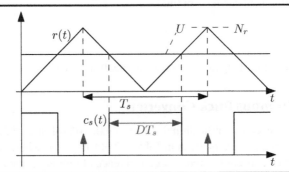

Discrete-Time Model

$$\Phi = e^{A_0 \frac{T_s}{2}(1-D)} e^{A_1 DT_s} e^{A_0 \frac{T_s}{2}(1-D)},$$

$$\gamma = \frac{T_s}{2N_r} e^{A_0 \frac{T_s}{2}(1-D)} \left[F_\downarrow + e^{A_1 DT_s} F_\uparrow\right],$$

$$\delta = C_0.$$

($F_\uparrow$ and $F_\downarrow$ as defined for the trailing-edge and leading-edge modulations)

Jordan form [134]. Once the matrix exponential is evaluated, one can go back to the original basis via the inverse basis transformation.

Matrix inversion is another operation involved in the discrete-time modeling formulas. For matrices greater than 2×2, hand calculation of the inverse quickly becomes a computationally intensive task.

If one seeks a closed-form expression for the transfer functions of interest, software tools for scientific computing such as Mathematica® can certainly bypass bothersome and error-prone hand calculations. Closed-form expressions, however, are only useful to the extent that they provide clear insights into the effects of various circuit parameters. In the z-domain formulation of the converter transfer functions, however, the parameters appear as part of exponential or trigonometric function arguments, making the interpretation difficult.

One possible, yet approximate, approach that retains a closed-form expression of the converter transfer functions consists in the expansion [36]

$$e^{A_c T} \approx I + A_c T, \tag{3.37}$$

which removes exponential or trigonometric functions from the analysis and is based on the assumption that the converter natural time constants are much longer than a switching period.

The approach followed in this section—and in the remaining chapters of the book—is a *numerical* one. The transfer functions of interest are evaluated numerically via systematic Matlab® scripting. A few simple Matlab® commands provide a fast, reliable, and overall better way to deal with discrete-time modeling in the design practice.

Unless otherwise stated, every design example discussed in this and in Chapter 4 assumes a normalized carrier amplitude

$$N_r = 1. \tag{3.38}$$

With this choice, control signal $u[k]$ varies in the $[0, 1]$ range.

3.2.1 Synchronous Buck Converter

As the first example, consider the digital voltage-mode control case study discussed in Chapter 2, with the parameters given in Table 1.2 of Chapter 1, where analog control of the same converter is considered. For convenience of reference, the block diagram of the digitally controlled Buck converter is reported again in Fig. 3.3.

It is assumed that a hardware-based controller is implemented, including the A/D converter, the digital compensator, and the trailing-edge digital pulse width modulator. The timing diagram of such controller is illustrated in Fig. 3.4. Sampling of the output voltage occurs $t_{cntrl} = 400$ ns before the DPWM latches the control command at the beginning of every modulation cycle.

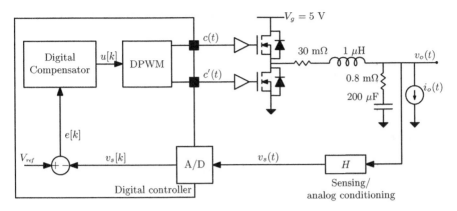

Figure 3.3   Synchronous Buck converter example: digital voltage-mode control.

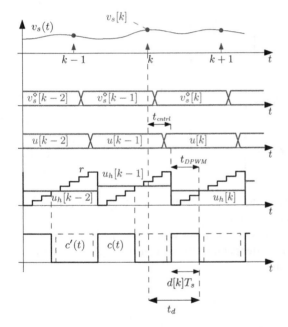

Figure 3.4   Synchronous Buck converter example: timing diagram.

A state-space representation of the converter is first required,

$$
\begin{bmatrix} \dfrac{di_L}{dt} \\[2mm] \dfrac{dv_C}{dt} \end{bmatrix} = \boldsymbol{A}_c \begin{bmatrix} i_L(t) \\[1mm] v_C(t) \end{bmatrix} + \boldsymbol{B}_c \begin{bmatrix} V_g \\[1mm] I_o \end{bmatrix},
$$

$$
\begin{bmatrix} i_L(t) \\[1mm] v_o(t) \end{bmatrix} = \boldsymbol{C}_c \begin{bmatrix} i_L(t) \\[1mm] v_C(t) \end{bmatrix} + \boldsymbol{E}_c \begin{bmatrix} V_g \\[1mm] I_o \end{bmatrix}.
$$

(3.39)

For the Buck converter under study, the state matrices A_1 and A_0 are the same,

$$A_1 = A_0 = \begin{bmatrix} -\dfrac{r_C + r_L}{L} & -\dfrac{1}{L} \\[2mm] \dfrac{1}{C} & 0 \end{bmatrix} \triangleq A, \tag{3.40}$$

whereas the matrices B_1 and B_0 are given by

$$B_1 = \begin{bmatrix} \dfrac{1}{L} & \dfrac{r_C}{L} \\[2mm] 0 & -\dfrac{1}{C} \end{bmatrix},$$

$$B_0 = \begin{bmatrix} 0 & \dfrac{r_C}{L} \\[2mm] 0 & -\dfrac{1}{C} \end{bmatrix}. \tag{3.41}$$

The output matrices C_c are as follows:

$$C_1 = C_0 = \begin{bmatrix} 1 & 0 \\[1mm] r_C & 1 \end{bmatrix} \triangleq C. \tag{3.42}$$

The matrices E_c do not need to be evaluated as, as anticipated, they do not enter into the control-to-output small-signal modeling.

The converter steady-state inputs are the input voltage V_g, the load current I_o, and the duty cycle D. Given the converter parameters and considering converter operation at full load $I_o = 5$ A, one has

$$V_g = 5 \text{ V},$$

$$I_o = 5 \text{ A},$$

$$D = \frac{V_o}{V_g} = \frac{1.8 \text{ V}}{5 \text{ V}} = 0.36. \tag{3.43}$$

From Table 3.1, the total loop delay t_d is

$$t_d = t_{cntrl} + t_{DPWM}$$

$$= t_{cntrl} + DT_s$$

$$= 400 \text{ ns} + 360 \text{ ns} = 760 \text{ ns}. \tag{3.44}$$

Inset 3.1 discusses the few Matlab® instructions needed to derive the discrete-time small-signal model of the converter based on the results from Section 3.1.

Inset 3.1 – Matlab® Example: Synchronous Buck Converter

This example shows how to develop the small-signal control-to-output transfer functions $G_{vu}(z)$ and $G_{iu}(z)$ using Matlab®. First, define the converter state-space matrices:

```
A1   =   [   -(rC+rL)/L   -1/L;    1/C 0    ];
A0   =   A1;
b1   =   [ 1/L    rC/L;    0    -1/C    ];
b0   =   [   0    rC/L;    0    -1/C];
c1   =   [1 0; rC 1];
c0   =   c1;
```

Next, evaluate $X_\downarrow$ according to Table 3.1. The expm method can be used to numerically evaluate the matrix exponential:

```
A1i  =   A1^-1;
A0i  =   A0^-1;
Xdown =    ((eye(2)-expm(A1*D*Ts)*expm(A0*Dprime*Ts))^-1)*...
              (-expm(A1*D*Ts)*A0i*(eye(2)-expm(A0*Dprime*Ts))*b0+...
                 -A1i*(eye(2)-expm(A1*D*Ts))*b1)*[Vg;Io]
```

The above Matlab® code assumes that matrices A_1 and A_0 are invertible, which is always the case as long as parasitic components (r_C, r_L) are included in the modeling step.

Then, construct small-signal model matrices Φ, γ, and δ as per Table 3.1:

```
Phi      =   expm(A0*(Ts-td))*expm(A1*D*Ts)*expm(A0*(td-D*Ts));
gamma    =   expm(A0*(Ts-td))*((A1-A0)*Xdown + (b1-b0)*[Vg;Io])*Ts;
delta    =   c0;
```

Finally, extract the control-to-output transfer functions $G_{vu}(z)$ and $G_{iu}(z)$ by converting the state-space representation (Φ, γ, δ) into Matlab® transfer function objects. First, build a state-space object using method ss and then the transfer function objects using tf:

```
sys    =   ss(Phi,gamma,delta(1,:),0,Ts);
Giuz   =   tf(sys);
sys    =   ss(Phi,gamma,delta(2,:),0,Ts);
Gvuz   =   tf(sys);
```

The magnitude and phase Bode plots of $G_{vu}(z)$ are shown in Fig. 3.5. For comparison, Bode plots of $G_{vu}(s) = G_{vd}(s)$ (recall that $N_r = 1$) obtained via conventional averaged small-signal modeling are also shown. The magnitude responses predicted by the two models are indeed quite similar, with a small departure visible

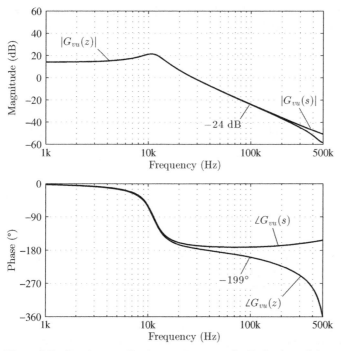

Figure 3.5 Synchronous Buck converter example: Bode plots of the control-to-output voltage transfer functions $G_{vu}(z)$ based on the discrete-time modeling approach and $G_{vu}(s)$ based on the standard averaged modeling approach.

in the proximity of the Nyquist rate $f_s/2 = 500$ kHz. On the other hand, the comparison between the *phase* responses of the two models reveals the additional phase lag caused by the loop delay t_d, correctly modeled by the discrete-time transfer function but absent in the s-domain averaged model.

Because of the very small ripple in the converter output voltage, the small-aliasing approximation discussed in Section 2.6.2 is indeed well satisfied in this example,

$$v_o[k] \approx \overline{v}_o(t_k). \tag{3.45}$$

On the basis of the considerations in Section 2.6.2, it is expected that the effective plant transfer function

$$G_{vu}^{\dagger}(s) \triangleq G_{vu}(s)e^{-st_d} \tag{3.46}$$

provides a valid correction to $G_{vu}(s)$ based on the averaged small-signal model. In other words, it is expected that $G_{vu}^{\dagger}(s)$ should closely approximate $G_{vu}(z)$. The Bode plots of $G_{vu}(z)$ and $G_{vu}^{\dagger}(s)$ compared in Fig. 3.6 confirm that this is indeed the case. A small departure between the z-domain model and the effective s-domain model is nonetheless visible close to the Nyquist rate, due to the fact that aliasing effects are not *entirely* absent.

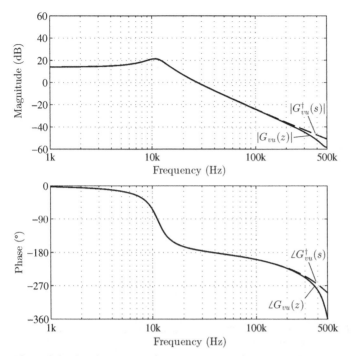

Figure 3.6 Synchronous Buck converter example: a comparison between the discrete-time model $G_{vu}(z)$ and the effective s-domain model $G_{vu}^{\dagger}(s) = G_{vu}(s)e^{-st_d}$ obtained by including the additional phase lag due to the total loop delay t_d.

Consider next the control-to-inductor current response, which would be of interest in digital current-mode control of the synchronous Buck converter. Suppose that a current-mode digital controller is implemented so that the Buck converter inductor current is sampled at the end of the turn-off interval. For simplicity, assume $t_{cntrl} = 0$ and that therefore the compensator updates the control command instantly. As a result, the total loop delay t_d equals the modulation delay $t_{DPWM} = DT_s$.

The Bode plots of $G_{iu}(z)$ versus $G_{iu}(s)$, reported in Fig. 3.7, compare the dynamics of the sampled inductor current $\hat{i}_L[k]$ and the averaged inductor current $\bar{\hat{i}}_L(t)$, respectively. As with the output voltage dynamics, an extra phase lag is predicted by the discrete-time model in the high-frequency range due to the modulation delay $t_{DPWM} = DT_s$. Even more importantly, significant differences between the responses predicted by the discrete-time and the continuous-time averaged modeling are found in Fig. 3.7 *at low frequencies*. The continuous-time averaged model $G_{iu}(s)$ predicts a *zero* at the origin $s = 0$,

$$G_{iu}(s) \triangleq \frac{\bar{\hat{i}}_L(s)}{\hat{u}(s)} = \frac{V_g}{N_r} \frac{sC}{1 + s\left(r_L + r_C\right)C + s^2 LC}, \tag{3.47}$$

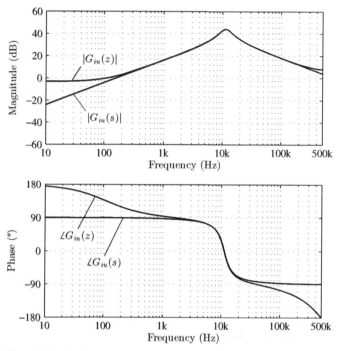

Figure 3.7 Synchronous Buck converter example: comparison between Bode plots of the control-to-inductor current transfer functions $G_{iu}(s)$ based on continuous-time averaged model and $G_{iu}(z)$ based on the discrete-time model.

while a finite, nonzero value is predicted by the discrete-time model. The difference is an example of aliasing effects due to sampling, which manifest themselves at dc.

To explain the different dynamics of the averaged current $\hat{\bar{i}}_L(t)$ and the current $\hat{i}_L[k]$ sampled at the end of the off-time interval, consider Fig. 3.8. If the converter is loaded with a constant current load I_o, as is the case in the example considered, a *static* variation $\hat{u}$ of the control command produces no variation in the average inductor current $\bar{i}_L = I_o$. This fact explains the zero in $G_{iu}(s)$ located at $s = 0$. On the other hand, the sampled current $i_L[k]$ undergoes a variation $\hat{i}_L[k]$ due to the change in the output voltage from $V_o = DV_g$ to $V_o + \hat{v}_o = (D + \hat{d})V_g$ and the corresponding variation of the inductor current ripple. Neglecting parasitic resistances r_L and r_C, one has

$$i_L[k] = I_o - \frac{V_o + \hat{v}_o}{L}(1 - D - \hat{d})\frac{T_s}{2} = I_o - \frac{T_s}{2L}(D + \hat{d})(1 - D - \hat{d})V_g. \quad (3.48)$$

The small-signal static variation $\hat{i}_L$ of the sampled current relative to the control perturbation $\hat{u}$ is, therefore,

$$G_{iu}(z = 1) \triangleq \left.\frac{\hat{i}_L(z)}{\hat{u}(z)}\right|_{z=1} = -\frac{T_s}{N_r}\frac{(1 - 2D)V_g}{2L}, \quad (3.49)$$

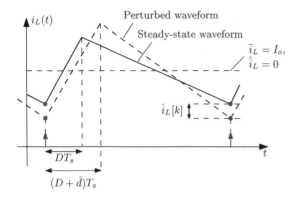

Figure 3.8   Effect of a static duty cycle perturbation on the inductor current sampled at the end of off-time interval.

which means that $G_{iu}(z)$ must present a nonzero value at dc. With the parameters of the considered Buck converter example,

$$G_{iu}(z = 1) \approx -0.7 = -3 \text{ dB}\angle{-180°}. \tag{3.50}$$

This is confirmed by the phase and the magnitude of $G_{iu}(z)$ in Fig. 3.7 at low frequencies.

The substantial difference between the discrete-time and the averaged dynamics at low frequencies is a result of the strong violation of the small-aliasing approximation. In this case, no simple corrections are available to improve predictions of the averaged small-signal model, and the design of a digital controller that relies on sampling of the signal of interest must be based on the discrete-time model that correctly considers the aliasing effects. To show that an attempt to correct for the delay effects is insufficient, suppose an effective control-to-inductor current transfer function is defined as

$$G_{iu}^{\dagger}(s) \triangleq G_{iu}(s)e^{-st_d}, \quad t_d = t_{DPWM} = DT_s, \tag{3.51}$$

with the purpose of modeling the additional small-signal phase delay introduced by the modulation. The Bode plots of the transfer functions $G_{iu}(z)$ and $G_{iu}^{\dagger}(s)$ are compared in Fig. 3.9. Although the high-frequency predictions of $G_{iu}^{\dagger}(s)$ are somewhat better compared to the high-frequency predictions of $G_{iu}(s)$ in Fig. 3.7, note that the aliasing effects are such that neither the high-frequency nor the low-frequency responses are modeled very well by $G_{iu}^{\dagger}(s)$. In particular, the responses of $G_{iu}^{\dagger}(s)$ and $G_{iu}(z)$ are qualitatively different at low frequencies.

3.2.2   Boost Converter

Consider now the digital version of the Boost converter average current control discussed in Section 1.5.2. The converter parameters are given in Table 1.3. A block diagram of the digital current-mode controller is illustrated in Fig. 3.10.

The inductor current is sampled at the middle of the turn-off interval, as illustrated in the control timing diagram of Fig. 3.11. Following the discussion in

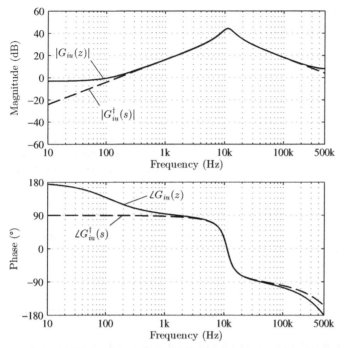

Figure 3.9 Synchronous Buck converter example: comparison between the Bode plots of the discrete-time model $G_{iu}(z)$ and the effective s-domain model $G_{iu}^{\dagger}(s)$.

Section 2.2.1, this sampling strategy is well suited when the goal is to regulate the average value of the current. The strategy is easily implemented by employing a symmetrical digital pulse width modulator and by synchronizing the sampling instants with the digital carrier peaking instants.

It is assumed that control calculations are carried out within a small fraction t_{cntrl} of the switching period immediately following the sampling event and that the DPWM immediately latches the updated control command. Such small control delay t_{cntrl} is also highlighted in Fig. 3.11, and the situation is precisely the one depicted in Table 3.3. It is important to note that, given the finite time t_{cntrl} available for control calculations, the control command must be suitably saturated in order not to produce a duty cycle larger than

$$D_{max} = 1 - \frac{2t_{cntrl}}{T_s}. \tag{3.52}$$

Determine first the discrete-time small-signal dynamics of the control-to-inductor current, described by transfer function $G_{iu}(z)$. Recall from Section 1.4.3 that the state-space matrices of the Boost converter during subtopologies S_0 and S_1

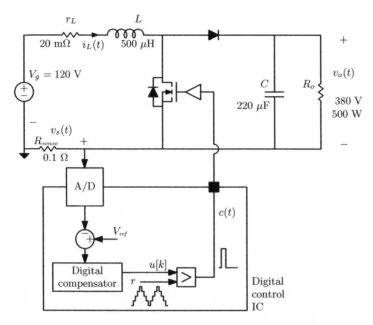

Figure 3.10   Boost digital current-mode control.

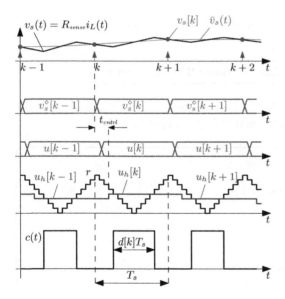

Figure 3.11   Boost converter current-mode control example: timing diagram.

are, respectively,

$$A_1 = \begin{bmatrix} -\dfrac{r_L + R_{sense}}{L} & 0 \\[3mm] 0 & -\dfrac{1}{R_o C} \end{bmatrix},$$

$$A_0 = \begin{bmatrix} -\dfrac{r_L + R_{sense}}{L} & -\dfrac{1}{L} \\[3mm] \dfrac{1}{C} & -\dfrac{1}{R_o C} \end{bmatrix},$$

(3.53)

while the input-to-state matrices are

$$B_0 = B_1 = \begin{bmatrix} \dfrac{1}{L} \\[3mm] 0 \end{bmatrix}.$$

(3.54)

The steady-state operating point of the converter, as determined by its operating conditions at full power $P_o = 500$ W, is

$$V_g = 120 \text{ V},$$
$$D = 0.68.$$

(3.55)

Matlab® modeling code for the Boost converter example is provided in Inset 3.2.

Inset 3.2 – Matlab® Example: Boost Converter

The following Matlab® lines implement the numerical calculations of the control-to-inductor current transfer function $G_{iu}(z)$ as per Table 3.3.

```
% Converter state-space matrices
A1 = [ -rL/L 0; 0 -1/Ro/C ];
A0  = [ -rL/L -1/L; 1/C -1/Ro/C ];
b1 = [ 1/L; 0];
b0 = b1;
c1  = [ 1 0; 0 1 ];
c0  = c1;

% Calculation of Xup and Xdown
A1i = A1^-1;
A0i = A0^-1;
Xdown = ((eye(2)-expm(A1*D*Ts)*expm(A0*Dprime*Ts))^-1)*...
                (-expm(A1*D*Ts)*A0i*(eye(2)-expm(A0*Dprime*Ts))*b0
        +...
```

```
                         -A1i*(eye(2)-expm(A1*D*Ts))*b1)*[Vg];
Xup  =    expm(A0*Dprime*Ts)*Xdown-A0i*(eye(2)-expm(A0*Dprime*Ts))
          *b2*[Vg];

Fdown   =    (A1-A0)*Xdown + (b1-b0)*[Vg;Iload;Vload];
Fup     =    (A1-A0)*Xup + (b1-b0)*[Vg;Iload;Vload];

% Small-signal model matrices
Phi     =    expm(A0*Dprime*Ts/2)*expm(A1*D*Ts)*expm(A0*Dprime*Ts/2);
gamma   =    (Ts/2)*expm(A0*Dprime*Ts/2)*(Fdown + expm(A1*D*Ts)*Fup);
delta = c0;

% Convert from state-space to transfer function object
sys  =    ss(Phi,gamma,delta(1,:),0,Ts);
Giuz =    tf(sys);
```

Figure 3.12 shows a comparison between the Bode plots of the discrete-time model $G_{iu}(z)$ found using the code in Inset 3.2, and the continuous-time averaged model $G_{iu}(s) = G_{id}(s)$ found in Section 1.5.2. As seen in the Buck example, the primary difference between the averaged and the discrete-time modeling is in the extra phase lag due to the control loop delays.

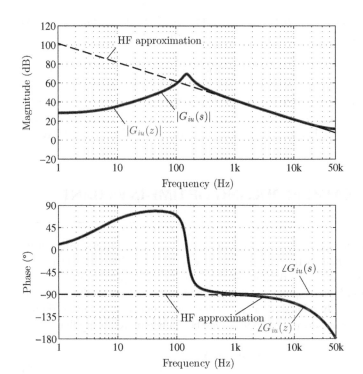

Figure 3.12 Boost converter example: Bode plots of the control-to-inductor current transfer functions $G_{iu}(z)$ based on the discrete-time model, and $G_{iu}(s)$ based on the averaged model.

In Section 1.5.2, the approximation

$$G_{id}(s) \approx \frac{V_o}{sL} \quad (\omega \gg \omega_0)$$

is mentioned as a useful way to quickly evaluate the high-frequency behavior of $G_{id}(s)$. A corresponding approximation exists for $G_{iu}(z)$,

$$G_{iu}(z) \approx \frac{T_s}{N_r} \frac{V_o}{L} \frac{z^{-1}}{1 - z^{-1}} \quad (\omega \gg \omega_0). \tag{3.56}$$

This high-frequency approximation is also depicted in Fig. 3.12. Equation (3.56) can be derived by neglecting the output voltage dynamics—an assumption well justified above the system resonance, where perturbations on v_o are well filtered. Under this assumption, the Boost converter becomes the same as the switched inductor example shown in Fig. 3.1(a), with $V_A = V_g$ and $-V_B = V_g - V_o$.

It should be noted that the inductor current ripple is by no means negligible with respect to the average value. However, the particular sampling strategy—symmetrical pulse width modulation combined with sampling the inductor current at the middle of the turn-off interval—*forces* the sampled current to closely match its averaged value, that is,

$$v_s[k] \approx \bar{v}_s(t_k) \Rightarrow i_L[k] \approx \bar{i}_L(t_k). \tag{3.57}$$

On the basis of the discussion in Section 2.6.2, the small-aliasing approximation is well satisfied. As a result, it is expected that the effective s-domain model

$$G_{iu}^{\dagger}(s) \triangleq G_{iu}(s)e^{-s\frac{T_s}{2}} \tag{3.58}$$

gives predictions close to the true small-signal dynamics predicted by $G_{iu}(z)$. This is confirmed by the comparison illustrated in Fig. 3.13, where only a minor discrepancy between the magnitude responses is observed in close proximity to the Nyquist frequency.

3.3   DISCRETE-TIME MODELING OF TIME-INVARIANT TOPOLOGIES

When a converter topology is *time-invariant*, it is possible to straightforwardly derive the control-to-output transfer function of the converter as a suitable discretization of its averaged small-signal model. By time-invariant topology, one refers to a converter circuit where subtopologies S_0 and S_1 are equal once the inputs are set to zero, and where outputs are linearly related to the converter inputs and state variables. Equivalently, the notion of time-invariant topology can be defined by requiring that

$$A_0 = A_1(\triangleq A),$$
$$C_0 = C_1(\triangleq C). \tag{3.59}$$

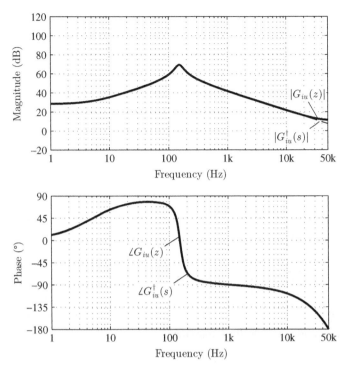

Figure 3.13   Boost converter example: comparison between the discrete-time model $G_{iu}(z)$ and the effective s-domain model $G_{iu}^{\dagger}(s)$.

When such conditions are satisfied and the system inputs are constant, the converter is *large-signal linear* with respect to $c(t)$. Notable examples of such converters are Buck topologies—see Fig. 3.14—as far as the inductor current and the output voltage are concerned.

Consider a switched-mode power converter satisfying (3.59). Figure 3.15 illustrates the system response to a control perturbation $\hat{u}$ superimposed on the steady-state command U. Owing to system linearity, the perturbed converter output voltage $\hat{v}_o(t)$ can be expressed as

$$\hat{v}_o(t) = \int_0^{+\infty} \hat{c}(\tau)g(t-\tau)\,d\tau = \int_0^t \hat{c}(\tau)g(t-\tau)\,d\tau, \qquad (3.60)$$

with $g(t)$ representing the impulse response of the converter to the control input $\hat{c}$. The last equality follows from the system causality, that is, from $g(t) = 0$ for all $t < 0$. It is important to stress that the concept of impulse response is applicable here *because the system is linear*.

In Section 2.5.2, it is recognized that, in the small-signal limit, $\hat{c}(t)$ is well approximated by a modulated train of Dirac pulses delayed by DT_s with respect to

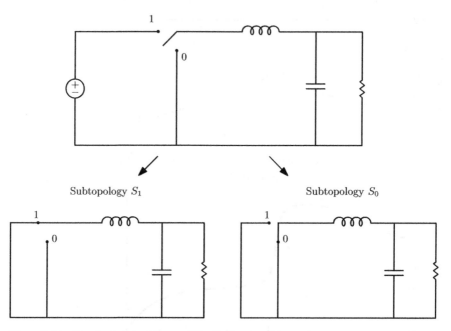

Figure 3.14   Topological invariance of the Buck converter.

the beginning of the switching interval,

$$\hat{c}(t) \approx \frac{T_s}{N_r} \sum_{n=0}^{+\infty} \hat{u}[n]\delta(t - nT_s - DT_s).$$ (3.61)

Therefore, one has

$$\hat{v}_o(t) = \frac{T_s}{N_r} \sum_{n=0}^{+\infty} \hat{u}[n] \int_0^{+\infty} \delta(\tau - nT_s - DT_s)g(t - \tau)\, d\tau$$

$$= \frac{T_s}{N_r} \sum_{n=0}^{+\infty} \hat{u}[n]g(t - nT_s - DT_s),$$

which, once sampled, becomes

$$\hat{v}_o[k] = \frac{T_s}{N_r} \sum_{n=0}^{+\infty} \hat{u}[n]g(kT_s - nT_s - DT_s).$$ (3.62)

Hence, the sampled output voltage $\hat{v}_o[k]$ is the discrete convolution between the control command $\hat{u}$ and $g(kT_s - DT_s)$, that is, the *delayed and sampled version of the converter impulse response*. The modulator delay $t_{DPWM} = DT_s$ naturally appears in the result.

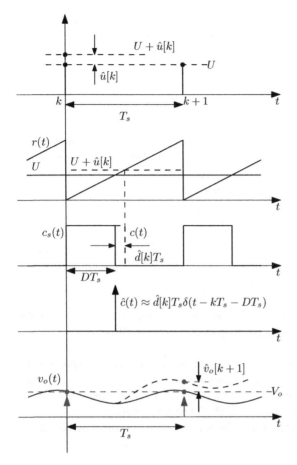

Figure 3.15 Small-signal discrete-time converter response to an input perturbation $\hat{u}$.

It is convenient, at this point, to introduce a notation for the above-mentioned discretization. If $G(s)$ is the transfer function of a continuous-time system and $g(t)$ the corresponding impulse response, define

$$Z_{T_s}\left[G(s)\right] \triangleq \sum_{n=0}^{+\infty} g(nT_s)z^{-n}. \tag{3.63}$$

The above-mentioned operation calculates the Z-transform of the sampled version of $g(t)$ and is commonly known as *impulse-invariant discretization*. The result of the operation is a z-domain transfer function $G(z)$ whose impulse response is the sampled version of $g(t)$.

With this definition, and taking the Z-transform of both sides of (3.62), one has

$$G_{vu}(z) = \frac{\hat{v}_o(z)}{\hat{u}(z)} = \frac{T_s}{N_r} Z_{T_s}\left[G(s)e^{-sDT_s}\right], \tag{3.64}$$

where $G(s)$ represents the Laplace transform of $g(t)$, that is, the transfer function between the PWM signal $c(t)$ and the converter output voltage $v_o(t)$. As the system is assumed to be large-signal linear, $G(s)$ must coincide with the averaged, small-signal control-to-output transfer function $G_{vd}(s)$. Therefore,

$$G_{vu}(z) = \frac{\hat{v}_o(z)}{\hat{u}(z)} = \frac{T_s}{N_r} Z_{T_s} \left[G_{vd}(s) e^{-sDT_s} \right]. \tag{3.65}$$

More generally, both control and modulation delays are present to yield an overall loop delay t_d. Analog sensing and conditioning, modeled by transfer function $H(s)$, can be included as well. In summary, one can use the above-mentioned arguments to evaluate the z-domain uncompensated loop gain as [131]

$$\boxed{\begin{aligned} T_u(z) &= \frac{T_s}{N_r} Z_{T_s} \left[T_u(s) e^{-st_d} \right] \\ &= \frac{T_s}{N_r} Z_{T_s} \left[T_u^\dagger(s) \right] \end{aligned}}, \tag{3.66}$$

where $T_u(s) \triangleq G_{vd}(s)H(s)$ is the Laplace-domain uncompensated loop gain of the system, evaluated assuming zero overall loop delay and unity DPWM carrier amplitude.

Equation (3.66) represents the main result of this section. It expresses the fact that *the small-signal discrete-time dynamics of a time-invariant topology is the impulse-invariant discretization of the effective small-signal s-domain dynamics*. It should be emphasized again that the result (3.66) holds only for time-invariant converters, that is, the converters such as the Buck converter, where the conditions (3.59) are met.

3.3.1 Equivalence to Discrete-Time Modeling

The objective of this section is to show that the exact discrete-time modeling framework discussed in Section 3.1 is equivalent the impulse-invariant discretization (3.66) when applied to time-invariant topologies. The approach is to start from the state-space description $(\Phi', \gamma', \delta')$ of (3.66) and to show that the small-signal matrices (Φ, γ, δ) of the discrete-time model indeed coincide with $(\Phi', \gamma', \delta')$ when the conditions (3.59) are met.

The system state-space equations are

$$\frac{d\boldsymbol{x}}{dt} = \boldsymbol{A}_c \boldsymbol{x}(t) + \boldsymbol{B}_c \boldsymbol{V},$$
$$\boldsymbol{y}(t) = \boldsymbol{C}_c \boldsymbol{x}(t) + \boldsymbol{E}_c \boldsymbol{V}. \tag{3.67}$$

From (3.59), the converter state-space equation becomes

$$\frac{d\boldsymbol{x}}{dt} = \boldsymbol{A}\boldsymbol{x}(t) + \boldsymbol{B}_c \boldsymbol{V}. \tag{3.68}$$

Equation (3.68) can now be integrated assuming that the PWM command $\hat{c}(t)$ consists of a train of Dirac pulses delayed by the modulator delay, in agreement with (3.61),

$$\hat{c}(t) \approx \frac{T_s}{N_r} \sum_{n=0}^{+\infty} \hat{u}[n]\delta(t - nT_s - t_d), \tag{3.69}$$

where the overall loop delay t_d has been considered to obtain the most general result. With such control input, the sampled dynamics of the state perturbation $\hat{x}[k]$ assumes the form

$$\hat{x}[k + 1] = \Phi'\hat{x}[k] + \gamma'\hat{u}[k],$$

$$\hat{y}[k] = \delta'\hat{x}[k], \tag{3.70}$$

where

$$\Phi' = e^{AT_s},$$

$$\gamma' = \frac{T_s}{N_r} e^{A(T_s-t_d)} (B_1 - B_0) V, \tag{3.71}$$

$$\delta' = C.$$

This result can be equivalently obtained by discretizing the converter state-space averaged, small-signal equations (1.34), provided that control input $\hat{d}(t)$ is replaced by the modulated train of delayed Dirac pulses (3.69). Therefore, under the above-mentioned conditions, (3.70) is the impulse-invariant discretization of the state-space averaged small-signal model. To complete the proof, then, one simply needs to show that $\Phi' = \Phi$ and $\gamma' = \gamma$ (observe that $\delta' = C = \delta$).

From matrix algebra, recall that if square matrices X and Y commute, then their exponentials commute too and their product is the exponential of $X + Y$,

$$XY = YX \quad \Rightarrow \quad e^X e^Y = e^Y e^X = e^{(X+Y)}. \tag{3.72}$$

As $A_0 = A_1 = A$, matrices A_0 and A_1 commute. This is therefore a sufficient condition for (3.32) to be written as

$$\Phi = e^{A(T_s-t_d+DT_s+t_d-DT_s)} = e^{AT_s} = \Phi'. \tag{3.73}$$

Furthermore, $A_0 = A_1$ implies that γ is independent of $X_\downarrow$ and therefore

$$\gamma = \frac{T_s}{N_r} e^{A(T_s-t_d)} [(B_1 - B_0) V] = \gamma', \tag{3.74}$$

completing the proof.

3.3.2 Relationship with the Modified Z-Transform

Discretization (3.66) can be rewritten in terms of the *modified* Z-*transform* of the analog process $T_u(s)$. The modified Z-transform of a continuous-time signal $g(t)$ is defined as

$$G(z;m) \triangleq \sum_{n=0}^{+\infty} g(nT_s + mT_s)z^{-n} = Z_{T_s}\left[G(s)e^{smT_s}\right], \qquad (3.75)$$

with $0 < m < 1$. It is therefore the Z-transform of the sequence $g(nT_s + mT_s)$ obtained from $g(t)$ by anticipating the signal by mT_s seconds and then sampling it with period T_s.

If $t_d = LT_s - mT_s$, $L \in \mathbb{N}$, (3.66) can be put in the form

$$T_u(z) = \frac{T_s}{N_r}z^{-L}Z_{T_s}\left[T_u(s)e^{smT_s}\right] = \frac{T_s}{N_r}T_u(z;m)z^{-L}, \qquad (3.76)$$

and the modeling task reduces to the evaluation of the modified Z-transform of $T_u(s)$ with parameter m.

3.3.3 Calculation of $T_u(z)$

Analytical calculation of $T_u(z)$ can be performed in several ways. Following (3.76), one approach is to anti-transform $T_u(s)$ back into the time domain, anticipate it by mT_s seconds, sample it with period T_s, evaluate the Z-transform $T_u(z;m)$ of the resulting sequence, and plug it into (3.76).

Consider, for instance, a first-order $T_u(s)$ such as

$$T_u(s) = \frac{T_{u0}}{1 + \dfrac{s}{\omega_P}}. \qquad (3.77)$$

Its inverse Laplace transform is the causal signal

$$g(t) = \begin{cases} \omega_P T_{u0}e^{-\omega_P t}, & t \geq 0, \\ 0 & t < 0. \end{cases} \qquad (3.78)$$

Anticipating $g(t)$ by mT_s and sampling yields

$$g(kT_s + mT_s) = \begin{cases} \omega_P T_{u0}e^{-\omega_P mT_s}e^{-\omega_P kT_s}, & k \geq 0, \\ 0 & k < 0, \end{cases} \qquad (3.79)$$

whose Z-transform is

$$T_u(z; m) = \sum_{k=0}^{+\infty} g(kT_s + mT_s) z^{-k}$$

$$= \omega_P T_{u0} e^{-\omega_P m T_s} \sum_{k=0}^{+\infty} (e^{-\omega_P T_s} z^{-1})^k \qquad (3.80)$$

$$= \frac{\omega_P T_{u0} e^{-\omega_P m T_s}}{1 - e^{-\omega_P T_s} z^{-1}}, \qquad |z| > e^{-\omega_P T_s}.$$

From (3.76)

$$T_u(z) = \frac{T_s}{N_r} \frac{\omega_P T_{u0} e^{-m\omega_P T_s}}{1 - e^{-\omega_P T_s} z^{-1}} z^{-L}. \qquad (3.81)$$

An alternative calculation approach, based on the residue theory, allows direct transformation of $T_u(s)$ into $T_u(z)$. Consider the expansion of $T_u(s)$ into partial fractions,

$$T_u(s) = \sum_{i=1}^{N} \frac{A_i}{s - s_i}, \qquad (3.82)$$

where, for simplicity, poles s_i of $T_u(s)$ are assumed to be all of order one. By antitransforming (3.82), anticipating it by mT_s seconds, sampling, and finally Z-transforming the resulting sequence, one finds

$$T_u(z; m) = \sum_{i=1}^{N} \frac{A_i e^{s_i m T_s}}{1 - e^{s_i T_s} z^{-1}}. \qquad (3.83)$$

Coefficients A_i are the residues of $T_u(s)$ associated with poles s_i. For simple poles, the A_i's are given by

$$A_i = \lim_{s \to s_i} (s - s_i) T_u(s). \qquad (3.84)$$

Plugging (3.83) into (3.76) yields

$$\boxed{T_u(z) = \frac{T_s}{N_r} z^{-L} \sum_{i=1}^{N} \frac{A_i e^{s_i m T_s}}{1 - e^{s_i T_s} z^{-1}}.} \qquad (3.85)$$

Note that this expansion also holds when there are complex conjugate pairs among the s_i's. In such case, corresponding A_i's would appear in complex conjugate pairs as well.

In the example above, the only residue of $T_u(s)$ is

$$A_P = \lim_{s \to -\omega_P} (s + \omega_P) T_u(s) = \omega_P T_{u0}, \qquad (3.86)$$

and one immediately obtains (3.81).

TABLE 3.4 Expressions of $T_u(z)$ for Three Different $T_u(s)$'s.
$t_d = LT_s - mT_s, L \in \mathbb{N}, 0 < m < 1$

$T_u(s)$	$T_u(z) = \dfrac{T_s}{N_r} T_u(z; m) z^{-L}$
$\dfrac{K}{s}$	$\dfrac{T_s}{N_r} \dfrac{K}{1 - z^{-1}} z^{-L}$
$\dfrac{T_{u0}}{1 + \dfrac{s}{\omega_P}}$	$\dfrac{T_s}{N_r} \dfrac{\omega_P T_{u0} e^{-m\omega_P T_s}}{1 - e^{-\omega_P T_s} z^{-1}} z^{-L}$
$\dfrac{T_{u0}}{1 + \dfrac{s}{\omega_0 Q} + \dfrac{s^2}{\omega_0{}^2}}$	$\dfrac{T_s}{N_r} \dfrac{\omega_0{}^2 T_{u0}}{\omega_d} \dfrac{\left(e^{m\alpha T_s} \sin(m\omega_d T_s) + e^{(1+m)\alpha T_s} \sin((1-m)\omega_d T_s) z^{-1} \right)}{1 - 2e^{\alpha T_s} \cos(\omega_d T_s) z^{-1} + e^{2\alpha T_s} z^{-2}} z^{-L}$ $\left(\alpha \triangleq -\dfrac{\omega_0}{2Q}, \omega_d \triangleq \omega_0 \sqrt{1 - \dfrac{1}{4Q^2}}, Q \geq \dfrac{1}{2} \right)$

Closed-form expressions of $T_u(z)$ for three different $T_u(s)$'s are reported in Table 3.4. Observe that the first two cases coincide with the results derived in Section 3.1.1 in the discussion of the switched inductor example. It should be noted that the switched inductor is also an example of time-invariant topology.

In the third example, a second-order plant is considered,

$$T_u(s) = \frac{T_{u0}}{1 + \dfrac{s}{\omega_0 Q} + \dfrac{s^2}{\omega_0{}^2}}, \tag{3.87}$$

with $Q \geq 0.5$. Poles s_1 and s_2 are complex conjugates and equal

$$s_{1,2} = \alpha \pm j\omega_d, \tag{3.88}$$

with

$$\alpha \triangleq -\frac{\omega_0}{2Q}, \tag{3.89}$$
$$\omega_d \triangleq \omega_0 \sqrt{1 - \frac{1}{4Q^2}}.$$

Following the residue approach, first write the continuous-time plant as

$$T_u(s) = \frac{A}{s - s_1} + \frac{A^*}{s - s_1^*}, \tag{3.90}$$

with

$$A = \lim_{s \to s_1} (s - s_1) T_u(s) = \frac{\omega_0^2 T_{u0}}{2j\omega_d}. \tag{3.91}$$

The expression for $T_u(z)$ reported in Table 3.4 is then obtained from (3.85) after a few algebraic manipulations.

Hand evaluation of $T_u(z)$ in more complex cases quickly becomes impractical and yields expressions that are difficult to manipulate for design purposes. A quicker alternative is to resort to software tools for scientific computing and evaluate $G(z)$ numerically. Matlab® provides dedicated commands for continuous-to-discrete-time conversion of a given transfer function that can be readily employed to evaluate (3.66).

As a final remark, observe from (3.85) that s-domain poles transform into z-domain poles according to

$$\boxed{s_i \rightarrow e^{s_i T_s}}. \tag{3.92}$$

On the other hand, no simple expressions for the locations of the z-domain zeros are available.

Inset 3.3 – Impulse-Invariant Discretization of $G_{vd}(s)$ Using Matlab®

Going back to the Buck converter example of Section 3.2.1, suppose an s-domain expression of $G_{vd}(s)$ is available in the Matlab® workspace as a transfer function object named Gvds. For instance, following (1.20), Gvds can be obtained by the following code:

```
s = tf('s');
Gvds = Vg*(1+s*rC*C)/(1+s*(rs+rC)*C+s^2*L*C);
```

To define a transport delay for Gvds, Matlab® allows the user to define an outputdelay property representing a transport delay block cascaded to Gvds. Setting such property to the intended small-signal loop delay td is accomplished as follows:

```
Gvds.outputdelay   =   td;
```

Variable td must be evaluated from the system timing diagram for the particular operating point around which the converter is linearized.

In the Buck example in Section 3.2.1, we have $t_d = 760$ ns.

Next, discretization (3.66) is accomplished via the c2d command:

```
Gvuz   =   (1/Nr)*c2d(Gvds,Ts,'imp');
```

where the 'imp' setting specifies the impulse-invariant discretization. This command produces a z-domain transfer function object Gvuz with a sampling time equal to the switching period Ts of the converter. This transfer function is equal to $G_{vu}(z)$ as defined by (3.66).

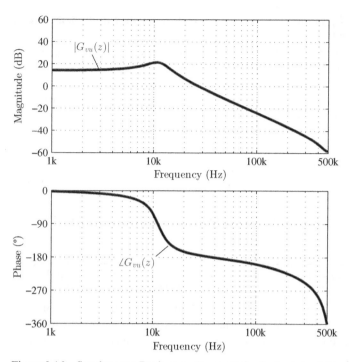

Figure 3.16   Synchronous Buck converter example: Bode plots of $G_{vd}(z)$ calculated using the general discrete-time modeling approach and using the impulse-invariant discretization.

3.3.4   Buck Converter Example Revisited

As the Buck converter examined in Section 3.2.1 is an example of time-invariant topology, one can verify that the same $G_{vu}(z)$ is predicted by the general discrete-time modeling framework—see Inset 3.1—and by the impulse-invariant approach—see Inset 3.3.

The comparison, illustrated in Fig. 3.16, confirms that for time-invariant topologies the two approaches yield the same result. The same verification can be carried out for the control-to-inductor current transfer function $G_{id}(z)$.

3.4   MATLAB® DISCRETE-TIME MODELING OF BASIC CONVERTERS

In this section, a simple Matlab® framework for numerically deriving the control-to-output small-signal model of a switched-mode power converter is presented. The framework is equipped with models for the Buck, Boost, and inverting Buck–Boost topologies and also allows for user-defined converters to be specified.

The system template considered by the Matlab® code is depicted in Fig. 3.17. The converter is assumed to operate from a stiff input voltage V_g, and with a load

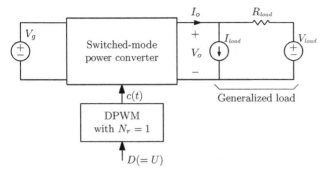

Figure 3.17   System template of the proposed Matlab® modeling framework.

modeled by an independent current sink I_{load}, in parallel with a Thévenin source (V_{load}, R_{load}). Such combination allows load modeling in a variety of situations, including a constant current load, a resistive load, or a Thévenin load, which is useful when the converter operates on a regulated dc bus.

The digital pulse width modulator can be a trailing-edge, a leading-edge, or a symmetrical DPWM. It is assumed that $N_r = 1$, so that the control input U coincides with the converter duty cycle D. For nonunity N_r values, the user simply needs to multiply the Matlab®-generated transfer functions by $1/N_r$.

The sampling strategy is defined by a single parameter t_d, which represents the total loop delay and defines the position of the sampling instant with respect to the position of the steady-state PWM modulated edge—refer to Figs 3.1 and 3.2. For the symmetrical modulator case, the sampling instant is always assumed to be located at the carrier peak, as shown in Fig. 3.3, and the parameter t_d is ignored.

The modeling code is implemented as the `extract_models` Matlab® script reported in Inset 3.4. The inputs in the script are as follows:

- `params`: structure object defining converter parameters. When modeling predefined topologies (Buck, Boost, or Buck–Boost), `params` contains the following fields:

  o `params.L`: converter inductance L

  o `params.rL`: converter inductor series resistance r_L

  o `params.C`: converter capacitance C

  o `params.rC`: converter capacitor equivalent series resistance r_C

  When a user-defined converter is to be modeled, `params` must specify matrices (A_c, B_c, C_c) relative to the converter subtopologies as follows:

  o `params.A1` for matrix A_1

  o `params.A0` for matrix A_0

  o `params.b1` for matrix B_1

  o `params.b0` for matrix B_0

  o `params.c1` for matrix C_1

  o `params.c0` for matrix C_0

- Vg: Converter input voltage V_g
- D: Converter steady-state duty cycle D
- Iload, Rload, Vload: Load parameters I_{load}, R_{load}, and V_{load}.
- td: Total loop delay t_d
- Ts: Converter switching period T_s
- converter: String defining the converter type. Default supported commands are 'buck', 'boost', and 'buckboost'. String 'custom' allows for user-defined models to be specified via the params structure.
- modulator: String defining the PWM carrier type. Supported types are 'te', 'le', or 'sym' for trailing-edge, leading-edge, or symmetrical modulation, respectively.

The output of the script is a transfer function matrix object Wz representing the discrete-time small-signal model transfer matrix $W(z)$ and transfer function matrix object Ws of the averaged small-signal model $W(s)$.

Inset 3.4 – Matlab® Modeling Script

```
function [Wz,Ws]    = ...
     extract_models(Vg,params,Rload,Iload,Vload,D,td,Ts,conv,
     modulator)

Dprime  =    1-D;

s   =   tf('s');
z   =   tf('z',Ts);

switch conv

   case 'buck'
       L    =   params.L;
       rL   =   params.rL;
       C    =   params.C;
       rC   =   params.rC;

       rpar   =   (rC)/(1+rC/Rload);
       A1   =   [-(rpar+rL)/L -1/(1+rC/Rload)/L; ...
                            1/(1+rC/Rload)/C -1/(Rload+rC)
                            /C];
       b1   =   [   1/L      rpar/L          -1/(1+Rload/rC)/L;
                      0    -1/(1+rC/Rload)/C  1/(Rload+rC)/C];
       c1   =   [1 0; rpar 1/(1+rC/Rload)];

       A0   =   A1;
       b0   =   [   0    rpar/L           -1/(1+Rload/rC)/L;
                      0   -1/(1+rC/Rload)/C  1/(Rload+rC)/C];
       c0   =   c1;
```

```
case 'boost'
     L    =    params.L;
     rL   =    params.rL;
     C    =    params.C;
     rC   =    params.rC;

     rpar    =    (rC)/(1+rC/Rload);
     A1   =    [-rL/L 0; 0 -1/(Rload+rC)/C];
     b1   =         [1/L      0                   0;
                      0     -1/(1+rC/Rload)/C  1/(Rload+rC)/C];
     c1   =    [1 0; 0 1/(1+rC/Rload)];

     A0   =    [-(rpar+rL)/L -1/(1+rC/Rload)/L; ...
                              1/(1+rC/Rload)/C -1/(Rload+rC)
                              /C];
     b0   =    [   1/L      rpar/L         -1/(1+Rload/rC)/L;
                      0     -1/(1+rC/Rload)/C  1/(Rload+rC)/C];
     c0   =    [1 0; rpar 1/(1+rC/Rload)];

case 'buckboost'
     L    =    params.L;
     rL   =    params.rL;
     C    =    params.C;
     rC   =    params.rC;

     rpar    =    (rC)/(1+rC/Rload);
     A1   =    [-rL/L 0; 0 -1/(Rload+rC)/C];
     b1   =    [   1/L      0                   0;
                      0     -1/(1+rC/Rload)/C  1/(Rload+rC)/C];
     c1   =    [1 0; 0 1/(1+rC/Rload)];

     A0   =    [-(rpar+rL)/L -1/(1+rC/Rload)/L; ...
                              1/(1+rC/Rload)/C -1/(Rload+rC)
                              /C];
     b0   =    [   0      rpar/L         -1/(1+Rload/rC)/L;
                      0     -1/(1+rC/Rload)/C  1/(Rload+rC)/C];
     c0   =    [1 0; rpar 1/(1+rC/Rload)];

case 'custom'
     A1   =    params.A1;
     A0   =    params.A0;
     b1   =    params.b1;
     b0   =    params.b0;
     c1   =    params.c1;
     c0   =    params.c0;
end;

%    *****************************************
%    Steady-state OP determination --- Continuous
%    *****************************************
X    =    -((D*A1+Dprime*A0)^-1)*(D*b1+Dprime*b0)*[Vg;Iload;Vload];
```

```
%      ****************************************
%      Steady-state OP determination --- Discrete
%      ****************************************
dim  =    min(size(A1));
A1i  =    A1^-1;
A2i  =    A0^-1;
Xdown    =    ((eye(dim)-expm(A1*D*Ts)*expm(A0*Dprime*Ts))^-1)*...
                  (-expm(A1*D*Ts)*A2i*(eye(dim)-expm(A0*Dprime*Ts))*
                  b0+...
                  -A1i*(eye(dim)-expm(A1*D*Ts))*b1)*[Vg;Iload;Vload];
Xup      =    expm(A0*Dprime*Ts)*Xdown-A2i*...
                              (eye(dim)-expm(A0*Dprime*Ts))*b0*...
                              [Vg;Iload;Vload];

Fdown    =    (A1-A0)*Xdown + (b1-b0)*[Vg;Iload;Vload];
Fup      =    (A1-A0)*Xup + (b1-b0)*[Vg;Iload;Vload];

%      ****************************************
%      Small-signal model---Continuous
%      ****************************************
A    =    D*A1+Dprime*A0;
F    =    (A1-A0)*X + (b1-b0)*[Vg;Iload;Vload];
C    =    D*c1+Dprime*c0;

sys  =    ss(A,F,C,0);
Ws   =    tf(sys);

%      ********************************
%      Small-signal model---Discrete
%      ********************************
switch modulator

    case 'te'
        Phi      =    expm(A0*(Ts-td))*expm(A1*D*Ts)*expm(A0*(td-D*
                      Ts));
        gamma    =    expm(A0*(Ts-td))*Fdown*Ts;
        delta    =    c0;

    case 'le'
        Phi      =    expm(A1*(Ts-td))*expm(A0*Dprime*Ts)*expm(A1*
                      (td-Dprime*Ts));
        gamma    =    expm(A1*(Ts-td))*Fup*Ts;
        delta    =    c1;

    case 'sym'
        Phi      =    expm(A0*Dprime*Ts/2)*expm(A1*D*Ts)*expm(A0*D
                      prime*
                          Ts/2);
        gamma    =    (Ts/2)*expm(A0*Dprime*Ts/2)*(Fdown + expm(A1*D*
                      Ts)
                          *Fup);
        delta    =    c0;
```

```
end;


sys  =   ss(Phi,gamma,delta,0,Ts);
Wz   =   tf(sys);

return;
```

3.5   SUMMARY OF KEY POINTS

- In the discrete-time modeling approach, the nonlinear *sampled* dynamics of the converter is linearized around the system's operating point. The result is a state-space discrete-time model of the power converter, which correctly accounts for sampling effects and delays present in the feedback loop.

- When the small-aliasing approximation introduced in Section 2.6.2 is justified, the effect of the total loop delay on the system's phase response can be approximated by attributing a transport delay to the standard averaged small-signal models. In other cases, no simple corrections are available to improve predictions of the averaged small-signal models, and digital control design must be based on discrete-time models that correctly consider sampling, aliasing, and delay effects.

- For time-invariant topologies, such as the Buck converter, the exact discrete-time model can be obtained via impulse-invariant discretization of the conventional averaged small-signal model, provided that the overall delay is included as an equivalent transport delay.

- Discrete-time modeling can be straightforwardly implemented via Matlab® scripting.

CHAPTER 4

DIGITAL CONTROL

The discrete-time modeling framework developed in Chapter 3 enables the direct-digital z-domain synthesis of the compensator transfer function based on the familiar frequency-domain specifications in terms of the desired crossover frequency, phase margin, and gain margin. This chapter is devoted to this subject.

The compensator design is presented in Section 4.1. Among many textbook approaches for direct-digital compensator design, the one emphasized here is based on the so-called *bilinear transform*, an effective tool for direct z-domain synthesis. The approach is outlined in Section 4.1.1, followed by a discussion of the digital proportional-integral-derivative (PID) compensator in Section 4.1.2. A number of design examples are then presented in Section 4.2, and closed-form expressions for the determination of the compensator coefficients are worked out analytically. In the design practice, it is often necessary to evaluate the impact of disturbances, such as input voltage or load current variations, acting on the closed-loop system. Section 4.3 is devoted to finding the transfer functions relevant for such evaluations in digitally controlled converters. Finally, Section 4.4 addresses important practical issues and mitigation strategies related to control saturation, when the closed-loop system is exposed to large-signal transients.

Practical application of the design methods developed in this chapter is best and more rapidly performed in the Matlab® design environment. Systematic Matlab® scripting allows the designer to bypass the rather lengthy calculations characterizing digital design, without losing analytical insights. Because of its practical importance, throughout this chapter, Matlab®-aided design is emphasized and developed alongside the analytical formulation.

4.1 SYSTEM-LEVEL COMPENSATOR DESIGN

The objective of a digital compensator design procedure is to determine the compensator z-domain transfer function $G_c(z)$ so that the closed-loop system meets certain design specifications. The approach developed in this section follows the usual frequency-domain approach where the compensator transfer function is shaped so that the system loop gain $T(z)$ achieves a target control bandwidth specification, with desired stability margins.

Digital Control of High-Frequency Switched-Mode Power Converters, First Edition.
Luca Corradini, Dragan Maksimović, Paolo Mattavelli, and Regan Zane.
© 2015 The Institute of Electrical and Electronics Engineers, Inc. Published 2015 by John Wiley & Sons, Inc.

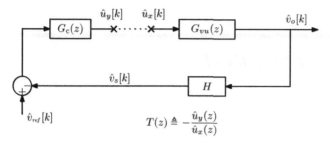

Figure 4.1   Definition of the z-domain loop gain $T(z)$.

Similar to the loop gain in analog controlled converters addressed in Section 1.3.3, the z-domain loop gain $T(z)$ of a single-loop digitally controlled system, such as the one sketched in Fig. 4.1, is defined as

$$T(z) \triangleq -\frac{\hat{u}_y(z)}{\hat{u}_x(z)} = G_c(z)HG_{vu}(z).$$ (4.1)

Likewise, the *uncompensated loop gain* $T_u(z)$ corresponds to the system loop gain with a unity compensation $G_c(z) = 1$,

$$T_u(z) = HG_{vu}(z).$$ (4.2)

Note that a wide-bandwidth sensing transfer function is assumed in Fig. 4.1, modeled by a static gain H.

A number of textbook approaches exist to synthesize $G_c(z)$ [7]. The one discussed here relies on the *bilinear transform* and belongs to a broader class of *mapping-based* approaches. The uncompensated loop gain $T_u(z)$, as well as the to-be-designed compensator transfer function $G_c(z)$, is first mapped into an equivalent continuous-time p-domain, where the actual design is performed. The compensator designed in p-domain is then back-mapped into the z-domain and implemented in the digital controller. As the actual design occurs in terms of the equivalent continuous-time system, all the techniques familiar to the analog designer can be reused without conceptual difficulties. Nevertheless, as the design starts and ends with exact z-domain transfer functions, including the converter discrete-time models developed in Chapter 3, the approach is not subject to any of the limitations associated with the use of averaged models discussed in Section 2.6.1.

4.1.1   Direct-Digital Design Using the Bilinear Transform Method

The bilinear map,

$$z(p) = \frac{1 + p\dfrac{T_s}{2}}{1 - p\dfrac{T_s}{2}},$$ (4.3)

provides a way to move from z-domain to an equivalent continuous-time p-domain, so that the digital compensator design can be performed entirely using the familiar Laplace-domain frequency response techniques. The inverse map,

$$p(z) = \frac{2}{T_s} \frac{1 - z^{-1}}{1 + z^{-1}}, \tag{4.4}$$

then allows the design to be moved back into the z-domain.

On the basis of (4.3), mapping of a z-domain transfer function $G(z)$ into the p-plane results in a p-domain function $G'(p)$ defined as

$$G'(p) \triangleq G(z(p)). \tag{4.5}$$

The bilinear map has the following properties:

1. If the transfer function $G_c(z)$ is rational in the variable z, its transform $G'_c(p) = G_c(z(p))$ is rational in the variable p.
2. The unit disk $|z| < 1$ is mapped into the left half-plane $\Re[p] < 0$, whereas the unstable portion of the z-plane $|z| > 1$ is mapped into the right half-plane $\Re[p] > 0$.
3. As illustrated in Fig. 4.2, the unit circle $z = e^{j\omega T_s}$ is mapped into the imaginary axis $p = j\omega'$:

$$z = e^{j\omega T_s} \quad \rightarrow \quad p = j\omega', \tag{4.6}$$

where ω and ω' represent the z-domain and p-domain angular frequencies, respectively. The relationship between ω and ω' is given by

$$\omega' = \frac{2}{T_s} \tan\left(\omega \frac{T_s}{2}\right). \tag{4.7}$$

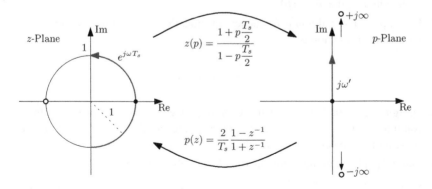

Figure 4.2   Bilinear mapping between the z-domain and the p-domain.

From the above-mentioned properties, it follows that the p-plane has the characteristics of a Laplace domain and that $G'_c(p)$ can be interpreted as the transfer function of a continuous-time system whose frequency response $G'_c(j\omega')$ is related to the original frequency response $G_c(e^{j\omega T_s})$ by

$$G'_c(j\omega') = G_c(e^{j\omega T_s}), \tag{4.8}$$

where ω' and ω are related through (4.7). Notably, as the z-domain frequency axis $z = e^{j\omega T_s}$ is mapped into the p-domain frequency axis $p = j\omega'$, the system frequency response is preserved by the transformation in both magnitude and phase. The frequency axis, however, undergoes a distortion as expressed by (4.7), which is commonly referred to as *frequency warping*. From (4.7), the distortion amounts to 1% at $\omega = \omega_s/18$ and 10% at $\omega = \omega_s/6$.

Application of the bilinear transform allows formulation of the digital design problem in the equivalent continuous-time domain and applications of the usual concepts and techniques commonly used in analog compensator designs. The result is then brought back to the z-domain by application of the inverse map (4.4). The frequency distortion can be easily compensated for by suitably *prewarping* all the specifications expressed in terms of a z-domain frequency.

It is important to note that frequency ω in the z-domain represents the actual frequency of interest in the system. All frequency-domain specifications, for example, crossover frequency or control-loop bandwidth, are expressed in terms of ω. On the other hand, the p-domain frequency ω' is distorted according to (4.7). This is why a new symbol p, as opposed to the standard symbol s, is used in the bilinear mapping for the equivalent continuous-time Laplace domain.

The design flow based on the bilinear transform method can now be summarized as follows:

1. Start with a z-domain model $T_u(z)$ of the uncompensated loop gain obtained via the discrete-time modeling method discussed in Chapter 3 and identify a template structure $G_c(z)$ of the compensator to be designed, for example, a digital PID compensator discussed in Section 4.1.2.

2. Apply (4.3) to both $T_u(z)$ and $G_c(z)$ and obtain the equivalent p-domain formulations $T'_u(p)$ and $G'_c(p)$.

3. Prewarp all the z-domain frequency specifications ω_{spec} into corresponding p-domain specifications ω'_{spec},

$$\omega'_{spec} = \frac{2}{T_s} \tan\left(\omega_{spec} \frac{T_s}{2}\right). \tag{4.9}$$

4. Design $G'_c(p)$ in the p-domain according to well-known approaches from continuous-time feedback theory, for example, as described in [1] in the context of analog control of switched-mode power converters.

5. Map $G'_c(p)$ back in the z-domain via (4.4) to obtain $G_c(z)$.

With a design of $G_c(z)$ completed, one can proceed to the compensator implementation, which is addressed in Chapter 6.

4.1.2   Digital PID Compensators in the z- and the p-Domains

The PID compensator is an important case, as it presents a suitable template in many practical designs. Two possible digital PID realizations, Euler and Tustin, are obtained from a continuous-time PID via discretization in Section 2.3. In this section, an opposite direction is taken, starting from a well-defined PID structure as a given template in the z-domain, without necessarily thinking of it as the result of a discretization process. The particular realization considered here, because of its inherent simplicity, is shown in Fig. 4.3. This is the Euler structure in its parallel—sometimes referred to as additive or noninteracting—form governed by the equations

$$
\begin{aligned}
u_p[k] &= K_p e[k], \\
u_i[k] &= u_i[k-1] + K_i e[k], \\
u_d[k] &= K_d(e[k] - e[k-1]), \\
u[k] &= u_p[k] + u_i[k] + u_d[k].
\end{aligned}
\tag{4.10}
$$

The compensator coefficients K_p, K_i, and K_d are the proportional, integral, and derivative gains, respectively.

Direct Z-transform of (4.10) yields the standard additive form of the digital PID transfer function

$$
G_{PID}(z) = K_p + \frac{K_i}{1 - z^{-1}} + K_d(1 - z^{-1}).
\tag{4.11}
$$

This transfer function, shown in a block-diagram form in Fig. 4.3, has two poles located at $z = 1$ and $z = 0$ and two zeros determined by the three coefficients (K_p, K_i, K_d).

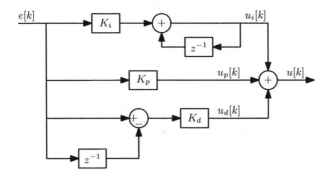

Figure 4.3   Block diagram of a digital PID compensator in the parallel form.

Once mapped into the p-domain via (4.3), (4.11) becomes

$$G'_{PID}(p) = \underbrace{K_p}_{\text{Proportional term}} + \underbrace{\frac{K_i}{T_s}\frac{1 + \dfrac{p}{\omega_p}}{p}}_{\text{Integral term}} + \underbrace{K_d T_s \frac{p}{1 + \dfrac{p}{\omega_p}}}_{\text{Derivative term}}, \quad (4.12)$$

where

$$\omega_p \triangleq \frac{2}{T_s} = \frac{\omega_s}{\pi}. \quad (4.13)$$

Figure 4.4 shows the asymptotic Bode plots of $G'_c(p)$, highlighting the proportional, integral, and derivative contributions.

Frequency ω_p, which originates from the finite switching frequency of the converter, is the zero frequency in the integral term, and the pole frequency in the derivative term of $G'_c(p)$. The reason such pole and zero appear in the p-domain is that the Nyquist angular frequency $\omega_N = \omega_s/2 = \pi/T_s$, represented in the z-plane by the $z = -1$ point, is mapped to infinity by (4.4). As both the integral and the derivative terms have finite gains at the Nyquist rate in the z-domain, they must have finite gains at $\omega' = +\infty$ in the p-domain. One can easily see that this is indeed the case in (4.12). Observe that ω_p is located at slightly less than one-third of the switching rate ω_s, which is relatively high compared to the crossover frequencies in typical control loops around switched-mode converters.

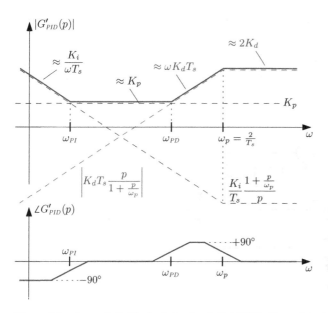

Figure 4.4   Asymptotic Bode plots of a digital PID in the p-domain.

For design purposes, it is useful to work with the multiplicative form of (4.12)

$$G'_{PID}(p) = \underbrace{G'_{PI\infty}\left(1 + \frac{\omega_{PI}}{p}\right)}_{PI}\underbrace{G'_{PD0}\frac{1 + \dfrac{p}{\omega_{PD}}}{1 + \dfrac{p}{\omega_p}}}_{PD}, \tag{4.14}$$

with $G'_{PI\infty} > 0$ and $G'_{PD0} > 0$. Once the p-domain parameters $(G'_{PI\infty}, \omega_{PI}, G'_{PD0}, \omega_{PD})$ are determined in the design process, the z-domain PID gains K_p, K_i, and K_d are calculated as

$$K_p = G'_{PI\infty}G'_{PD0}\left(1 + \frac{\omega_{PI}}{\omega_{PD}} - \frac{2\omega_{PI}}{\omega_p}\right),$$

$$K_i = 2G'_{PI\infty}G'_{PD0}\frac{\omega_{PI}}{\omega_p}, \tag{4.15}$$

$$K_d = \frac{G'_{PI\infty}G'_{PD0}}{2}\left(1 - \frac{\omega_{PI}}{\omega_p}\right)\left(\frac{\omega_p}{\omega_{PD}} - 1\right).$$

For the above-mentioned equations to yield valid PID coefficients $K_p \geq 0$, $K_i \geq 0$, and $K_d \geq 0$, one must have

$$0 \leq \omega_{PI} \leq \omega_p,$$

$$0 \leq \omega_{PD} \leq \omega_p. \tag{4.16}$$

Letting $K_d = 0$ in (4.12) yields a simpler proportional-integral (PI)—or *lag*—compensation

$$G'_{PI}(p) = K_p + \frac{K_i}{T_s}\frac{1 + \dfrac{p}{\omega_p}}{p}$$

$$= G'_{PI\infty}\left(1 + \frac{\omega_{PI}}{p}\right), \tag{4.17}$$

where the parameters of the parallel and multiplicative forms are related by

$$K_p = G'_{PI\infty}\left(1 - \frac{\omega_{PI}}{\omega_p}\right),$$

$$K_i = 2G'_{PI\infty}\frac{\omega_{PI}}{\omega_p}. \tag{4.18}$$

These equations are special cases of (4.15) when $\omega_{PD} \to \omega_p$ and $G'_{PD0} \to 1$.

In another limiting case where the integral gain is zero, the proportional-derivative (PD)—or *lead*—compensator is obtained,

$$G'_{PD}(p) = K_p + K_d T_s \frac{p}{1 + \dfrac{p}{\omega_p}}$$

$$= G'_{PD0} \frac{1 + \dfrac{p}{\omega_{PD}}}{1 + \dfrac{p}{\omega_p}},$$

(4.19)

with

$$K_p = G'_{PD0},$$

$$K_d = \frac{G'_{PD0}}{2} \left(\frac{\omega_p}{\omega_{PD}} - 1 \right).$$

(4.20)

This form is a limit case of (4.15) when $\omega_{PI} \to 0$ and $G'_{PI\infty} \to 1$.

4.2 DESIGN EXAMPLES

The objectives in this section are to illustrate applications of the compensator design flow presented in Section 4.1 in several practical examples.

4.2.1 Digital Voltage-Mode Control of a Synchronous Buck Converter

The first example considered is the example of digital voltage-mode control of the synchronous buck converter introduced in Section 2.6. The converter exact discrete-time model and the control-to-output voltage transfer function $G_{vu}(z)$ are derived in Section 3.3.4. The magnitude and phase Bode plots of the uncompensated loop gain of the system,

$$T_u(z) = H G_{vu}(z),$$

(4.21)

are shown in Fig. 4.5. The design specifications are to achieve a crossover frequency equal to $f_c = f_s/10 = 100$ kHz with a phase margin $\varphi_m = 45°$. From the Bode plots in Fig. 4.5, it is clear that a suitable phase boost must be introduced by the control action. The PID compensator template is selected for the compensator $G_c(z)$.

Following the approach similar to the standard analog control design exemplified in Section 1.5.1, the compensator design is carried out in two successive steps: first, a PD compensator is designed that achieves the target crossover frequency and phase margin. Such PD compensator has two parameters and is uniquely determined by the f_c and φ_m specifications. Secondly, an integral action is introduced in the compensation. The rationale behind this two-step approach is that the proportional and derivative terms exploit their action at frequencies close to f_c, whereas the main

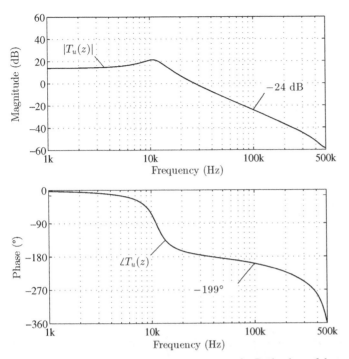

Figure 4.5 Synchronous Buck converter example: Bode plots of the uncompensated loop gain $T_u(z)$.

purpose of the integral term is to increase the low-frequency gain and thus ensure good static regulation.

The overall PID transfer function in the p-domain has the general form (4.14). The crossover frequency specification in the p-domain becomes, according to (4.9),

$$\omega_c' = \frac{2}{T_s}\tan\left(\omega_c\frac{T_s}{2}\right) \approx 2\pi \cdot (103.4 \text{ kHz}). \tag{4.22}$$

As expected, $\omega_c' \approx \omega_c$. Furthermore, one has

$$\omega_p = \frac{2}{T_s} \approx 2\pi \cdot (318 \text{ kHz}). \tag{4.23}$$

Following the method presented in Section 4.1.1, the p-domain transform $T_u'(p)$ of $T_u(z)$ should be evaluated first. It should be noted that the magnitude and phase values of $T_u(z)$ are required only *at the target crossover frequency* ω_c. Therefore, it is not necessary to perform the formal step of the z-to-p mapping of the entire uncompensated loop gain. In this example, the magnitude and the phase of T_u

at ω_c—or, equivalently, the magnitude and the phase of T_u' at ω_c'—are

$$
\begin{aligned}
|T_u(e^{j\omega_c T_s})| &= |T_u'(j\omega_c')| \approx 63.1 \cdot 10^{-3} \Rightarrow -24 \text{ dB}, \\
\angle T_u(e^{j\omega_c T_s}) &= \angle T_u'(j\omega_c') \approx -199°.
\end{aligned}
\tag{4.24}
$$

as illustrated in Fig. 4.5. The design objectives for the PD portion of the compensation can be expressed by the complex constraint that requires the $T'(j\omega_c')$ to have unity magnitude and $-\pi + \varphi_m$ phase at ω_c',

$$
T'(j\omega_c') \triangleq G_{PD}'(j\omega_c')T_u'(j\omega_c') = e^{j(-\pi+\varphi_m)}.
\tag{4.25}
$$

The complex constraint (4.25) corresponds to the magnitude and phase constraints

$$
|T_u'(j\omega_c')||G_{PD}'(j\omega_c')| = 1,
\tag{4.26}
$$

$$
\angle T_u'(j\omega_c') + \angle G_{PD}'(j\omega_c') = -\pi + \varphi_m.
\tag{4.27}
$$

From (4.27), using (4.19), one has

$$
\angle T_u'(j\omega_c') + \arctan\left(\frac{\omega_c'}{\omega_{PD}}\right) - \arctan\left(\frac{\omega_c'}{\omega_p}\right) = -\pi + \varphi_m,
\tag{4.28}
$$

which yields

$$
\boxed{\omega_{PD} = \frac{\omega_c'}{\tan\left(\varphi_m - \varphi_{m,u} + \arctan\left(\dfrac{\omega_c'}{\omega_p}\right)\right)},}
\tag{4.29}
$$

where the *uncompensated phase margin*

$$
\boxed{\varphi_{m,u} \triangleq \pi + \angle T_u'(j\omega_c') = \pi + \angle T_u(e^{j\omega_c T_s})}
\tag{4.30}
$$

represents the system phase margin when a purely proportional compensation is designed to obtain the crossover frequency f_c.

From (4.20), a valid solution $K_d > 0$ is found if and only if $0 < \omega_{PD} < \omega_p$. Applying this constraint to (4.29) yields an upper and a lower bound for the achievable phase margin φ_m,

$$
\boxed{\varphi_{m,u} < \varphi_m < \varphi_{m,u} + \frac{\pi}{2} - \arctan\left(\frac{\omega_c'}{\omega_p}\right).}
\tag{4.31}
$$

The phase margin bounds (4.31) have a straightforward interpretation. The lower bound for φ_m is equal to $\varphi_{m,u}$, which corresponds to the case when the compensation would not require a derivative action. The upper bound for φ_m expresses the maximum achievable phase boost by the digital PD compensator. If a phase margin greater than the upper bound is requested, it would not be possible to meet such specification using a valid, realizable PD compensator, and one would have to look for more complex compensator structures. Note that the effect of ω_p appears in the upper bound due to the phase-lagging effect of ω_p.

For the synchronous Buck converter under study, $\angle T_u'(j\omega_c') \approx -199°$ and therefore $\varphi_{m,u} \approx -19°$, while $\arctan(\omega_c'/\omega_p) = 18°$. Condition (4.31) becomes

$$-19° < \varphi_m < 53°, \tag{4.32}$$

which implies that the target $\varphi_m = 45°$ specification can be met. A valid PD compensation for the system then exists. From (4.29), ω_{PD} is equal to

$$\omega_{PD} = 2\pi \cdot (14.9 \text{ kHz}). \tag{4.33}$$

The PD gain G_{PD0}' is determined from the magnitude constraint (4.26),

$$G_{PD0}' = \frac{1}{|T_u'(j\omega_c')|} \frac{\sqrt{1 + \left(\dfrac{\omega_c'}{\omega_p}\right)^2}}{\sqrt{1 + \left(\dfrac{\omega_c'}{\omega_{PD}}\right)^2}}, \tag{4.34}$$

which yields

$$G_{PD0}' = 2.37. \tag{4.35}$$

Next, an integral action is introduced, the purpose of which is to null the steady-state regulation error and, more generally, increase the loop gain at low frequencies and therefore improve the in-band disturbance rejection.

The PI-related zero ω_{PI} should not appreciably alter the crossover frequency and the phase margin attained by the PD compensation. Similarly, the high-frequency PI gain $G_{PI\infty}'$ should not alter the loop gain magnitude in the neighborhood of ω_c. Choose then ω_{PI} to be one-twentieth of the crossover angular frequency ω_c,

$$\omega_{PI} = 2\pi \cdot (5 \text{ kHz}), \tag{4.36}$$

and

$$G_{PI\infty}' = 1. \tag{4.37}$$

The PID compensation transfer function (4.14) is now completely defined in p-domain, with

$$G_{PI\infty}' = 1,$$

$$G_{PD0}' = 2.37,$$

$$\omega_{PI} = 2\pi \cdot (5 \text{ kHz}),$$

$$\omega_{PD} = 2\pi \cdot (14.9 \text{ kHz}). \tag{4.38}$$

Finally, the z-domain PID gains are obtained from (4.15),

$$K_p = 3.09,$$

$$K_i = 74.52 \cdot 10^{-3}, \qquad (4.39)$$

$$K_d = 23.8.$$

Figure 4.6 illustrates the Bode plots of the final compensator transfer function before and after the introduction of the integral term, whereas Fig. 4.7 reports both the final compensated loop gain and the loop gain with PD compensation only. As expected, both the crossover frequency and the phase margin are essentially unaltered after the integral term is included, whereas the low-frequency magnitude of the loop gain is substantially improved. As a verification, Fig. 4.7 also reports the simulated system loop gain, obtained following the Middlebrook's injection approach [78], similar to the simulation performed in the analog controlled converter in Section 1.5.1. Figure 4.8 depicts the setup used to determine the z-domain loop gain and verify the modeling and design results by simulation. Note that a discrete-time perturbation $u_{pert}[k]$, sampled at the converter switching frequency, is used in Fig. 4.8, rather

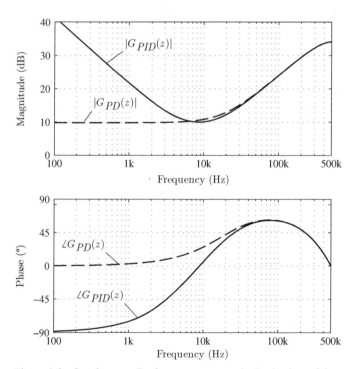

Figure 4.6 Synchronous Buck converter example: Bode plots of the compensator transfer function $G_c(z)$: proportional-derivative (PD) part (dashed line) and the overall proportional-integral-derivative (PID) compensator (solid line).

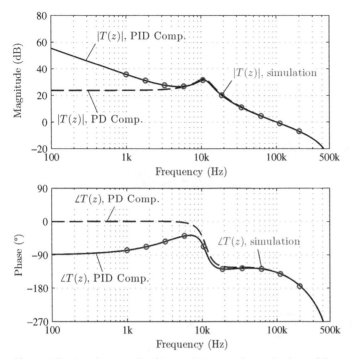

Figure 4.7   Synchronous Buck converter example: Bode plots of the compensated voltage loop gain $T(z)$ with proportional-derivative compensation (dashed line) and after the introduction of the integral term (solid line).

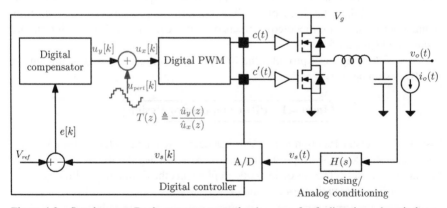

Figure 4.8   Synchronous Buck converter example: the setup for finding the z-domain loop gain $T(z)$ by simulation, which emulates operation of a network analyzer.

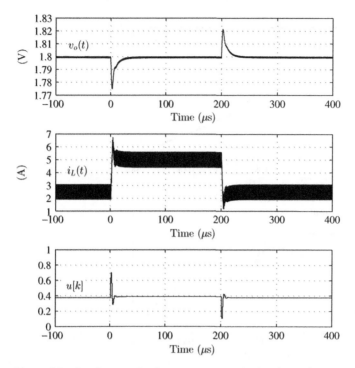

Figure 4.9 Synchronous Buck converter example: 2.5 A ↔5 A step-load response.

than a continuous-time sinusoidal perturbation. Figure 4.9 illustrates the simulated closed-loop response of the converter following a 5 to 2.5 A step-up/step-down load current transient. An expanded view of the step-up case is reported in Fig. 4.10. The simulations are carried out according to the timing diagram of Fig. 3.4 and therefore account for the total delay of the digital control loop. Amplitude quantization, on the other hand, is not modeled—effects of finite A/D and DPWM resolutions are discussed in Chapter 5.

The following inset provides the Matlab® instructions required to implement the PID design described earlier.

Inset 4.1 – PID Compensator Design

Assume the transfer function object Tuz has been calculated (refer to Inset 3.3) and that variables wc and mphi represent the target crossover angular frequency and phase margin, respectively. The following code implements the design of the PID compensator according to equations (4.29) and (4.36).

```
%   Target crossover frequency and phase margin
wc = 2*pi*100e3;
phm = (pi/180)*45; % In radians
```

```
%    Magnitude and phase of Tuz at the target crossover frequency
[m,p]    =    bode(Tuz,wc);
p        =    (pi/180)*p;
%    Prewarping on wc
wcp      =    (2/Ts)*tan(wc*Ts/2);
%    PD Design
wp  =    2*pi*fs/pi;
pw  =    atan(wcp/wp);
wPD =    1/(tan(-pi+mphi-p+pw)/wcp)
GPD0 =    sqrt(1+(wcp/wp)^2)/(m*(sqrt(1+(wcp/wPD)^2)))
%    PI zero and high-frequency gain
wPI =    wc/20;
GPIinf = 1;
%    Proportional, Integral and Derivative Gains
Kp  =    GPIinf*GPD0*(1+wPI/wPD-2*wPI/wp);
Ki  =    2*GPIinf*GPD0*wPI/wp;
Kd  =    GPIinf*GPD0/2*(1-wPI/wp)*(wp/wPD-1);
%    PID Transfer function
z   =    tf('z',Ts);
Gcz =    Kp + Ki/(1-z^-1) + Kd*(1-z^-1);
```

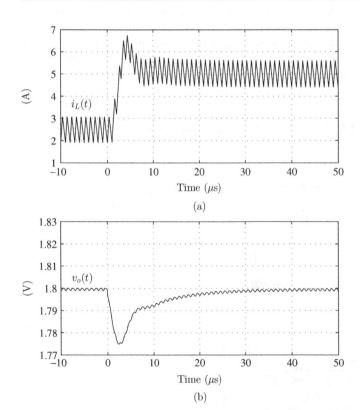

(a)

(b)

Figure 4.10   Synchronous Buck converter example: expanded view of the 2.5 A→5 A
step-load response: (a) inductor current and (b) output voltage.

4.2.2   Digital Current-Mode Control of a Boost Converter

The Boost converter with digital current-mode control introduced in Section 3.2.2 is reported in Fig. 4.11 for convenience. The system parameters are listed in Table 1.3, the same as in the analog current-mode control example in Section 1.5.2. Modeling of the Boost converter, including a derivation of the control-to-inductor current transfer function $G_{iu}(z)$ using the discrete-time modeling approach, is presented in Section 3.2.2. Recall that a symmetrical PWM carrier is employed, with sampling instants located at the midpoints of the turn-off intervals. As discussed in Section 2.2.1, this enables sampling of the true average value of $i_L(t)$.

The Bode plots of the uncompensated current loop gain $T_u(z)$,

$$T_u(z) = R_{sense}G_{iu}(z), \qquad (4.40)$$

are illustrated in Fig. 4.12. Not surprisingly, the phase response remains well-behaved, only mildly lagging below $-90°$ due to the delays in the control loop. A PI controller is therefore sufficient to compensate the system dynamics.

With the PI compensation designed for an $f_c = f_s/10 = 10$ kHz crossover frequency and $\varphi_m = 50°$ phase margin, the Bode plots of the compensated loop gain $T(z)$ and the values obtained by simulations (for verification purposes) are superimposed in Fig. 4.12.

Design of the PI compensator proceeds along the same lines as in Section 4.2.1, except that a unique solution is found given that the two design constraints (crossover frequency and phase margin) are used to determine the two

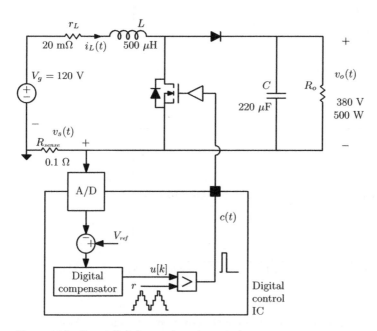

Figure 4.11   Boost digital current-mode control.

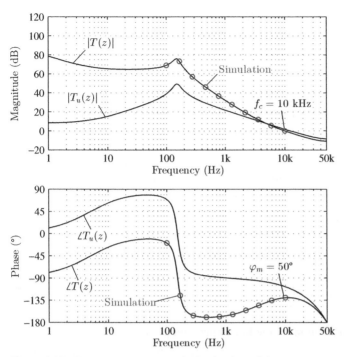

Figure 4.12 Boost converter example: Bode plots of the uncompensated loop gain $T_u(z)$ and the loop gain $T(z)$.

parameters of the PI compensator (the proportional and the integral gain). In this example, $\omega_p = 2\pi \cdot (31.8 \text{ kHz})$ and $\omega_c' \approx 2\pi \cdot (10.3 \text{ kHz})$.

First, using (4.40), express the magnitude and phase of the uncompensated loop gain at the desired crossover frequency as

$$|T_u(e^{j\omega_c T_s})| = |T_u'(j\omega_c')| \approx 1.23 \Rightarrow 1.78 \text{ dB},$$
$$\angle T_u(e^{j\omega_c T_s}) = \angle T_u'(j\omega_c') \approx -108°. \tag{4.41}$$

The uncompensated phase margin $\varphi_{m,u}$, defined as in (4.30), equals $\varphi_{m,u} = 72°$.

Consider next the PI transfer function in the p-domain, as per (4.17),

$$G_{PI}'(p) = G_{PI\infty}'\left(1 + \frac{\omega_{PI}}{p}\right), \tag{4.42}$$

and impose the complex design constraint

$$T'(j\omega_c') \triangleq G_{PD}'(j\omega_c')T_u'(j\omega_c') = e^{j(-\pi+\varphi_m)}. \tag{4.43}$$

Solving for ω_{PI} and $G_{PI\infty}'$ yields

$$\boxed{\omega_{PI} = \omega_c' \tan(\varphi_{m,u} - \varphi_m)}, \tag{4.44}$$

$$G'_{PI\infty} = \frac{1}{|T'_u(j\omega'_c)|} \frac{1}{\sqrt{1 + \left(\dfrac{\omega_{PI}}{\omega'_c}\right)^2}}. \tag{4.45}$$

Notice again that bounds can be found to determine whether a valid PI compensator exists. From (4.18), conditions $K_p > 0$ and $K_i > 0$ translate into $0 < \omega_{PI} < \omega_p$. Applied to (4.44), these constraints become

$$\varphi_{m,u} - \arctan\left(\frac{\omega_p}{\omega'_c}\right) < \varphi_m < \varphi_{m,u}, \tag{4.46}$$

which states the fact that a PI compensation cannot boost the phase beyond $\varphi_{m,u}$, nor can reduce the uncompensated phase margin below a certain limit. In the considered example, $\arctan(\omega_p/\omega'_c) \approx 72°$. Hence

$$\varphi_{m,u} - \arctan\left(\frac{\omega_p}{\omega'_c}\right) = 0° < \varphi_m = 50° < \varphi_{m,u} = 72°, \tag{4.47}$$

and therefore a valid solution for the PI compensator exists,

$$\omega_{PI} \approx 2\pi \cdot (4.2 \text{ kHz}),$$
$$G'_{PI\infty} \approx 0.754 \Rightarrow -2.45 \text{ dB}. \tag{4.48}$$

Finally, solving for the proportional and integral gains of the digital compensator via (4.18) yields

$$K_p \approx 0.6543,$$
$$K_i \approx 0.2. \tag{4.49}$$

Figure 4.13 illustrates the simulated closed-loop response of the system to a 4.2 A→2.1 A step of the current setpoint, corresponding to a 500 W→250 W power transient. The transient response is consistent with the frequency-domain design specifications.

Various other approaches to digital current-mode control have been discussed in the literature, including predictive current-mode control [135, 136] and digital current-mode control based on low-resolution current sensing [137].

4.2.3 Multiloop Control of a Synchronous Buck Converter

In analog controlled dc–dc converters, a multiloop control approach is often applied, where a wide-bandwidth current control loop is nested inside an outer, lower-bandwidth loop regulating the output voltage. Advantages of this approach include the ability to directly control and limit the current during transients as a protection feature, as well as the ability to ensure current sharing in systems where

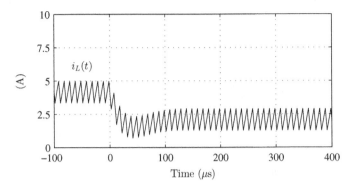

Figure 4.13   Boost converter example: 500 W→250 W step reference response.

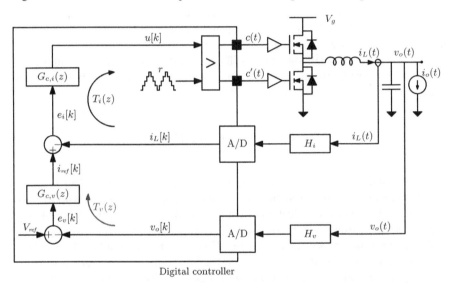

Figure 4.14   Digital multiloop control of a synchronous Buck converter.

multiple converters are operated in parallel. Furthermore, compensation of the system is overall less critical and more robust than in the pure voltage-mode control scheme.

A digital multiloop controller of the synchronous buck converter examined previously is shown in Fig. 4.14. The converter is driven by a symmetrical digital PWM modulator, and both the inductor current and the output voltage are sampled at the carrier peak, that is, at the middle of the turn-off interval. For the inductor current, this implies that a digital average current-mode control of $i_L(t)$ is implemented. It is assumed that a timing diagram such as the one illustrated in Fig. 3.11 holds for this system as well, with the same constraints regarding control calculations and consequent duty cycle limitation already discussed in Section 4.2.2.

As reported in Fig. 4.14, the inner control loop regulating the inductor current receives its setpoint $i_{ref}[k]$ from the voltage loop regulator. The basic idea behind multiloop control is to decouple the dynamics of current from the voltage by making

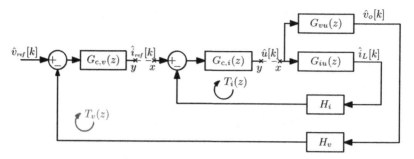

Figure 4.15   Small-signal block diagram of the digital multiloop control.

the current control loop much faster. This philosophy directly reflects upon the design procedure, so that the current loop design and the voltage loop design are carried out in a two-step sequence, rather than simultaneously.

The equivalent small-signal block diagram of the control system is illustrated in Fig. 4.15. Consider the current loop compensator design first. Assuming that the voltage loop is *open*, the current loop gain $T_i(z)$ is

$$T_i(z) \triangleq -\frac{\hat{u}_y(z)}{\hat{u}_x(z)} = G_{c,i}(z)H_iG_{iu}(z), \qquad (4.50)$$

where H_i is the inductor current sensing gain and $G_{c,i}(z)$ is the current loop compensator transfer function to be designed. It is assumed that the sensing gain is normalized to $H_i = 1$. The control-to-inductor current transfer function $G_{iu}(z)$ can be obtained straightforwardly from the converter models derived in Section 3.2.1. The Bode plots of the uncompensated loop gain $T_{u,i}(z) = H_iG_{iu}(z)$, reported in Fig. 4.16, exhibit well-behaved dynamics up to a significant fraction of the Nyquist rate, a situation that only differs from the familiar analog control case by the additional phase lag associated with the overall loop delay

$$t_d = t_{DPWM} = \frac{T_s}{2} = 500 \text{ ns}. \qquad (4.51)$$

Below resonance, $T_{u,i}(z)$ has a slope of 20 dB/decade as a result of a zero located in close proximity of $z = 1$. Notice that the sampling strategy considered in this example is such that the sampled current very closely follows the actual average current. As a result, the low-frequency behavior of $G_{iu}(z)$ accurately resembles its s-domain counterpart $G_{id}(s)$, and a zero at dc is predicted.[1] On the basis of the above-mentioned

---

[1] A closer examination of the location of the zero shows a value slightly different from $z = 1$. This is due to the fact that the inductor current waveform, regardless of the sampling strategy, is never *exactly* triangular, and therefore the sampled current slightly departs from the average current. The z-domain model correctly accounts for this aliasing effect, which nonetheless remains unimportant from a practical standpoint.

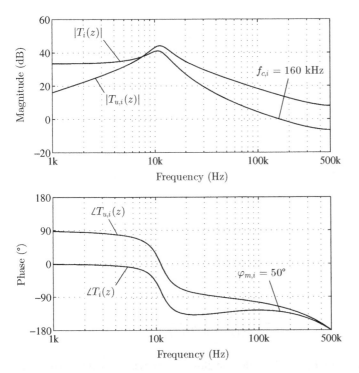

Figure 4.16   Multiloop control example: Bode plots of the current loop gain.

considerations, a PI compensation

$$G_{c,i}(z) = K_{p,i} + \frac{K_{i,i}}{1 - z^{-1}} \qquad (4.52)$$

can be designed for the inner current loop. Following the same steps as in the Boost current-mode controller of Section 4.2.2 and assuming a crossover frequency $\omega_{c,i} = 2\pi \cdot (160 \text{ kHz})$ and a phase margin $\varphi_{m,i} = 50°$, one derives the parameters $K_{p,i}$ and $K_{i,i}$ of the current loop compensator,

$$K_{p,i} \approx 0.1637,$$
$$K_{i,i} \approx 0.0468. \qquad (4.53)$$

The Bode plots of the compensated current loop gain $T_i(z)$ are also reported in Fig. 4.16. The finite loop gain value in the low-frequency range is due to the cancellation between the PI integrator pole and the zero of $T_{u,i}(z)$ located at $z \approx 1$. A finite low-frequency loop gain is a rather common situation in average current-mode controllers, which may be addressed by embedding a double integrator in the compensator transfer function. In the multiloop control system considered here, however, it is important to realize that any regulation error in the inner current loop is entirely offset by the outer voltage loop, which adjusts the current setpoint accordingly. For this reason, the finite low-frequency current loop gain is not a concern.

Referring again to Fig. 4.15, the voltage loop gain is evaluated with the current loop *closed*,

$$T_v(z) \triangleq -\frac{\hat{i}_{ref,y}(z)}{\hat{i}_{ref,x}(x)} = G_{c,v}(z)H_vG_{vu}(z)\frac{G_{c,i}(z)}{1+T_i(z)}, \tag{4.54}$$

where $G_{c,v}(z)$ is the voltage loop compensator transfer function to be designed. In this numerical example, it is assumed that $H_v = 1$. The uncompensated voltage loop gain $T_{u,v}(z)$ can be manipulated as follows:

$$\begin{aligned} T_{u,v}(z) &= H_vG_{vu}(z)\frac{G_{c,i}(z)}{1+T_i(z)} \\ &= \frac{H_v}{H_i}\frac{G_{vu}(z)}{G_{iu}(z)}\frac{G_{c,i}(z)H_iG_{iu}(z)}{1+T_i(z)} \\ &= \frac{H_v}{H_i}\frac{G_{vu}(z)}{G_{iu}(z)}\frac{T_i(z)}{1+T_i(z)}. \end{aligned} \tag{4.55}$$

This exact expression for $T_{u,v}(z)$ includes the closed-loop dynamics of the inner current loop expressed by the term $T_i(z)/(1+T_i(z))$. In the limit of very large current loop bandwidth—or, more precisely, for $\omega \ll \omega_{c,i}$—one has

$$\frac{T_i(z)}{1+T_i(z)} \approx 1 \Rightarrow T_{u,v}(z) \approx \frac{H_v}{H_i}\frac{G_{vu}(z)}{G_{iu}(z)}. \tag{4.56}$$

The quantity

$$Z_{vi}(z) \triangleq \frac{G_{vu}(z)}{G_{iu}(z)} = \frac{\hat{v}_o(z)}{\hat{i}_L(z)} \tag{4.57}$$

is *not* a converter transfer function because $\hat{i}_L$ is not an input for the control system. Nonetheless, it is a property of the converter, as it expresses the small-signal relationship between the sampled inductor current and the sampled output voltage. As the voltage loop crossover frequency $\omega_{c,v}$ is usually selected so that $\omega_{c,v} \ll \omega_{c,i}$, approximation (4.56) is usually well satisfied and the design of $G_{c,v}(z)$ can be carried out by taking

$$T_{u,v}(z) \approx \frac{H_v}{H_i}Z_{vi}(z) \tag{4.58}$$

as an approximate expression for $T_{u,v}(z)$. Figure 4.17 illustrates the exact Bode plots of $T_{u,v}(z)$ along with the approximation (4.58). Not surprisingly, the frequency response of $T_{u,v}(z)$ resembles that of a capacitive impedance $1/sC$.

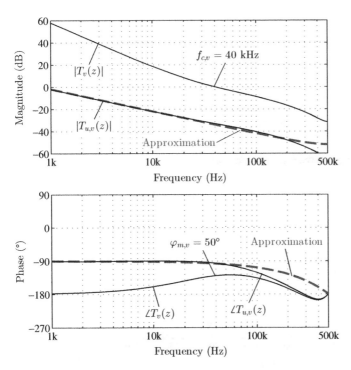

Figure 4.17 Multiloop control example: Bode plots of the voltage loop gain.

Given the frequency response of $T_{u,v}(z)$, a digital PI compensator

$$G_{c,v}(z) = K_{p,v} + \frac{K_{i,v}}{1 - z^{-1}} \tag{4.59}$$

can be designed for $\omega_{c,v} = 2\pi \cdot (40 \text{ kHz})$ and $\varphi_{m,v} = 50°$, leading to

$$
\begin{aligned}
K_{p,v} &\approx 33.81, \\
K_{i,v} &\approx 6.34.
\end{aligned}
\tag{4.60}
$$

The Bode plots of the compensated voltage loop gain $T_v(z) = G_{c,v}(z)T_{u,v}(z)$ are reported in Fig. 4.17 as well. The closed-loop transient performance of the designed multiloop controlled converter is illustrated in Fig. 4.18, which shows the response to a 0 A→5 A→0 A step-load sequence.

4.2.4 Boost Power Factor Corrector

Single-phase power factor correctors (PFCs) are ac–dc rectifiers widely employed as input stages of electrical and electronic appliances, for the purpose of taking power

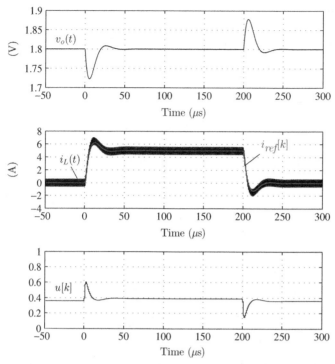

Figure 4.18 Multiloop control example: closed-loop response to a 0 A→5 A→0 A step-load sequence.

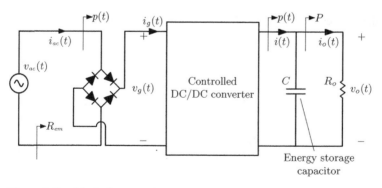

Figure 4.19 Block diagram of a PFC system.

from the electric power grid ideally at unity power factor, and with low harmonic content of the input ac current. More details about operation, analysis, design, modeling, and analog control of PFC rectifiers can be found in power electronics textbooks [1, 2, 4, 5], with only a brief summary provided here before addressing discrete-time modeling and digital control design details.

A general block diagram of a PFC rectifier is reported in Fig. 4.19, with steady-state waveforms illustrating operating principles common to all PFC

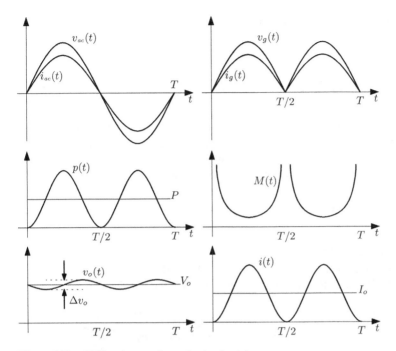

Figure 4.20   PFC main waveforms in steady state.

realizations illustrated in Fig. 4.20. Let $v_{ac}(t)$ and $i_{ac}(t)$ be the ac line voltage and current, respectively. These are first rectified by a diode bridge and then processed by a dc–dc converter controlled so as to (i) emulate a resistive load on the input side and (ii) regulate the dc voltage V_o at the output, which supplies dc power P to a load or a downstream power conversion stage modeled as a resistive load R_o.

With the objective of taking power from the ac line at unity power factor and assuming sinusoidal ac line voltage, one has

$$v_{ac}(t) = \sqrt{2}V_{ac,rms}\sin(\omega t)$$
$$i_{ac}(t) = \sqrt{2}\frac{V_{ac,rms}}{R_{em}}\sin(\omega t),$$

(4.61)

where $\omega = 2\pi/T$ is the grid angular frequency and the *emulated resistance* R_{em} related to the active power P demanded by the dc load. Assuming that PFC losses can be neglected, we have

$$\frac{V_{ac,rms}^2}{R_{em}} = P.$$

(4.62)

The rectified voltage $v_g(t)$ and current $i_g(t)$ at the input of the dc–dc stage are

$$v_g(t) = |v_{ac}(t)| = \sqrt{2}V_{ac,rms}|\sin(\omega t)|,$$

$$i_g(t) = |i_{ac}(t)| = \sqrt{2}\frac{V_{ac,rms}}{R_{em}}|\sin(\omega t)|. \tag{4.63}$$

Observe that $v_g(t)$ and $i_g(t)$ have a fundamental frequency equal to twice the line frequency. One consequence of the above-mentioned equations is that the dc–dc converter in the PFC rectifier must operate with a voltage conversion ratio $M(t) = V_o/v_g(t)$ that varies over time with a period equal to one-half of the ac line period,

$$M(t) \triangleq \frac{V_o}{v_g(t)} = \frac{V_o}{\sqrt{2}V_{ac,rms}}\frac{1}{|\sin(\omega t)|}, \tag{4.64}$$

as illustrated in Fig. 4.20.

Examine now the *instantaneous* power $p(t)$ absorbed from the line. It consists of the dc term P plus a fluctuating component at twice the line frequency,

$$p(t) \triangleq v_{ac}(t)i_{ac}(t) = P(1 - \cos 2\omega t) = \underbrace{P}_{Active\ power} - \underbrace{P\cos(2\omega t)}_{Fluctuating\ power}. \tag{4.65}$$

As the PFC output power is constant, the fluctuating power must be reactively exchanged between the PFC and the grid. In other words, an energy storage element—most commonly the output capacitor, as in Fig. 4.19—must be appropriately sized in order to filter the effect of the fluctuating power on the output voltage ripple. The PFC output voltage $v_o(t)$ therefore consists of the regulated dc component V_o, plus a small ripple at twice the line frequency originated by the fluctuating power exchanged by the storage element. A simplified—although accurate enough for practical design purposes—design equation for the energy storage element based on the target peak-to-peak voltage ripple Δv_o can be derived from (4.65),

$$\Delta v_o \approx \frac{P}{\omega C V_o}. \tag{4.66}$$

On the other hand, in consequence of the large energy exchange between the PFC converter and the energy storage element, current $i(t)$ output by the dc–dc converter exhibits a large ripple component at twice the line frequency, superimposed to the dc current I_o absorbed by the load. Under the small-ripple approximation for the output voltage, one has

$$i(t) = \frac{p(t)}{v_o(t)} \approx \frac{p(t)}{V_o} = I_o - I_o\cos(2\omega t), \tag{4.67}$$

which is also depicted in Fig. 4.20.

In a practical PFC implementation, the dc–dc converter introduces additional harmonic components due to its switching activity. In this sense, the above-mentioned quantities must be interpreted as the *averaged* versions of the instantaneous signals over the converter switching period T_s, that is,

$$p(t) \rightarrow \overline{p}(t),$$
$$i_g(t) \rightarrow \overline{i}_g(t),$$
$$v_o(t) \rightarrow \overline{v}_o(t),$$
$$i(t) \rightarrow \overline{i}(t).$$

(4.68)

The PFC control example discussed here is based on the Boost converter. A block diagram of the digitally controlled Boost PFC is shown in Fig. 4.21, with the system parameters given in Table 4.1. The power stage parameters are the same as in the Boost dc–dc converter with digital current-mode control considered in Section 4.2.2, which provides a starting point in the application of digital control design principles to the PFC example.

In the case study scenario, the typical US household ac voltage (120 V, 60 Hz) is rectified and elevated to supply a 380 V, 500 W dc load. At this power level, the RMS input current and the dc output current are

$$I_{ac,rms} = \frac{P}{V_{ac,rms}} = \frac{500 \text{ W}}{120 \text{ V}} \approx 4.2 \text{ A},$$
$$I_o = \frac{P}{V_o} = \frac{500 \text{ W}}{380 \text{ V}} \approx 1.3 \text{ A}.$$

(4.69)

With the selected output capacitor, the expected peak-to-peak output voltage ripple is, according to (4.66),

$$\Delta v_o \approx 16 \text{ V},$$

(4.70)

TABLE 4.1   Boost PFC Example Parameters

Parameter	Value
Input voltage RMS $V_{ac,rms}$	120 V
Line frequency ω	$2\pi \cdot (60 \text{ Hz})$
Output voltage V_o	380 V
Output power P	500 W
Switching frequency f_s	100 kHz
Inductance L	500 µH
Inductor series resistance r_L	20 mΩ
Filter capacitance C	220 µF
Output voltage sensing gain H_v	1 V/V
Current sensing gain R_{sense}	0.1 V/A

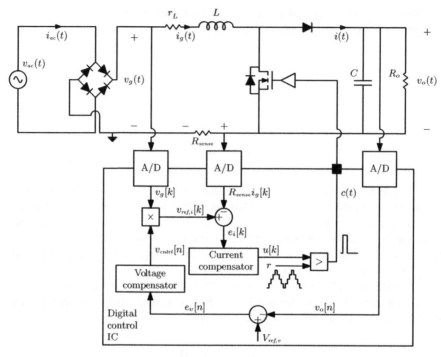

Figure 4.21 Digitally controlled Boost power factor corrector.

or $\approx 4\%$ of the dc value. The inductor peak-to-peak current ripple,

$$\Delta i_L = \frac{v_g}{L} dT_s \leq \frac{\sqrt{2}V_{ac,rms}}{f_s L} \left(1 - \frac{\sqrt{2}V_{ac,rms}}{V_o} \right) \approx 1.88 \text{ A}, \qquad (4.71)$$

has the maximum value occurring around the peak of the line voltage. The Boost converter operates in CCM throughout the entire line cycle as long as [1]

$$R_{em} \leq 2Lf_s \Leftrightarrow P \geq 144 \text{ W}. \qquad (4.72)$$

The PFC digital controller considered here closely follows the well-known multiplier-based analog implementation [1]. As shown in Fig. 4.21, it is a multiloop control system with two distinct control loops:

- A *wide-bandwidth current loop* is employed to control the rectified current $\bar{i}_g(t)$ so as to closely follow the rectified input voltage waveshape $v_g(t)$, so that the input port of the rectifier behaves as the emulated resistance R_{em}. A variable setpoint $v_{ref}[k]$ for the inner current loop is obtained by multiplying the rectified voltage $v_g(t)$ by a controllable signal v_{cntrl},

$$v_{ref,i}[k] \triangleq k_x v_g[k] v_{cntrl}, \qquad (4.73)$$

where k_x is a proportionality constant of the digital multiplier. The scale factor between $v_g(t)$ and $\bar{i}_g(t)$ represents the emulated resistance R_{em}, and its value is determined by v_{cntrl}. In steady-state one has $v_{cntrl} = V_{cntrl}$. Assuming that $R_{sense}\bar{i}_g \approx v_{ref,i}$ due to the wide-bandwidth current loop, one has

$$R_{sense}\bar{i}_g \approx k_x v_g V_{cntrl} \Rightarrow \boxed{R_{em} \triangleq \frac{v_g}{\bar{i}_g} = \frac{R_{sense}}{k_x V_{cntrl}}}, \quad (4.74)$$

which yields the dependence of R_{em} on the voltage control command v_{cntrl}. Furthermore, plugging the above-mentioned equation into (4.62) yields

$$\boxed{P = k_x \frac{V_{ac,rms}^2}{R_{sense}} V_{cntrl}}. \quad (4.75)$$

Therefore, v_{cntrl} directly controls the average power drawn from the line.

- A *low-bandwidth voltage loop* is established to regulate the converter output voltage at a constant value $V_{ref,v}$. When the output voltage falls below $V_{ref,v}$, the voltage feedback increases v_{cntrl}, forcing a higher ac power absorption and restoring the regulation. The opposite mechanism occurs when the output voltage exceeds the desired setpoint. In this sense, the voltage loop acts as a *power balancing* loop.

A large bandwidth separation between the fast current loop and the slow voltage loop is essential to guarantee proper operation of the PFC system. The current loop bandwidth $f_{c,i}$ is normally designed to be as large as possible, typical values being around one-seventh to one-tenth of the converter switching rate f_s. On the other hand, an excessively rapid adjustment of R_{em} by the voltage loop would inevitably distort the input current and ultimately compromise the power factor correction function. The voltage loop bandwidth $f_{c,v}$ is typically limited to much less than twice the line frequency. In summary,

$$\underbrace{f_{c,v}}_{\text{Voltage loop bandwidth}} \quad \ll \quad \underbrace{2f}_{\text{Twice the line frequency}} \quad \ll \quad \underbrace{f_{c,i}}_{\text{Current loop bandwidth}} \quad \ll \quad \underbrace{f_s}_{\text{Switching frequency}} .$$
$$(4.76)$$

One additional control provision, not shown in Fig. 4.21 but frequently adopted in practical PFC implementations, is to make the multiplier gain k_x inversely proportional to the squared input RMS voltage $V_{ac,rms}^2$. Referred to as *input voltage feedforward*, this technique allows to suppress disturbances in $v_{ac}(t)$ from the converter feedback [1]. As the input voltage feedforward does not alter the control-to-output dynamics of the converter, it is not considered in this example.

Consider first the design of the inner, wide-bandwidth current loop. The purpose of the inner loop is simply to track the current setpoint $v_{ref,i}[k]$. In this regard, the modeling technique and the design approach are the same as for the dc–dc case examined in Section 4.2.2. An additional issue to be addressed, however, is that the converter input voltage $v_g(t)$ is now slowly varying between

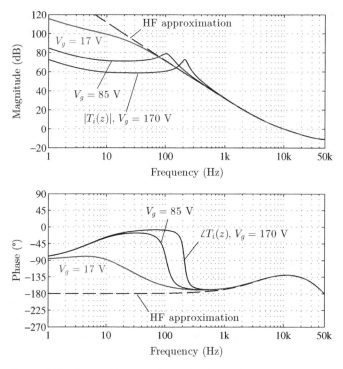

Figure 4.22 Boost PFC example: Bode plots of the current loop gain for three different V_g's.

0 V and $\sqrt{2}V_{ac,rms} \approx 170$ V because of the variations in the ac line voltage. As the fundamental frequency of $v_g(t)$ is $2f \ll f_{c,i}$, its motion is relatively slow compared with the dynamics of the wide-bandwidth current loop. The variation of $v_g(t)$ is therefore approximated as a *quasi-static* change of the converter operating point. Figure 4.22 illustrates the current loop gain $T_i(z)$, designed for $f_{c,i} = 10$ kHz and $\varphi_{m,i} = 50°$ at the peak input voltage $V_g = 170$ V, together with the same loop gain evaluated at two other intermediate values, $V_g = 17$ V and $V_g = 85$ V. In plotting these latter two cases, the steady-state duty cycle D is recalculated to keep $V_o = 380$ V. Small-signal-wise, under the quasi-static approximation, the change in the operating point due to the change in $v_g(t)$ affects the low-frequency portion of the current loop gain but keeps unaltered both the crossover frequency and the phase margin. A high-frequency approximation of $T_i(z)$, derived from (3.56), is also depicted in Fig. 4.22 to illustrate that the crossover frequency and the phase margin of $T_i(z)$ can be designed independent from the value of $v_g(t)$.[2] In summary, the design of the wide-bandwidth inner current loop involves the same design tools and methodologies seen for the dc–dc case.

---

[2] It is important to mention that while this property holds for the Boost converter, it is not generally true for other converter topologies.

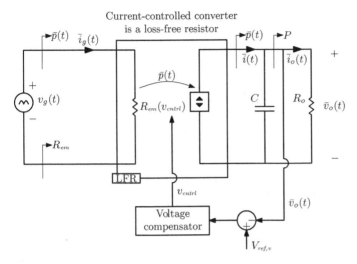

Figure 4.23   Loss-free resistor as the core of the voltage loop dynamics.

Consider now the problem of designing the low-bandwidth outer voltage loop responsible for regulating the PFC output voltage. As anticipated, such loop acts on the value of R_{em} by means of v_{cntrl} in order to maintain the power balance between the input and the output. In doing so, a dynamic response much slower than twice the line frequency is desired not to excessively distort the input current.

In the framework of continuous-time modeling, the uncompensated dynamics of the outer voltage loop is studied with the aid of the *loss-free resistor* (LFR) model, illustrated in Fig. 4.23 [1]. Assuming the inner current loop operates ideally, the input port of the converter behaves as a resistance R_{em} controlled by v_{cntrl}. On the other hand, the output port becomes a *controlled power source*, which transmits to the output the average power absorbed from the line. Equations of the LFR are

$$v_g(t) = R_{em}(v_{cntrl})\bar{i}_g(t),$$

$$\bar{v}_o(t)\bar{i}(t) = \frac{v_g^2(t)}{R_{em}(v_{cntrl})}. \tag{4.77}$$

The LFR model yields nonlinear, time-varying equations due to the nonlinearity of the controlled power source element and due to the input voltage variation, which produces a power oscillation at twice the line frequency. To handle this, the LFR model is first averaged over half the line period, an operation that removes the time-varying nature of the equations in the same manner as the conventional averaging operation over a switching period is used to obtain a time-invariant nonlinear model of a switched-mode converter. Denote with

$$\bar{\bar{v}}_o(t) \triangleq \langle \bar{v}_o(t) \rangle_{T_{2L}} \tag{4.78}$$

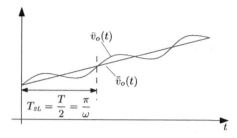

Figure 4.24 Removal of the components of $v_o(t)$ at the harmonics of the line frequency, by averaging over one-half of the line period T_{2L}.

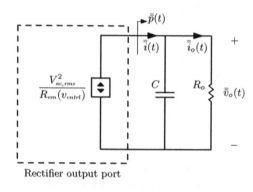

Rectifier output port

Figure 4.25 Model of the LFR output port averaged over one-half of the line period.

the output voltage averaged first over T_s, then over the half line period $T_{2L} \triangleq \pi/\omega$ (Fig. 4.24). Applying the above-mentioned averaging operation to the instantaneous power $\bar{p}(t)$ removes the fluctuating term,

$$\bar{\bar{p}}(t) = \frac{V_{ac,rms}^2}{R_{em}(v_{cntrl}(t))} = k_x \frac{V_{ac,rms}^2}{R_{sense}} v_{cntrl}(t). \qquad (4.79)$$

The averaged version of the LFR controlled power source is illustrated in Fig. 4.25. This model is still nonlinear, but is time-invariant, and can be linearized around its steady-state operating point to find the small-signal dynamics between the control input v_{cntrl} and the PFC output voltage.

In developing a digital version of the voltage regulation loop, the first item to be addressed is sampling, which brings in the same issues already discussed in Section 2.2.1. The main difference is that relevant harmonics responsible of aliasing effects are in this case those originating from the fluctuating component of the instantaneous power $\bar{p}(t)$ at twice the line frequency and its multiples. Following the considerations developed in Section 2.2.1, a suitable approach is to sample the PFC output voltage at twice the line frequency. Denote then the sampled output voltage as

$$v_o[n] \triangleq v_o(t = nT_{2L}), \qquad (4.80)$$

index n counting the number of half line periods T_{2L} rather than the number of converter switching periods. With this sampling strategy, one inherently eliminates the baseband spectral components originated by sampling, and the residual aliasing effect only manifests at dc. Furthermore, because of the small ripple Δv_o superimposed to the output voltage, it is appropriate to invoke the small-aliasing approximation on $v_o[n]$ and assume that the residual dc aliasing effects remain negligibly small as well. In conclusion,

$$\boxed{v_o[n] \approx \bar{\bar{v}}_o(nT_{2L})}. \tag{4.81}$$

Qualitative spectra of the PFC output voltage $\bar{v}_o(t)$, of its averaged version $\bar{\bar{v}}_o(t)$ and of $v_o[n]$, are shown in Fig. 4.26.

The small-aliasing approximation (4.81) allows formulation of the discrete-time dynamics of the uncompensated voltage loop as a discretization of its continuous-time averaged dynamics. Start from the nonlinear equation governing the LFR dynamics over half the line period. As shown in Fig. 4.25, one has

$$\frac{d\bar{\bar{v}}_o}{dt} = \frac{1}{C}\left(\bar{\bar{i}}(t) - \bar{\bar{i}}_o(t)\right)$$

$$= \frac{1}{C}\left(\frac{\bar{\bar{p}}(t)}{\bar{\bar{v}}_o} - \frac{\bar{\bar{v}}_o}{R_o}\right) \tag{4.82}$$

$$= -\frac{\bar{\bar{v}}_o(t)}{R_o C} + k_x \frac{V_{ac,rms}^2}{R_{sense}C}\frac{v_{cntrl}(t)}{\bar{\bar{v}}_o(t)}.$$

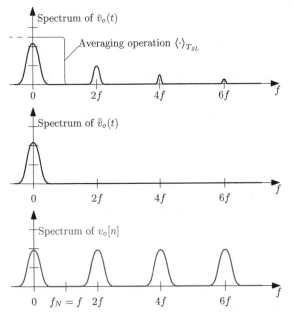

Figure 4.26   Qualitative spectra of (top) the PFC output voltage, (middle) its averaged version over half line period, and (bottom) the sampled PFC output voltage.

Integration of this equation from an arbitrary initial condition $\overline{\overline{v}}_o(0)$ yields[3]

$$\overline{\overline{v}}_o(t) = \sqrt{e^{-\frac{2t}{R_o C}}\overline{\overline{v}}_o^2(0) + 2k_x \frac{V_{ac,rms}^2}{R_{sense}C} \int_0^t e^{-2\frac{t-\tau}{R_o C}} v_{cntrl}(\tau)\, d\tau}. \qquad (4.83)$$

This equation can be discretized in order to obtain the sampled nonlinear dynamics of $v_o[n]$. Considering that v_{cntrl} is constant between two consecutive sampling instants, one has

$$v_o[n+1] = \sqrt{e^{-\frac{2T_{2L}}{R_o C}} v_o^2[n] + k_x \frac{V_{ac,rms}^2}{R_{sense}} R_o \left(1 - e^{-\frac{2T_{2L}}{R_o C}}\right) v_{cntrl}[n]}$$

$$= f(v_o[n], v_{cntrl}[n]). \qquad (4.84)$$

Perturbation and linearization of the above-mentioned equation proceeds in a straight-forward manner, with

$$v_o[n] \rightarrow V_o + \hat{v}_o[n],$$

$$v_{cntrl}[n] \rightarrow V_{cntrl} + \hat{v}_{cntrl}[n] \qquad (4.85)$$

representing the decomposition into dc and perturbation terms. Note that V_{cntrl} is related to the steady-state power level drawn from the grid by (4.75), and therefore

$$k_x \frac{V_{ac,rms}^2}{R_{sense}} R_o = \frac{P}{V_{cntrl}} R_o = \frac{V_o^2}{V_{cntrl}}. \qquad (4.86)$$

The result of the linearization step is then

$$\hat{v}_o[n+1] = e^{-\frac{2T_{2L}}{R_o C}}\hat{v}_o[n] + \frac{V_o}{2V_{cntrl}} \left(1 - e^{-\frac{2T_{2L}}{R_o C}}\right)\hat{v}_{cntrl}[n]. \qquad (4.87)$$

In the z-domain, the above-mentioned equation allows derivation of the uncompensated voltage loop gain

$$\boxed{T_{u,v}(z_{2L}) \triangleq \frac{\hat{v}_o(z_{2L})}{\hat{v}_{cntrl}(z_{2L})} = \frac{V_o}{2V_{cntrl}} \frac{\left(1 - e^{-\frac{2T_{2L}}{R_o C}}\right) z_{2L}^{-1}}{1 - e^{-\frac{2T_{2L}}{R_o C}} z_{2L}^{-1}}}, \qquad (4.88)$$

where the subscript $2L$ to the complex variable z serves as an indicator that the voltage loop is sampled at twice the line frequency rather than at the switching frequency.

[3]The equation is of the type

$$y\frac{dy}{dt} + \frac{y^2}{\tau_0} = Ax,$$

which can be integrated with the substitutions $u \triangleq y^2$ and $\frac{du}{dt} = 2y\frac{dy}{dt}$.

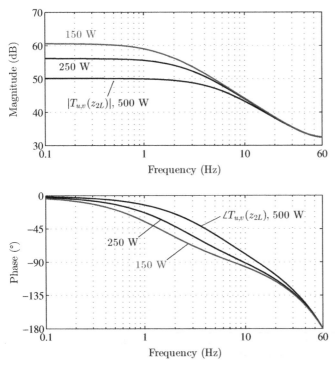

Figure 4.27   Boost PFC example: Bode plots of the uncompensated voltage loop gain $T_{u,v}(z_{2L})$ for $P = 500$, 250, and 150 W.

It should be noted that sampling of the output voltage could also be performed at a rate faster than T_{2L}. For instance, $v_o(t)$ could be sampled at the switching rate, with the voltage loop controller clocked accordingly. In this case, the voltage controller would process not only the average component of the output voltage but also aliased spectral images. This case is not considered further here.

As in the continuous-time case, the uncompensated voltage dynamics is of the first order and determined by the load time constant R_oC. Bode plots of $T_{u,v}(z_{2L})$ are depicted in Fig. 4.27 for the case study under consideration and relative to three distinct power levels. Figure 4.28 reports the compensated loop gain $T_v(z_{2L})$, designed to achieve $f_{c,v} = 6$ Hz and $\varphi_{m,v} = 70°$ at $P = 500$ W, with a PI compensation and using the methodologies developed in this chapter. Simulation results, obtained using the Middlebrook injection method, are also reported in order to verify the accuracy of the model. Loop gain variations under different power levels are illustrated as well.

Time-domain simulated waveforms reporting the system closed-loop response to a $P = 500$ to 250 W step load are depicted in Fig. 4.29.

Quantization effects due to A/D conversion and digital modulation, neglected in this and the other examples considered in this chapter, pose additional issues and design constraints. These considerations are addressed in Chapter 5. Other digital PFC control architectures and control approaches have been reported in the literature,

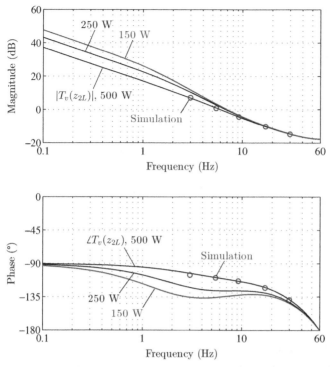

Figure 4.28 Boost PFC example: Bode plots of the voltage loop gain $T_v(z_{2L})$ for $P = 500$, 250, and 150 W.

leading to faster voltage regulation [73], elimination of the need to sense the input voltage [138, 139], or the need to sense the input current [140].

4.3 OTHER CONVERTER TRANSFER FUNCTIONS

In the design practice it is usually necessary to assess the impact of *disturbances* of various nature acting on the converter system. These disturbances are collectively grouped into the input vector $v(t)$ at the beginning of Chapter 3. Although $v(t)$ has been assumed constant up to this point, it is important now to make the situation more general and assume that a generic small-signal perturbation is superimposed to it. In the dc–dc application represented by the voltage-mode control example, for instance, the disturbances are usually represented by the converter input voltage and load current,

$$v(t) = \begin{bmatrix} v_g(t) \\ i_o(t) \end{bmatrix}, \tag{4.89}$$

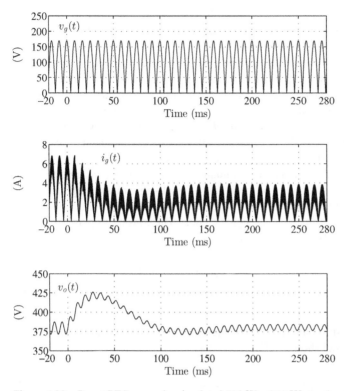

Figure 4.29   Boost PFC example: simulated 500 W→250 W step load response.

and their small-signal effects on the converter output voltage are quantified by the converter open-loop and closed-loop audiosusceptibility and output impedance, respectively. The open-loop and the closed-loop definitions of these quantities are

$$G_{vg}(s) \triangleq \left. \frac{\hat{\bar{v}}_o(s)}{\hat{\bar{v}}_g(s)} \right|_{\hat{u}=0,\hat{\bar{i}}_o=0} \qquad \text{(Open-loop)},$$

$$G_{vg,cl}(s) \triangleq \left. \frac{\hat{\bar{v}}_o(s)}{\hat{\bar{v}}_g(s)} \right|_{\hat{\bar{v}}_{ref}=0,\hat{\bar{i}}_o=0} \qquad \text{(Closed-loop)},$$

$$\text{(4.90)}$$

$$Z_o(s) \triangleq -\left. \frac{\hat{\bar{v}}_o(s)}{\hat{\bar{i}}_o(s)} \right|_{\hat{u}=0,\hat{\bar{v}}_g=0} \qquad \text{(Open-loop)},$$

$$Z_{o,cl}(s) \triangleq -\left. \frac{\hat{\bar{v}}_o(s)}{\hat{\bar{i}}_o(s)} \right|_{\hat{\bar{v}}_{ref}=0,\hat{\bar{v}}_g=0} \qquad \text{(Closed-loop)}.$$

$$\text{(4.91)}$$

One may observe that the above-mentioned transfer functions are defined in the continuous-time *Laplace* domain, the same way they are defined in the context of analog control. Unlike control transfer functions, which relate two discrete-time signals—the control command $u[k]$ and the sampled output vector $y[k]$—the disturbance transfer functions relate quantities that are inherently continuous-time. Furthermore, *averaged* values are considered. An important motivation to retain this definition even in the context of digital control is that $G_{vg}(s)$ and $Z_o(s)$ are expected to model the frequency responses *measured* by a network analyzer. During a frequency sweep, a network analyzer filters out all frequency components except for a narrow band around the frequency of the injected perturbation.

Regarding *open-loop* dynamics, evaluation of any disturbance transfer functions can be performed according to the averaged small-signal modeling theory [1]. There are no conceptual difficulties in this case because the control action is constant and does not appear in the small-signal models. The converter open-loop output impedance has already been determined in (1.69):

$$Z_o(s) = r_L \frac{(1 + sr_C C)\left(1 + s\dfrac{L}{r_L}\right)}{1 + s(r_C + r_L)C + s^2 LC}. \tag{4.92}$$

The open-loop audiosusceptibility can be determined from the averaged small-signal equivalent circuit model shown in Fig. 1.11(a) as

$$G_{vg}(s) = D\frac{1 + sr_C C}{1 + s(r_C + r_L)C + s^2 LC}. \tag{4.93}$$

On the other hand, an analytical difficulty arises in the calculation of *closed-loop* disturbance dynamics because of the sampled-data nature of the feedback system. In particular, although input and output quantities are analog, the control system responds to the *sampled* disturbance as seen at the output.

In this context, it is appropriate to clarify when and how continuous-time averaged small-signal models can still be employed to obtain meaningful *approximations* of $G_{vg,cl}(s)$ and $Z_{o,cl}(s)$. As anticipated in Section 2.6.2, the *averaged* uncompensated converter dynamics can be described by the effective uncompensated loop gain

$$T_u^\dagger(s) \triangleq T_u(s)e^{-st_d}, \tag{4.94}$$

which accounts for the total loop delay. Under the small-aliasing approximation already discussed in Section 2.6.2,

$$\boxed{v_s(t_k) \approx \overline{v}_s(t_k)}\,, \tag{4.95}$$

$T_u^\dagger(s)$ becomes a good description of the dynamics seen by the digital compensator. This suggests to define the effective loop gain $T^\dagger(s)$ as

$$T^\dagger(s) \triangleq G_c^\dagger(s)T_u^\dagger(s), \tag{4.96}$$

closed-loop input impedances

$$Z_g(s) \triangleq \left(\frac{\hat{i}_L(s)}{\hat{v}_g(s)} \right)^{-1} \Bigg|_{\hat{u}=0},$$

$$Z_{g,cl}(s) \triangleq \left(\frac{\hat{i}_L(s)}{\hat{v}_g(s)} \right)^{-1} \Bigg|_{\hat{v}_{ref}=0} \tag{4.100}$$

$$= Z_g(s) \left(1 + T^\dagger(s) \right),$$

where $T^\dagger(s)$ is defined as in (4.96), whereas the effective uncompensated loop gain $T_u^\dagger(s)$ is

$$T_u^\dagger(s) \triangleq R_{sense} G_{iu}(s) e^{-s\frac{T_s}{2}}. \tag{4.101}$$

The Bode plots of $Z_g(s)$ and $Z_{g,cl}(s)$ are shown in Fig. 4.32, along with the simulated frequency responses.

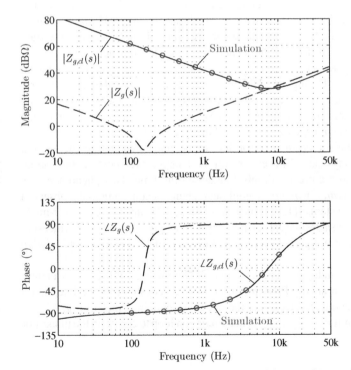

Figure 4.32   Boost converter example: Bode plots of the open-loop and closed-loop input impedances.

The Laplace-domain treatment based on the small-aliasing approximation is one among various approaches for modeling converter transfer functions of digitally controlled converters. Other approximated methods exist, for instance, in which the formulation is carried out in the z-domain [141].

4.4   ACTUATOR SATURATION AND INTEGRAL ANTI-WINDUP PROVISIONS

So far, only the controller behavior within its linear range of operation has been addressed. In practice, large transients may drive the control system into large-signal motions requiring, in general, specific provisions in order to be properly handled. Saturation of the pulse width modulator—which represents the actuator in a PWM-controlled converter—is an example of such nonlinear effects. It should be noted that saturation of duty cycle and other nonlinear effects due to various circuit limitations also arise in analog controllers where provisions such as clamping of voltages within the controller circuit are often applied. The purpose of this section is to discuss saturation effects and possible mitigation strategies in the context of digital controller implementation.

Suppose, for instance, that the digital voltage-mode control designed in Section 4.2.1 is set to $V_{ref} = 3.3$ V reference and consider the closed-loop response to a large 0 A→10 A step-load transient, as shown in Fig. 4.33 for two input voltages, $V_g = 5$ V and $V_g = 4$ V. The control command, and therefore the duty cycle, saturates to 100% for a significant fraction of the transient duration, the saturation becoming deeper at lower V_g because the steady-state control command becomes closer to 1. During the DPWM saturation interval, the integral term $u_i[k]$ of the control action increases due to the positive regulation error $e[k]$ being accumulated over time. By the time the DPWM goes back into its linear region of operation, $u_i[k]$ has reached a value that largely exceeds the steady-state value. As a result, a sufficient *negative* error has to be integrated before the system settles back to steady state. The end result is a larger overshoot in the output voltage response and a correspondingly longer settling time. This phenomenon is an example of *integrator windup*, a nonlinear phenomenon commonly observed in PID-controlled systems [127] (including analog controlled switched-mode converters), which degrades the large-signal system response. Any countermeasure aimed at preventing or mitigating this effect is usually referred to as an *anti-windup* provision.

One basic anti-windup provision, similar to clamping the voltage across an integrating capacitor in an analog PID circuit, consists of constraining the integration range of the PID accumulator. Consider the integrator building block of Fig. 4.34. A *saturation block* is introduced within the accumulation loop to constrain the integrator

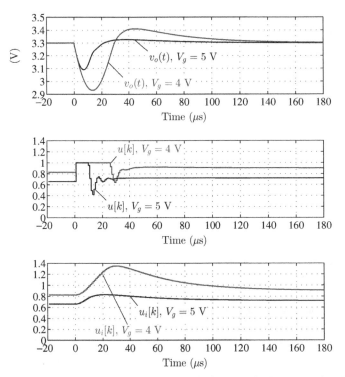

Figure 4.33   Synchronous Buck example: impact of PWM saturation during a 0 A→10 A step-load transient.

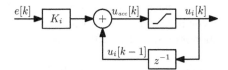

Figure 4.34   Saturated integrator.

state variable to within the range $[U_i^-, U_i^+]$. Equations of the integrator become

$$u_{acc}[k] = K_i e[k] + u_i[k-1],$$

$$u_i[k] = \begin{cases} U_i^+ & \text{if } u_{acc}[k] \geq U_i^+, \\ U_i^- & \text{if } u_{acc}[k] \leq U_i^-, \\ u_{acc}[k] & \text{otherwise.} \end{cases} \qquad (4.102)$$

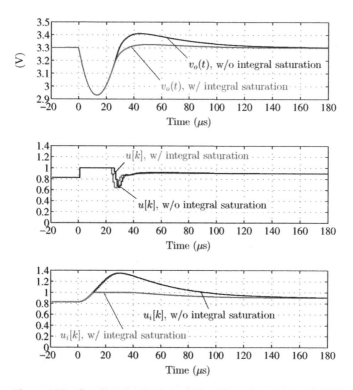

Figure 4.35 Synchronous Buck example: effect of saturation of the integrator state variable.

A common choice is $[U_i^-, U_i^+] = [0, 1]$, so that the integral term never exceeds the steady-state range of operation. Figure 4.35 compares the 0 A→10 A step-load responses with and without the integral saturation provision. The settling time of the transient is substantially improved.

Detrimental effects of the actuator saturation can further be exemplified in the multiloop control example of Section 4.2.3. The block diagram of Fig. 4.36 highlights two saturation blocks present in the system. One is, similar to the voltage-mode control case discussed earlier, the inherent saturation of the control command $u[k]$. The second saturation block is intentionally introduced to limit the current loop setpoint $i_{ref}[k]$, which provides a useful limitation of the inductor current. Note, however, that this is a *soft* inductor current limitation because $i_L(t)$ is only limited to the extent that it follows $i_{ref}[k]$, that is, within the current loop bandwidth.

In the example considered, positive and negative limits for $i_{ref}[k]$ are set to 6 A and -1 A, respectively. The system response to a 1.8 to 3.3 V step of the voltage setpoint V_{ref} is illustrated in Fig. 4.37. Such transient causes a steep increase in the inductor current demand and therefore a temporary saturation of the current loop setpoint $i_{ref}[k]$. Without any kind of anti-windup provision, the saturation of $i_{ref}[k]$ causes the integral state variable of the voltage-loop compensator to windup, driving the system into a nonlinear oscillating mode, which strongly degrades the transient

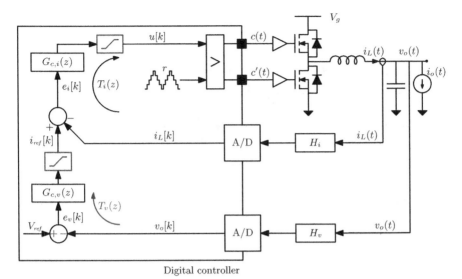

Figure 4.36 Multiloop control example: saturation of the current loop setpoint and the control command.

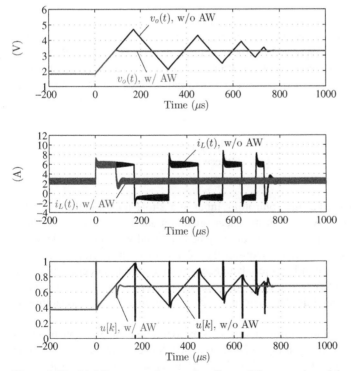

Figure 4.37 Multiloop control example: effects of the saturation of the current loop setpoint without ("w/o AW") and with anti-windup ("w AW").

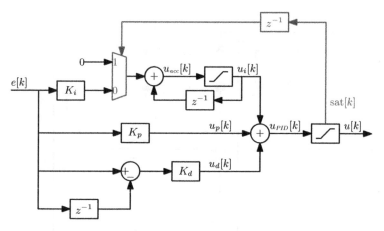

Figure 4.38   Anti-windup scheme based on conditional integration.

performance. On the other hand, if the voltage-loop compensator is equipped with a saturated integrator, the system gracefully exits the saturated condition and resumes control of the converter, as shown in the waveforms of Fig. 4.37 with anti-windup provision ('w/ AW').

An even more effective approach to anti-windup is *conditional integration*, which consists of freezing the integration process upon saturation of the *overall* control command $u[k]$. This approach is illustrated in Fig. 4.38. The overall PID output is subject to a saturation block that constrains $u[k]$ to the range $[0, 1]$, which corresponds to the 0–100% duty cycle limits. The saturation block outputs a digital flag $\text{sat}[k]$ signaling the saturation status,

$$\text{sat}[k] = \begin{cases} 0 & 0 \leq u_{PID}[k] \leq 1, \\ 1 & \text{otherwise} . \end{cases} \tag{4.103}$$

The signal $\text{sat}[k]$ is used as the controlling input of a 2-to-1 digital multiplexer that passes the scaled error $K_i e[k]$ during linear operation and the constant 0 when in saturation. Notice the saturation flag can only be employed with a one-cycle delay, as its determination during cycle k requires the calculation of the overall control signal. Conditional integration has the advantage, over the simple integral saturation provision, to *immediately* freeze the error accumulation upon PWM saturation and therefore to further speed up the recovery process when linear control operation is regained.

Conditional integration makes it unnecessary to limit the integral state by introducing a saturation block as in Fig. 4.34. In practice, however, the accumulation is always implemented in *saturated arithmetic*, which inherently implements saturation of the integral term. This is discussed in more detail in Chapter 6, along with other controller implementation issues.

The basic approaches highlighted in this section are relatively simple and effective and do not significantly impact the controller complexity. Other more sophisticated anti-windup techniques, which have been proposed for both analog and digital controllers can be found in the literature [127, 142].

4.5   SUMMARY OF KEY POINTS

- Once the z-domain model of the uncompensated dynamics is available, based on the techniques presented in Chapter 3, the controller design can proceed with the usual goals of ensuring the required control bandwidth and stability margins.

- In the bilinear transform approach presented in this chapter, the problem of compensator design is reformulated in an equivalent continuous-time domain (p-domain). This allows straightforward application of well-known frequency-domain analog control design techniques, and simple direct z-domain synthesis of digital compensators without introducing any approximations.

- Other converter transfer functions of interest, such as the closed-loop output impedance or audiosusceptibility, rely on standard s-domain formulation even when the controller is digital. An approximated way to do so, applicable under the small-aliasing approximation, is to make use of the effective s-domain control loop dynamics introduced in Section 2.6.

- Actuator saturation may lead to integral windup phenomena and consequently degraded transient responses. Mitigation of these effects is typically accomplished by embedding suitable anti-windup provisions into the digital compensator architecture.

CHAPTER 5

AMPLITUDE QUANTIZATION

Up to this point, the digital nature of the control has been modeled and discussed only in terms of *time quantization*. In this chapter, the picture is completed by examining *amplitude quantization* effects. Nonlinear interactions between analog to digital (A/D) and digital pulse width modulator (DPWM) quantizations may result in steady-state movements of the system state-space trajectory often referred to as *limit cycling* [37, 38, 143–147]. Limit cycling is a concern because it potentially affects regulation accuracy and performance of digitally controlled regulators.

In contrast to time quantization, amplitude quantization does not preserve the system linearity and produces effects that cannot be treated using standard linear system analysis tools. As a result, a comprehensive analytical treatment of quantization-related phenomena is more involved and more difficult. This chapter is devoted to explaining how limit cycling arises in relation to the existence of a steady-state solution in a digitally controlled converter with A/D and DPWM quantizations. The focus is on digitally controlled dc–dc converters. Quantization effects and limit cycling in digitally controlled single-phase PFC rectifiers are addressed in [145]. Quantization characteristics are summarized in Section 5.1, and the problem of finding a steady-state solution in a digitally controlled converter is discussed in Section 5.2. Necessary *no-limit-cycling* conditions aimed at preventing limit cycling are discussed in Section 5.3. Section 5.4 reviews some of the techniques targeting high-resolution DPWM and A/D implementations, whereas the key points are summarized in Section 5.5.

5.1 SYSTEM QUANTIZATIONS

For reference, Fig. 5.1 reiterates the block diagram of a digital voltage-mode controlled synchronous Buck converter studied, as a modeling and control loop design example, in Chapters 2–4. This section reviews the amplitude quantization characteristics of the A/D converter and the DPWM in the digital control loop.

5.1.1 A/D Converter

An introductory description of the A/D conversion process and associated quantization are provided in Section 2.2. For convenience, the A/D converter model and its

Digital Control of High-Frequency Switched-Mode Power Converters, First Edition.
Luca Corradini, Dragan Maksimović, Paolo Mattavelli, and Regan Zane.

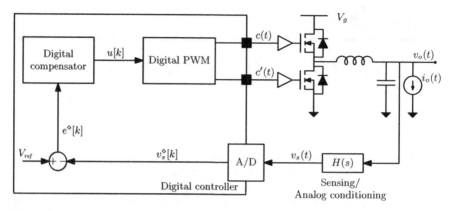

Figure 5.1   Digital voltage-mode control of a synchronous Buck converter.

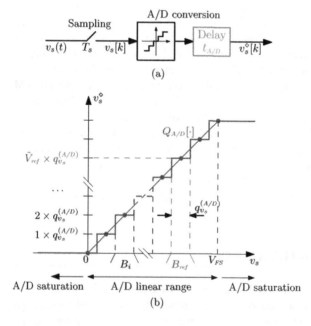

Figure 5.2   (a) A/D converter block diagram and (b) its quantization characteristic.

quantization characteristic $Q_{A/D}[\cdot]$ are shown again in Fig. 5.2. In this chapter, the focus is on the amplitude quantization effects. Therefore, the A/D conversion time and any other delays are neglected from here on, treating the A/D conversion process as essentially instantaneous. The quantization characteristic is shown in Fig. 5.2(b) for the case when the linear conversion range of the A/D converter spans from 0 V to a full-scale voltage $v_s = V_{FS}$.

Recall that the *zero-error bin* represents the particular quantization level at which the sensed signal v_s is to be regulated and that such a bin is identified by the controller digital setpoint V_{ref}. Assuming that a steady-state operating point exists in the closed-loop system such that $e^\circ = 0$, the quantized A/D output v_s° is equal to V_{ref}, which means that the sampled version of the sensed analog signal resides inside the zero-error bin B_{ref},

$$e^\circ = 0 \Leftrightarrow v_s^\circ[k] = V_{ref} \Leftrightarrow v_s[k] \in B_{ref}. \tag{5.1}$$

It should be noted that, *a priori*, the very existence of such an operating point where the error is zero ($e^\circ = 0$) is not automatically guaranteed once quantizations are included in the system. In addition to the issue of existence of the desired steady-state solution, *stability* of the steady-state operating point is another delicate matter when the quantizer nonlinearities are considered. Ideally, one wants such an operating point to exist and be stable in the sense described by (5.1): the controller must be able to reach the zero-error bin *and remain there indefinitely* in the absence of external disturbances.

Quantization of the sensed signal v_s translates into a corresponding quantization of the converter output voltage v_o. Denoting with $H_0 = H(s = 0)$ the dc gain of the voltage sensing circuitry, the width of the equivalent output voltage quantization bin is

$$q_{v_o}^{(A/D)} = \frac{q_{v_s}^{(A/D)}}{H_0} = \frac{V_{FS,o}}{2^{n_{A/D}}}, \tag{5.2}$$

where $n_{A/D}$ denotes the number of bits of A/D resolution and

$$V_{FS,o} = \frac{V_{FS}}{H_0} \tag{5.3}$$

defines the equivalent A/D conversion range on the output voltage. Analog output voltages within a zero-error bin of width $q_{v_o}^{(A/D)}$ produce a zero digital error signal $e^\circ = 0$, which implies that the LSB resolution $q_{v_o}^{(A/D)}$ determines how well the output voltage can be regulated by the digital control loop. Suppose that the regulation bin $q_{v_o}^{(A/D)}$ must be less than $\epsilon \%$ of the nominal output voltage V_{ref}/H_0. From 5.2, it follows that the A/D converter must have

$$n_{A/D} > \log_2\left(\frac{100}{\epsilon}\right) + \log_2\left(\frac{V_{FS}}{V_{ref}}\right) \tag{5.4}$$

bits of resolution. As an example, assuming $\epsilon = 1$ and $V_{FS}/V_{ref} = 2$, an A/D converter having at least $n_{A/D} = 8$ bits of resolution is required.

5.1.2 DPWM Quantization

As anticipated in Section 2.4, the DPWM can only produce pulses of quantized duty cycle. Such quantization can be equivalently modeled as a quantization on the control command $u[k]$. Given the duty cycle resolution bin, that is, the smallest duty cycle variation q_D, the smallest command variation q_u that the modulator is capable of resolving is

$$q_u = q_D N_r = \frac{N_r}{2^{n_{DPWM}}}. \tag{5.5}$$

Here, n_{DPWM} is the number of bits of DPWM resolution. Therefore, the DPWM behaves as an *infinite* resolution DPWM preceded by a quantizer $Q_{DPWM}[\cdot]$ on $u[k]$,

$$u^\diamond[k] = Q_{DPWM}[u[k]] \triangleq N_r Q_D \left[\frac{u[k]}{N_r} \right]. \tag{5.6}$$

This model is implicitly used throughout the previous chapters to represent the DPWM quantization in the control block diagrams. The most important implication of (5.6) is the corresponding quantization of the steady-state converter output voltage. If $M(D)$ is the converter conversion ratio,

$$M(D) \triangleq \frac{V_o}{V_g}, \tag{5.7}$$

then

$$V_o(D^\diamond) = M(D^\diamond)V_g \tag{5.8}$$

are the steady-state output voltages reproducible by a constant, quantized duty cycle $D^\diamond$. In general, the corresponding quantization of V_o is not uniform, as $M(D)$ depends on D and therefore on the converter operating point. In the neighborhood of a steady-state $D^\diamond$, the smallest duty cycle variation q_D produces a variation $q_{v_o}^{(DPWM)}$ approximately equal to

$$q_{v_o}^{(DPWM)} \approx \left. \frac{\partial M}{\partial D} \right|_{D^\diamond} q_D V_g. \tag{5.9}$$

In a Buck converter, for instance, $M(D) = D$ and therefore

$$q_{v_o}^{(DPWM)} = q_D V_g = \frac{q_u}{N_r} V_g \qquad \text{(Buck)}, \tag{5.10}$$

which is independent of $D^\diamond$. This situation is depicted in Fig. 5.3 for a 3-bit DPWM example. Note, however, that $q_{v_o}^{(DPWM)}$ depends on V_g because the input voltage affects the small-signal gain of the power stage. Larger input voltages lead to coarser DPWM quantization steps on V_o. In other words, the DPWM and the power converter operate as a digital-to-analog (D/A) converter, and DPWM quantization determines how precisely the converter output voltage can be positioned.

As another example, a Boost converter has

$$M(D) = \frac{1}{1-D} \Rightarrow \left. \frac{\partial M}{\partial D} \right|_{D^\diamond} = \frac{1}{(1-D^\diamond)^2}, \tag{5.11}$$

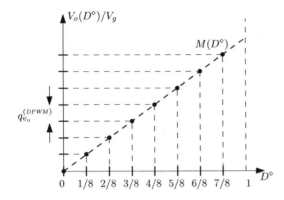

Figure 5.3   DPWM-induced quantization on the output voltage, Buck converter example.

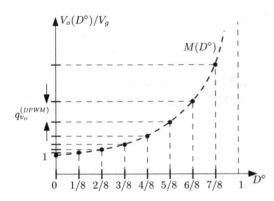

Figure 5.4   DPWM-induced quantization on the output voltage, Boost converter example.

and therefore

$$q_{v_o}^{(DPWM)} \approx q_D \frac{1}{(1 - D^\diamond)^2} V_g \quad \text{(Boost)}, \qquad (5.12)$$

which suggests a finer quantization of V_o as $D^\diamond$ decreases, as Fig. 5.4 confirms.

The DPWM-induced quantization of the output voltage can be expressed more compactly by noting that the slope of $M(D)$ times the input voltage V_g is the dc gain G_{vd0} of the control-to-output voltage small-signal transfer function,

$$\frac{\partial M}{\partial D} V_g = G_{vd}(s = 0) = N_r G_{vu}(z = 1), \qquad (5.13)$$

from which, in general, it follows that

$$\boxed{q_{v_o}^{(DPWM)} \approx G_{vd}(s = 0)q_D = G_{vu}(z = 1)q_u}, \qquad (5.14)$$

which is valid for every converter topology.

5.2   STEADY-STATE SOLUTION

Assuming that a stable control loop has been designed, a digitally controlled converter is expected to operate at an equilibrium, that is, steady-state operating point where all controller variables have constant values and where all converter waveforms are periodic, with the period equal to the switching period $T_s = 1/f_s$. To find the steady-state solution, consider a dc model of a digitally controlled converter, including A/D and DPWM quantization, as shown in Fig. 5.5. This is a static model, so the discrete-time compensator is represented by its dc gain G_{c0},

$$G_{c0} \triangleq G_c(z)\big|_{z \to 1}, \tag{5.15}$$

while H_0 is the sensor dc gain. Neglecting losses, the converter is represented by an ideal $1 : M(D^\circ)$ transformer, where $M(D) = V_o/V_g$ is the dc conversion ratio.

Suppose first that very high-resolution A/D and DPWM are employed, so that $q_{v_s}^{(A/D)} \approx 0$ and $q_u \approx 0$ or, equivalently, $V_s^\circ \approx V_s$ and $u^\circ \approx u$. An equilibrium solution in the model of Fig. 5.5 can be found using a graphical approach illustrated in Fig. 5.6, where the quantized sensed signal V_s° at the A/D converter output is shown as a function of the sensed analog signal V_s at the A/D converter input.

For the case when the A/D resolution is very high, in the linear region, the A/D quantization characteristic becomes simply

$$V_s^\circ = V_s. \tag{5.16}$$

Similarly, going from V_s° to V_s through the blocks around the loop, one has

$$V_s = \frac{H_0 V_g G_{c0}}{N_r} \left(V_{ref} - V_s^\circ \right), \tag{5.17}$$

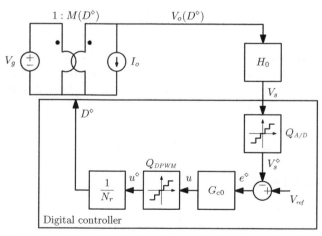

Figure 5.5   DC model of a digitally controlled converter, including A/D and DPWM quantization.

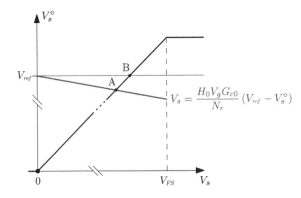

Figure 5.6   A graphical approach to finding a quiescent operating point in a digitally controlled converter with A/D converter and DPWM having very high resolutions. The expression for the sensed dc voltage V_s as a function of the quantized value $V_s^\diamond$ is shown for the synchronous Buck converter example.

assuming very high-resolution DPWM in the synchronous Buck converter example ($M(D) = D$). The steady-state solution, which is found at the intersection of (5.16) and (5.17), as shown in Fig. 5.6, in this case allows a simple algebraic solution. Eliminating $V_s^\diamond$ from (5.16) and (5.17), the dc output voltage V_s/H_0 is obtained,

$$V_o = \frac{V_{ref}}{H_0} \frac{\frac{H_0 V_g G_{c0}}{N_r}}{1 + \frac{H_0 V_g G_{c0}}{N_r}}, \tag{5.18}$$

where the quantity $\left(H_0 V_g G_{c0}\right)/N_r = T_0$ can be recognized as the dc value of the loop gain in the synchronous Buck converter. Assuming very high-resolution A/D and DPWM, the steady-state solution (5.18) is exactly the same as with analog control: a large, but finite, dc compensator gain G_{c0} results in a small, but not zero, dc regulation error, as illustrated by point A in Fig. 5.6. If, on the other hand, the compensator includes an integral action, $G_{c0} \to \infty$, (5.17) becomes a horizontal line $V_s^\diamond = V_{ref}$, and the equilibrium solution is at point B, which corresponds to zero dc error, $V_o = V_{ref}/H_0$.

Consider next a case when practical, finite-resolution A/D and DPWM are employed. A graphical solution is illustrated in Fig. 5.7. The A/D quantization characteristic is now highly nonlinear,

$$V_s^\diamond = Q_{A/D}[V_s], \tag{5.19}$$

with the widths of the A/D quantization bins equal to $q_{v_s}^{(A/D)}$. Furthermore, because of the DPWM quantization, the characteristic from $V_s^\diamond$ to V_s around the loop is also nonlinear,

$$V_s = \frac{H_0 V_g}{N_r} Q_{DPWM}\left[G_{c0}(V_{ref} - V_s^\diamond)\right], \tag{5.20}$$

assuming, again, the synchronous Buck converter example, with $M(D) = D$. The widths of the horizontal bins in the characteristic (5.20) around the loop are equal

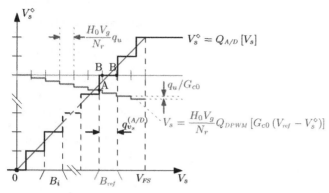

Figure 5.7 A graphical approach to finding a quiescent operating point in a digitally controlled converter with A/D converter and DPWM having finite resolution. The expression for the sensed dc voltage V_s as a function of the quantized value $V_s^\diamond$ is shown for the synchronous Buck converter example.

to $H_0 V_g q_D$ where $q_D = q_u/N_r = 1/2^{n_{DPWM}}$ is the bin width due to the DPWM quantization, as shown in (5.5). The height of a vertical step in the characteristic (5.20) is equal to q_u/G_{c0}.

If the compensator dc gain G_{c0} is large, but finite, the equilibrium solution is illustrated by point A in Fig. 5.7. This point is on a *vertical* segment of the A/D quantization characteristic. However, the A/D output $V_s^\diamond$ can only be equal to an integer multiple of $q_{v_s}^{(A/D)}$. Therefore, in contrast to point A in Fig. 5.6, the equilibrium point A in Fig. 5.7 is *not feasible*. In conclusion, given a large, but finite dc gain of the compensator, the digitally controlled converter does not have a fixed equilibrium point. Instead, the A/D converter output must bounce among two or more quantization steps, resulting in a persistent disturbance (limit cycling) in converter waveforms.

If the compensator includes an integral action, so that $G_{c0} \to \infty$, the widths of the vertical steps in the characteristic (5.20) vanish to zero, $q_u/G_{c0} \to 0$. The characteristic from $V_s^\diamond$ to V_s around the loop becomes a series of points, $H_0 V_g q_D$ apart, as shown in Fig. 5.7. In this case, multiple equilibrium solutions may exist, as illustrated by the two points B in Fig. 5.7. Each one of the two possible equilibrium solutions is inside the A/D converter zero-error bin B_{ref}.

It should be noted that the existence of multiple possible equilibrium solutions corresponding to $e^\diamond = 0$ is predicated upon the assumption that the compensator includes integral action, $K_i > 0$, and that the widths of the bins due to DPWM quantization are *shorter* than the A/D bins,

$$\frac{H_0 V_g q_u}{N_r} < q_{v_s}^{(A/D)}, \qquad (5.21)$$

as illustrated in Fig. 5.7. If the condition (5.21) is not satisfied, a steady-state solution may or may not exist, depending on whether there is a DPWM quantized sensed output voltage V_s inside the A/D zero-error bin or not. In the case where no such point

B exists, the control loop acts to bounce the output voltage through two or more bins around the zero-error bin, thus leading to limit cycling. To achieve a steady-state operating point where all controller variables have constant values and where all converter waveforms are periodic, with the period equal to the switching period $T_s = 1/f_s$, it is necessary to ensure that a dc solution exists inside the zero-error bin of the A/D converter. This conclusion leads to necessary no-limit-cycling conditions discussed in Section 5.3.

5.3   NO-LIMIT-CYCLING CONDITIONS

On the basis of the discussion in Section 5.2, the existence of a steady-state solution inside the A/D converter zero-error bin presents a necessary no-limit-cycling condition. This condition, which requires that the compensator must include integral action, $K_i > 0$, is developed further in this section, leading to formulation of necessary no-limit-cycling conditions in terms of DPWM and A/D resolution, as well as the value of the integral gain K_i.

5.3.1   DPWM versus A/D Resolution

Assuming $K_i > 0$, (5.21) is a necessary condition to guarantee existence of a dc solution inside the A/D converter zero-error bin in the digitally controlled synchronous Buck converter example. More generally, this condition can be interpreted in terms of the equivalent DPWM and A/D quantizations of the output voltage.

Consider Fig. 5.8, illustrating a set of possible output voltages due to DPWM quantization, as well as the output voltage bins due to A/D quantization on the V_o axis. For clarity, a 3-bit A/D converter is assumed where bin 011 represents the zero-error bin. In the situation depicted, no DPWM quantization level of V_o falls into A/D bin 011 (or bin 000). As the zero-error condition cannot be attained, the controller must continuously adjust the output voltage in the neighborhood of the setpoint in the futile attempt to null the regulation error. As a result, a limit cycle arises.

Consider next the situation depicted in Fig. 5.9, in which the DPWM resolution has been increased to the point that at least one DPWM quantization level must fall

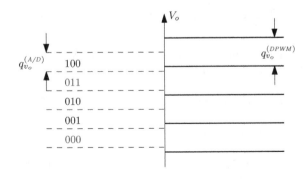

Figure 5.8   DPWM resolution is coarser than the A/D resolution in terms of output voltage quantization.

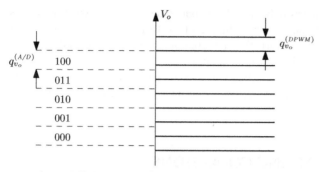

Figure 5.9   DPWM resolution is finer than the A/D resolution in terms of output voltage quantization.

into every A/D bin, which means that a stable quiescent operating point, void of limit cycling, is feasible.

The above-mentioned considerations, together with the assumption that the compensator employs an integral action, $(K_i > 0)$, lead to a general no-limit-cycling condition

$$\boxed{q_{v_o}^{(DPWM)} < q_{v_o}^{(A/D)}},$$ (5.22)

which states that the DPWM resolution must be finer than the A/D resolution when both are expressed in terms of the output voltage quantization. Using (5.9), this condition can be expressed in terms of the DPWM quantization bin q_u and the A/D quantization bin $q_{v_s}^{(A/D)}$,

$$H_0 V_g \left.\frac{\partial M}{\partial D}\right|_{D^\circ} \frac{q_u}{N_r} < q_{v_s}^{(A/D)},$$ (5.23)

which, for the Buck converter, reduces to (5.21).

As an example, consider applying these concepts to the digital control design of the synchronous Buck converter, assuming that an 8-bit A/D converter is employed, which operates on a full-scale range $V_{FS} = 2$ V. The sensor gain is $H_0 = 1$, and $V_g = 5$ V. The A/D quantization step on the output voltage V_o is then

$$q_{v_o}^{(A/D)} = \frac{q_{v_s}^{(A/D)}}{H_0} = \frac{2 \text{ V}}{2^8} \approx 7.8 \text{ mV}.$$ (5.24)

Suppose also an 8-bit DPWM is employed. As for this example $N_r = 1$, the duty cycle and control command resolutions coincide and

$$q_D = q_u \approx 0.39\%.$$ (5.25)

This translates into an equivalent quantization on V_o equal to

$$q_{v_o}^{(DPWM)} = V_g q_D \approx 19.5 \text{ mV}.$$ (5.26)

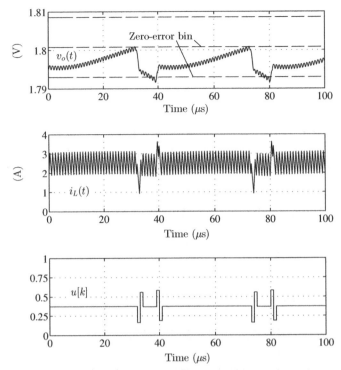

Figure 5.10   Simulated steady-state operation of the converter with a coarse DPWM resolution, so that the no-limit-cycling condition (5.22) is not met.

As seen, the no-limit-cycling condition (5.22) is not fulfilled. Figure 5.10 illustrates the simulated steady-state behavior of the controller under these conditions, confirming that periodic limit cycling affects converter operation.

If (5.22) is not satisfied, the equilibrium solution may or may not exist, depending on whether there is a DPWM quantization point inside the A/D converter zero-error bin or not. Another important observation is that limit cycling, if it does occur, is relatively small in amplitude, in the order of the quantization resolution $q_{v_o}^{(A/D)}$ of the A/D converter, as illustrated by the waveforms of Fig. 5.10.

Suppose that the DPWM resolution is increased to $n_{DPWM} = 10$ bits. In this case, the output voltage quantization shrinks down to $q_{v_o}^{(DPWM)} \approx 4.5$ mV, which is finer than $q_{v_o}^{(A/D)} \approx 7.8$ mV, so that the no-limit-cycling condition (5.22) is met. In this case, the limit cycle disappears, as shown in Fig. 5.11.

Note that (5.22) and equivalently (5.23) depend, in general, on the converter operating point. In other words, the DPWM quantization on V_o is not uniform, and both situations depicted in Figs 5.8 and 5.9 can occur in the same system as a result of a change in the operating condition. To guarantee feasibility of a no-limit-cycling steady-state solution, (5.22) must be fulfilled with sufficient margin over the entire envisioned operating range of the converter.

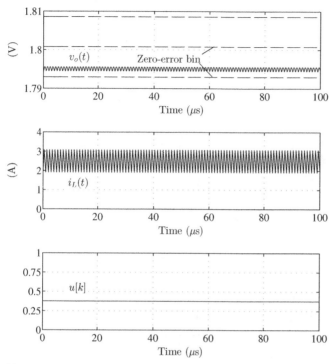

Figure 5.11 Simulated steady-state operation of the converter with a fine DPWM resolution, so that the no-limit-cycling condition (5.22) is met.

5.3.2 Integral Gain

The no-limit-cycling condition (5.22) suggests that a sufficiently fine DPWM resolution prevents the DPWM quantization to interact with the A/D converter and generate a limit cycle. Recall that the condition is predicated on the assumption that the compensator employs an integral action, $K_i > 0$, as detailed in Section 5.2. However, even when (5.22) is fulfilled, limit cycling can still arise if the integral gain K_i is too large because the A/D quantization, in combination with the integral action in the compensator, results in an effective steady-state quantization of the duty cycle command $u[k]$.

To see this, consider first the response of a simple integral compensator to a unit error impulse of amplitude equal to $q_{v_s}^{(A/D)}$, that is, the smallest possible disturbance at the compensator input. The integrator response to this unit impulse is a step, as shown in Fig. 5.12, where K_i is the integral gain. The step amplitude in $u[k]$ is equal to $K_i q_{v_s}^{(A/D)}$. In conclusion, because of the A/D quantization and the integral gain K_i in the compensator, the control command signal $u[k]$ is effectively quantized with a bin width equal to $K_i q_{v_s}^{(A/D)}$, *regardless of how high the DPWM resolution may be*. To generalize the above discussion, suppose that the system controlled by a digital PID compensator has reached a steady-state condition in which $e°[k] = 0$ from a certain

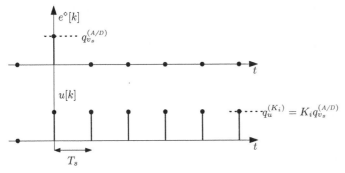

Figure 5.12 Waveforms illustrating quantization of the DPWM input signal $u[k]$ due to A/D quantization and integral action of the digital compensator: (top) an impulse in error $v_e^\diamond$ and (bottom) impulse response of an integral digital compensator with integral gain K_i.

k_0 on. The steady-state control action then consists solely of the steady-state integral term U_i:

$$e^\diamond[k] = 0 \Leftrightarrow u[k] = u_i[k] = U_i. \tag{5.27}$$

Signal $u_i[k]$, on the other hand, is the result of the accumulation of all past regulation errors up to instant k, times the integral gain K_i:

$$u_i[k] = K_i \sum_{n=-\infty}^{k} e^\diamond[n]. \tag{5.28}$$

Because of the quantization of the regulation error, one has

$$e^\diamond[k] = \tilde{e}[k]q_{v_s}^{(A/D)}, \qquad \tilde{e}[k] \in \mathbb{Z}, \tag{5.29}$$

and therefore

$$U_i = K_i \underbrace{\left(\sum_{n=-\infty}^{n=k} \tilde{e}[n] \right)}_{N[k] \in \mathbb{Z}} q_{v_s}^{(A/D)} = N[k] \left(K_i q_{v_s}^{(A/D)} \right). \tag{5.30}$$

Hence, the steady-state control command U_i embeds an inherent granularity that is determined by the A/D converter quantization step $q_{v_s}^{(A/D)}$, multiplied by the integral gain K_i,

$$q_u^{(K_i)} \triangleq K_i q_{v_s}^{(A/D)} = K_i H_0 q_{v_o}^{(A/D)}. \tag{5.31}$$

Whatever the steady-state condition, the integrator is only capable of positioning U_i to within a quantization step $q_u^{(K_i)}$ wide. If $q_u \ll q_u^{(K_i)}$ so that the hardware DPWM quantization is much finer than $q_u^{(K_i)}$, the integral term quantization becomes the dominant one, and the equivalent quantization step on the output voltage due to (5.31) is

$$q_{v_o}^{(K_i)} \approx G_{vd}(s = 0)\frac{q_u^{(K_i)}}{N_r} = G_{vd}(s = 0)\frac{K_i H_0 q_{v_o}^{(A/D)}}{N_r}, \qquad (5.32)$$

where

$$G_{vd}(s = 0) = N_r G_{vu}(z = 1) = \left.\frac{\partial M}{\partial D}\right|_{D^\circ} \qquad (5.33)$$

is the dc control-to-output gain.

Such equivalent quantization of V_o interacts with the A/D quantization in a very similar way as the hardware DPWM quantization does. By assumption, $e^\circ[k] = 0$ so *at least* one quantization level (5.32) must reside in the zero-error bin:

$$q_{v_o}^{(K_i)} < q_{v_o}^{(A/D)}, \qquad (5.34)$$

which leads to the following no-limit-cycling condition involving the integral gain:

$$\boxed{G_{vd}(s = 0)\frac{K_i H_0}{N_r} < 1}. \qquad (5.35)$$

As sketched in Fig. 5.13, when the foregoing constraint is not satisfied, a situation similar to the one previously depicted in Fig. 5.8 is obtained, in which the absence of a modulation level inside the zero-error A/D bin triggers limit cycling. The situation depicted in Fig. 5.13 can be prevented by decreasing the integral gain K_i until every A/D bin can be reached by a modulation level—this is the condition illustrated in Fig. 5.14.

For the Buck converter control of Fig. 5.1, one has $G_{vd}(s = 0) \approx V_g$ and $N_r = 1$, therefore the no-limit-cycling condition (5.35) becomes

$$H_0 V_g K_i < 1. \qquad (5.36)$$

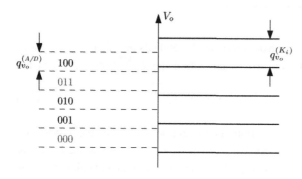

Figure 5.13 Integral term resolution is coarser than the A/D resolution in terms of output voltage quantization.

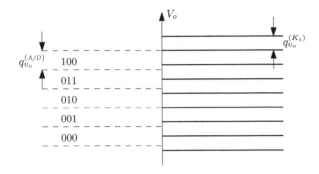

Figure 5.14   Integral term resolution is finer than the A/D resolution in terms of output voltage quantization.

In the considered synchronous Buck design example, from the value of K_i in (4.39), one has

$$H_0 V_g K_i \approx 0.37 \qquad (5.37)$$

and the no-limit-cycling condition (5.35) is fulfilled with large margin.

5.3.3   Dynamic Quantization Effects

The no-limit-cycling conditions developed in this section require that the controller employs an integral action $K_i > 0$ and are based on guaranteeing the existence of a steady-state dc operating point. The first condition, given by (5.22) or equivalently (5.23), requires that a sufficiently high-resolution DPWM be employed so as to prevent the DPWM and the A/D quantizations to interact with each other and generate a limit cycle. The second condition, given by (5.34) or equivalently (5.35) requires that the controller integral gain $K_i > 0$ must be sufficiently small so that the controller is able to position the regulated signal into the zero-error bin of the A/D converter. These two basic no-limit-cycling conditions are, in the majority of cases, sufficient for a first-cut design.

If the two no-limit-cycling conditions discussed in this section are met, a digitally controlled converter is guaranteed to have at least one equilibrium solution in the zero-error bin of the A/D converter, $e^\circ = 0$. It should be understood, however, that the existence of an equilibrium solution is in general *not sufficient* to guarantee a steady-state operation without limit cycling. With quantization effects, the converter is a complex nonlinear dynamic system, and limit-cycling disturbances can sometimes be observed even when the loop is design for stable operation, and when the basic no-limit-cycling conditions are met. On the other hand, for a stable, well-designed loop with high-resolution A/D and DPWM components, the amplitudes of any such limit-cycling disturbances in the output voltage tend to be relatively small, in the order of $q_{v_o}^{(A/D)}$. In practice, such small-amplitude disturbances may be tolerated.

More rigorous dynamic stability and performance analysis of quantization effects in digitally controlled switched-mode converters is a subject of ongoing investigations. An approximate describing function-based analysis suggests that, in addition to meeting the two basic no-limit-cycling conditions developed in this section, it is desirable to design a control loop with sufficiently large gain margin [38]. To mitigate the need to perform extensive time-domain simulations, a statistical analysis approach is presented in [143], with design guidelines relating the control loop bandwidth to the probability of limit-cycle oscillations. An energy-based approach for predicting limit-cycle oscillations is presented in [144], together with design guidelines related to the control bandwidth, positioning of PID zeros, and system damping. A more comprehensive set of no-limit-cycling conditions can be found in [146, 147] for the case of the synchronous Buck converter with a PI compensator.

5.4 DPWM AND A/D IMPLEMENTATION TECHNIQUES

As discussed in Sections 5.1–5.3, high-resolution DPWM and A/D are required to achieve precise regulation and mitigate limit cycling oscillations in digitally controlled converters. The purpose of this section is to briefly review some of the DPWM and A/D implementation techniques in the context of high-frequency digitally controlled switched-mode power converters.

5.4.1 DPWM Hardware Implementation Techniques

A block diagram and operating waveforms of a standard counter-based DPWM introduced in Section 2.4 are shown again in Fig. 5.15. The counter-based DPWM replicates an analog pulse width modulator by replacing a saw-tooth or a triangular analog carrier waveform by the digital ramp $r[nT_{clk}]$ at the output of a counter clocked at a clock frequency f_{clk}. A digital comparator outputs the modulated waveform by comparing the counter output with a latched digital control command u_h. The counter-based DPWM of resolution n_{DPWM} requires a clock frequency $f_{clk} = 2^{n_{DPWM}} f_s$, where f_s is the switching frequency. As an example, in Section 5.3, it is found that a 10-bit DPWM is required in order to meet no-limit-cycling conditions in the synchronous Buck converter operating at $f_s = 1$ MHz. To achieve this resolution with the counter-based DPWM, a clock frequency greater than 1 GHz would be required.

  Various alternative DPWM architectures have been proposed to achieve practical realizations of high-resolution DPWM's for high-frequency switched-mode power converters [148]. The main idea behind these architectures is to achieve high-resolution time quantization using a tapped string of delay cells, commonly referred to as a *delay line*, instead of the very high-frequency clock. A block diagram and simplified operating waveforms of a basic *delay-line* DPWM are shown in Fig. 5.16. A clock signal at a frequency equal to the switching frequency, $f_{clk} = f_s$, sets the output latch at the start of a switching period. The same clock signal is propagated through a delay line so that the output at tap m_k is delayed with respect to the output at tap m_{k-1} by a cell delay $\Delta t_{DPWM} = t_c$. The latched digital control command u_h selects that tap resets the output signal using a digital $2^{n_{DPWM}} : 1$ multiplexer. In the example waveforms shown in Fig. 5.16(b), $u_h = u = 2$, tap m_2 is

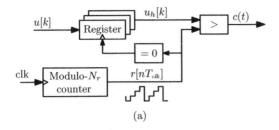

(a)

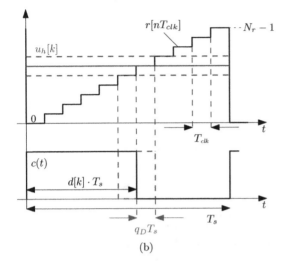

(b)

Figure 5.15   (a) Block diagram of a counter-based digital pulse width modulator and (b) associated waveforms.

selected, which results in the DPWM output pulse width equal to $dT_s = 3t_c$. A clear advantage of this architecture is that the time resolution in the output signal is set by a cell delay t_c as opposed to the clock period T_{clk}. As a result, the clock frequency equals the switching frequency. Other practical difficulties relate to the fact that the total delay may be too short or too long compared to the desired switching period T_s and that adjusting the switching frequency is not as simple as in the counter-based architecture. These issues can be addressed in one of two ways: (1) by closing the delay line into a self-oscillating ring that serves as a clock generator or (2) by employing a delay-locked loop capable of adjusting the cell delay t_c to secure a lock between a clock generator and the delay line.

A disadvantage of the delay-line approach is that the length of the delay line and the size of the multiplexer grow exponentially with the number of bits n_{DPWM}. In hybrid DPWM architectures, counter and delay-line approaches are combined, as illustrated in Fig. 5.17. The latched control command u having n_{DPWM} bits is split into two parts as shown in Fig. 5.18, the least significant m-bit long u_{LS} and the most-significant $(n_{DPWM} - m)$-bit long u_{MS}. The most-significant part u_{MS} acts as the control command for the counter-based portion of the hybrid DPWM, denoted as the "synchronous modulator" in Fig. 5.17. As in the counter-based DPWM of Fig. 5.15, a zero count ($r = 0$) initiates the output pulse $c(t)$. At the time when the ramp r reaches the most-significant part u_{MS}, the synchronous modulator generates

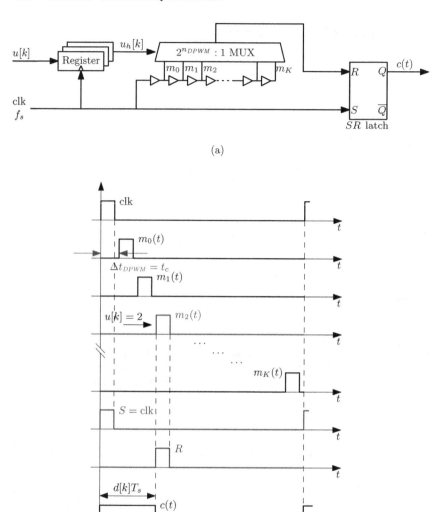

(a)

(b)

Figure 5.16 Delay line DPWM: (a) architecture and (b) operating waveforms.

a pulse $m_0(t)$, as shown in Fig. 5.17(b). This pulse, instead of resetting the output as in the pure counter-based DPWM, acts as the input to the delay-line portion of the hybrid modulator. Along the delay line, a tap is selected based on the least-significant command part u_{LS}, which extends the output pulse by $u_{LS}t_c$, where t_c is the propagation delay of a delay cell. In the example shown in Fig. 5.17(b), the high-resolution extension is $2t_c$. The duty cycle of the output pulse $c(t)$ is equal to

$$d[k] = \left(u_{MS} + \frac{u_{LS}}{2^m}\right)\frac{T_{clk}}{T_s}, \tag{5.38}$$

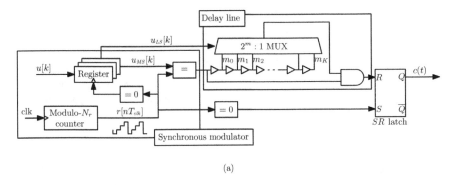

(a)

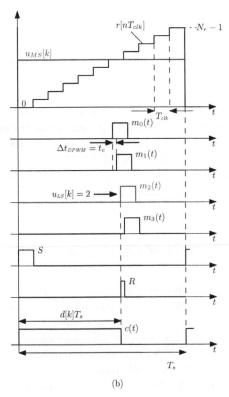

(b)

Figure 5.17   Hybrid DPWM: (a) architecture and (b) operating waveforms.

where

$$\frac{T_{clk}}{T_s} = \frac{f_s}{f_{clk}} = \frac{1}{2^{(n_{DPWM} - m)}}.$$ (5.39)

It should be noted that the hybrid architecture requires a cell delay t_c such that $2^m t_c = T_{clk}$. As in delay-line architectures, this can be accomplished using delay-locked loop or other techniques.

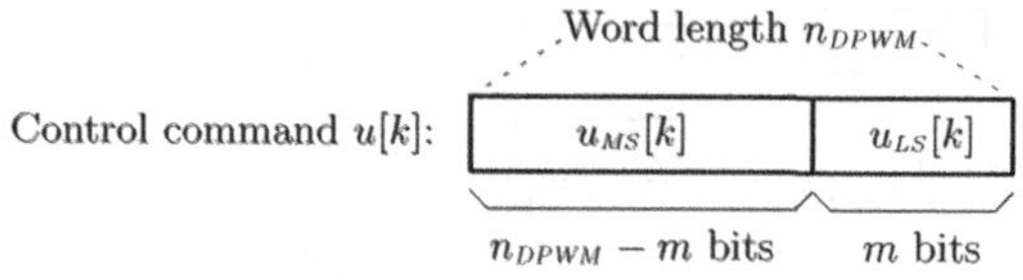

Word length n_{DPWM}.

Control command $u[k]$:

$u_{MS}[k]$	$u_{LS}[k]$

$n_{DPWM} - m$ bits m bits

$$u[k] = u_{MS}[k] + \frac{u_{LS}[k]}{2^m}$$

Figure 5.18   The control command u in the hybrid DPWM of Fig. 5.17.

On the basis of (5.38) and (5.39), it is clear that the length m of the least-significant part provides a way to achieve a desirable trade-off between the DPWM size and the required clock rate, that is, a trade-off between the delay-line DPWM and the counter-based DPWM. A larger m implies a longer delay line and a larger multiplexer, while the required clock rate is reduced. In the limit, for $m = n_{DPWM}$, the hybrid DPWM becomes a pure delay-line DPWM with $f_{clk} = f_s$. On the other hand, with $m = 0$, the hybrid DPWM becomes a pure counter-based DPWM with $f_{clk} = 2^{n_{DPWM}} f_s$.

In both delay-line and hybrid architectures, implementation issues related to clock and delay-line lock, delay matching, and circuit layout require attention. High-resolution DPWM designs based on these architectures have been demonstrated both in custom-integrated circuits [22, 24, 25, 27, 29, 148–151] and in field programmable gate arrays (FPGAs) [152–156]. Other approaches to high-resolution digital pulse width modulation, including multiphase modulators, can be found in [157–164].

5.4.2   Effective DPWM Resolution Improvements via $\Sigma\Delta$ Modulation

A digital modulator in combination with the converter power stage can be viewed as a power D/A converter, taking digital command u as an input and producing converter voltages or currents as analog outputs. The power-D/A view has led to a number of DPWM developments based on techniques adopted from the signal D/A conversion area. In particular, $\Sigma\Delta$ modulation techniques, which have been used in signal processing, data converter, and digital audio applications [165], have been applied to achieve effective DPWM resolution improvements in digitally controlled converters [31].

Figure 5.19 shows a general architecture of a $\Sigma\Delta$ modulator following the "error-feedback" architecture [165], which has an advantage of including no delays in the forward path from the high-resolution n_{HR}-bit command $u_{HR}[k]$ to the low-resolution command $u[k]$ provided to the n_{DPWM}-bit hardware DPWM unit. The truncation block takes the n_{DPWM} most-significant bits, while remaining $n_{HR} - n_{DPWM}$ represent the quantization noise. The quantization noise is filtered by a digital filter $1 - NTF(z)$, where $NTF(z)$ is the noise transfer function. The $\Sigma\Delta$ modulator shifts the quantization noise to high frequencies, where it is filtered by the low-pass action of the switched-mode power converter, thus leading to effective

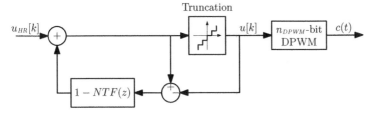

Figure 5.19 $\Sigma\Delta$ modular placed in front of a hardware digital pulse width modulator.

resolution improvements. In the second-order $\Sigma\Delta$ modulator,

$$NTF(z) = \left(1 - z^{-1}\right)^{2},$$ (5.40)

$$1 - NTF(z) = 2z^{-1} - z^{-2}.$$ (5.41)

To illustrate an application of the second-order $\Sigma\Delta$ modulator, consider the example discussed in Section 5.3.1, where it is found that the resolution of the 8-bit DPWM is not high enough, leading to limit cycle oscillations, as shown in Fig. 5.10. When a second-order $\Sigma\Delta$ modulator is placed in front of this $n_{DPWM} = 8$-bit DPWM, the command resolution is increased to $n_{HR} = 10$ bits, and the resulting operating waveforms of the synchronous Buck converter are shown in 5.20.

Given the high-resolution command u_{HR}, the $\Sigma\Delta$ modulator generates a pattern in the lower resolution command u at the input of the hardware DPWM. The pattern is such that the low-frequency *average* value of u equals the high-resolution command u_{HR}, while the spectrum of variations in u is shifted to high frequencies so that it is filtered by the low-pass action of the power converter. As a result of the variations in u, a jitter can be observed in the signal $c(t)$, which controls switches in the power converter. This jitter is low-pass filtered, and its effects are essentially invisible in the converter current and voltage waveforms shown in Fig. 5.20. The output voltage $v_o(t)$ is positioned inside the zero-error bin and no limit cycles are observed. The effect of the effective resolution improvement is very similar to the effect of using a higher resolution DPWM exemplified in Section 5.3.1, in the waveforms of Fig. 5.9.

The effective resolution improvements based on $\Sigma\Delta$ modulation have been studied in [166], in relation to the hardware DPWM resolution, and the power-stage filter corner frequency relative to the switching frequency. It is found that the second-order $\Sigma\Delta$ modulation easily offers several bits of effective resolution improvement.

5.4.3   A/D Converters

In the context of digital control of switched-mode power converters, the specifications driving A/D converter implementations include the conversion time, which must be much shorter than a switching period, and the resolution, which must be high enough to achieve precise regulation. On the other hand, linearity or wide conversion range may be compromised in order to reduce the A/D complexity. It should be noted that

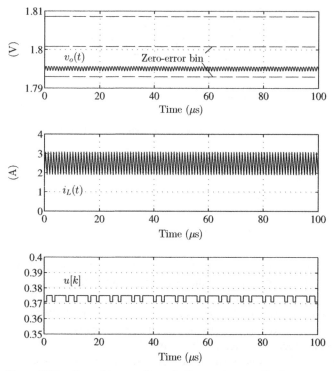

Figure 5.20 Operating waveforms in the synchronous Buck converter example in Section 5.3.1 using 8-bit DPWM and a second-order $\Sigma\Delta$ modular to increase the effective resolution to 10 bits.

these specifications differ from the requirements in standard A/D converters targeting signal-processing, open-loop sensing, or slow control system applications [167], which is why various A/D realizations targeting digital control of switched-mode converters have received attention.

In all examples considered so far, a standard A/D characteristic is assumed, where a signal is sampled and converted over a wide linear range, and the quantized signal is then compared to a digital setpoint reference V_{ref}, as shown in Figs. 5.1 and 5.2. As an alternative, Fig. 5.21 shows a windowed-flash A/D converter that meets the converter control-loop requirements using a small number of analog comparators [26]. An analog signal v_o is compared to a set of levels $q^{(A/D)}$ apart and centered around an analog reference V_{ref}. The comparator outputs represent the error signal in what is commonly referred to as the "thermometer code." A digital encoder then outputs a standard binary representation of the error, which is then sampled by a system clock to produce the digital error signal $e[k]$. The conversion time is short, as it consists only of propagation delays of the comparators and the encoder. The width of the linear conversion range is centered around V_{ref} is determined by the number of comparators employed. The conversion range can be restricted to as few as three A/D output levels ($+1, 0,$ and -1), which allows the windowed-flash implementation with

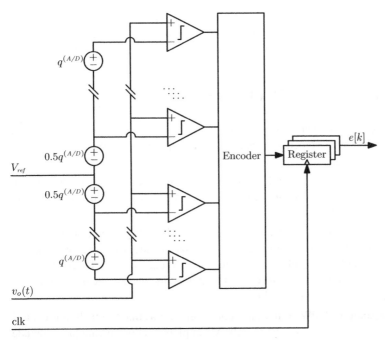

Figure 5.21   Windowed-flash A/D converter.

only two comparators [24]. More generally, the conversion range is designed to be at least as wide as the expected deviations in the regulated signal when the closed-loop controlled converter is exposed to expected disturbances and transients.

A number of variations in the windowed-flash architecture have been explored. For example, a nonuniform windowed-flash A/D characteristic can be programmed by setting the comparison levels in order to improve the converter dynamic responses [71]. Targeting custom-integrated circuit implementations in digital CMOS processes, delay-line-based A/D converters with windowed characteristics, are described in [12, 21, 22, 32]. Instead of analog comparators, the voltage-dependent delay characteristic of logic gates is used to perform voltage-to-delay and delay-to-digital conversion. A similar approach, using a ring-oscillator A/D and targeting very low-power converters for mobile applications, has been described in [27]. As another alternative for A/D circuit realization, threshold inverter quantization (TIQ) has been proposed in [35]. In the TIQ A/D approach, logic inverters with programmed thresholds replace analog comparators of the windowed-flash approach, which enables fast conversion and asynchronous sampling in a high-performance digital hysteretic controller.

A/D implementation approaches have also been developed based on the specific ripple characteristics of converter waveforms to simplify the A/D conversion or improve the effective resolution. For example, the digital controller in [168] employs a single comparator and a counter, whereas the digital controller architecture in [66] is based on single-comparator sensing in combination with D/A converters used to set the comparison thresholds. In power-factor correction (PFC) applications, a single-comparator voltage A/D has been proposed, based on the output voltage ripple [169]. The concept has been extended to inductor current sensing in [170].

5.5 SUMMARY OF KEY POINTS

- A/D and DPWM quantizations introduce nonlinear effects, which may result in steady-state movements of the system state-space trajectory often referred to as *limit cycling*.
- In the presence of A/D and DPWM quantizations, a dc steady-state solution may not exist in a digitally controlled converter, a situation that results in limit-cycle oscillations.
- The basic no-limit-cycling conditions are formulated to guarantee the existence of a dc steady-state solution. The conditions include a requirement that the equivalent DPWM resolution is better than the equivalent A/D resolution, as well as a requirement that the control loop must include an integral action with an upper limit on the integral gain.
- High-resolution DPWM and A/D converters are available for digital control applications. High-resolution DPWM implementation techniques include delay-line and hybrid approaches, as well as effective resolution improvements using $\Sigma\Delta$ modulation.

CHAPTER 6

COMPENSATOR IMPLEMENTATION

In previous chapters, the compensator design is formulated from a *system-level* perspective. Focusing on PID compensators, the main outcome of the design procedure in Chapter 4 is the z-domain transfer function of the compensator

$$G_c(z) \triangleq \frac{\hat{u}(z)}{\hat{e}(z)} = K_p + \frac{K_i}{1 - z^{-1}} + K_d \left(1 - z^{-1} \right), \qquad (6.1)$$

or, equivalently, of the associated difference equations governing the controller behavior

$$u_p[k] = K_p e[k],$$

$$u_i[k] = u_i[k-1] + K_i e[k],$$

$$u_d[k] = K_d(e[k] - e[k-1]), \qquad (6.2)$$

$$u[k] = u_p[k] + u_i[k] + u_d[k].$$

Both (6.1) and (6.2) are completely defined once the vector of coefficients $K = \left[K_p, K_i, K_d \right]^T$ is specified.

This chapter is devoted to *compensator implementation*, which refers to translating (6.2) into a form suitable for its physical realization:

- In *software-based* controllers, (6.2) is converted into an algorithm, usually coded in a high-level programming language such as C.

- In *hardware-based* controllers, (6.2) is converted into a digital circuit consisting of a combination of digital arithmetic blocks and registers. The mainstream and the most portable way for low-level controller description is based on the use of a *hardware description language* (HDL) such as VHDL or Verilog.

This chapter covers essential aspects of implementation of *hardware-based* compensators, which are the solutions most commonly adopted in high-frequency dc–dc applications. To some extent, however, considerations developed in this chapter also apply to software-based compensator realizations.

Digital Control of High-Frequency Switched-Mode Power Converters, First Edition.
Luca Corradini, Dragan Maksimović, Paolo Mattavelli, and Regan Zane.

There are two phases of the implementation process, namely, the *coefficients quantization* and the *fixed-point implementation* of the controller:

- *Coefficients quantization* is the round-off process by which the vector of coefficients K is represented using a finite number of bits. This step has the effect of altering the compensator transfer function actually realized with respect to the one originally designed. As a result, the implemented system loop gain $\tilde{T}(z)$ differs from the target loop gain $T(z)$ determined from the system-level design step. Figure 6.1 exemplifies possible effects of coefficient quantization on the implemented system loop gain $\tilde{T}(z)$ for the synchronous Buck converter control design examined in the previous chapters. With a sufficiently fine quantization, the implemented loop gain closely resembles the one designed in Chapter 4, as exemplified by the $\tilde{T}_1(z)$ responses in Fig. 6.1. However, as quantizations of the compensator gains are made coarser, differences arise. An excessively coarse quantization on the proportional and derivative gains inevitably leads to severe bandwidth and phase margin errors, which impact closed-loop system dynamics and may even result in instability, as is the case for the $\tilde{T}_2(z)$ responses in Fig. 6.1. On the other hand, excessive quantization of the integral gain can lead to low-frequency deviations and can degrade controller regulation and disturbance rejection capabilities. In the situation depicted—$\tilde{T}_3(z)$ responses in Fig. 6.1—the relative error in the low-frequency gain is around 20%.

Selection of the required coefficient resolution is guided by suitable constraints aimed at preserving the designed small-signal loop gain responses. Two constraints are considered in this chapter, one quantifying the tolerable loop gain error at the target crossover frequency ω_c and the other related to the low-frequency region $\omega \to 0$. If $\tilde{T}(z)$ is the actual system loop gain and $T(z)$ represents the target—unquantized—loop gain, the crossover frequency constraint can be formulated as

$$\text{Constraint I:} \quad \boxed{\begin{array}{c} \left|\,|\tilde{T}(z)| - 1\,\right|_{\omega=\omega_c} < \varepsilon_c \\[2mm] \left|\angle\tilde{T}(z) - \angle T(z)\right|_{\omega=\omega_c} < \alpha_c \end{array}}, \tag{6.3}$$

where ε_c and α_c are design constraints on the tolerable magnitude and phase deviation, respectively. Similarly, the constraint on the low-frequency loop gain magnitude can be expressed as

$$\text{Constraint II:} \quad \boxed{\left|\,\frac{|\tilde{T}(z)| - |T(z)|}{|T(z)|}\,\right|_{\omega \to 0} < \varepsilon_0}, \tag{6.4}$$

where $\omega \to 0$ implies that the left-hand side of the constraint is to be taken in its limit value at dc.

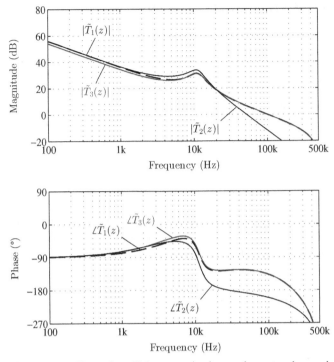

Figure 6.1   Effects of coefficient quantization on the system loop gain: adequate quantization ($\tilde{T}_1(z)$), excessively coarse K_p and K_d quantization ($\tilde{T}_2(z)$), and excessively coarse K_i quantization ($\tilde{T}_3(z)$).

- The *fixed-point implementation* step formulates the physical realization of the control law, where a finite number of bits is available to perform control operations. *Fixed-point arithmetic* is emphasized, in contrast to floating-point arithmetic, primarily because of its much wider applicability in both hardware-based and software-based implementations in digital power electronics applications. A goal of this step is to define a finite precision controller whose behavior negligibly departs from the ideal controller determined in the system-level design step. This entails making round-off and truncation effects negligible, as well as avoiding signal saturations.

Coefficient quantization is discussed in Section 6.2, with a design example in Section 6.3. The analysis developed in this section applies equally well to hardware-based and software-based compensators. Section 6.4 discusses fixed-point controller implementation for hardware-based compensators, which entails different considerations than those involved when dealing with software-based controllers.

The notations employed in this chapter and the basic aspects of fixed-point arithmetic and VHDL manipulation of binary two's complement (B2C) quantities are described in Appendix B.

6.1 PID COMPENSATOR REALIZATIONS

The issue of compensator implementation is closely related to how the compensation law is realized in hardware. As known from the field of digital filter design, any given transfer function has many *realizations* [8], each having a different hardware complexity and sensitivity to coefficient round-off.

Equation (6.1) directly leads to the *parallel* implementation of the PID structure depicted in Fig. 6.2. The transfer function of the parallel realization is denoted as $G_{PID}(z; K)$, K being the vector of the proportional, integral, and derivative gains,

$$G_{PID}(z; K) \triangleq K_p + \frac{K_i}{1 - z^{-1}} + K_d(1 - z^{-1}), \qquad K \triangleq [K_p, K_i, K_d]^T. \tag{6.5}$$

Along with the parallel form, two other important realizations of the PID structure are considered. The *direct* structure, or *direct-II* form as it is more precisely referred to in [8], is illustrated in Fig. 6.3 and denoted as

$$G_{PID}(z; b) \triangleq \frac{b_0 + b_1 z^{-1} + b_2 z^{-2}}{1 - z^{-1}}, \qquad b \triangleq [b_0, b_1, b_2]^T. \tag{6.6}$$

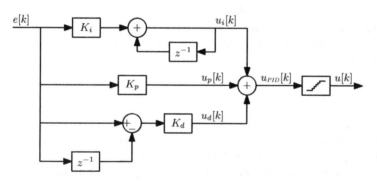

Figure 6.2   Parallel realization of a digital PID compensator.

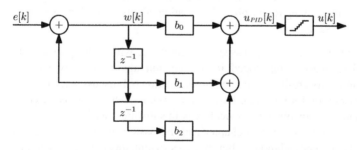

Figure 6.3   Direct realization of a digital PID compensator.

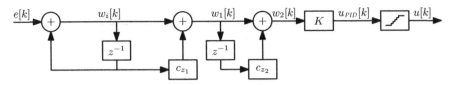

Figure 6.4   Cascade realization of a digital PID compensator.

In the time domain, the direct realization translates into the set of difference equations

$$w[k] = w[k-1] + e[k],$$
$$u[k] = b_0 w[k] + b_1 w[k-1] + b_2 w[k-2].$$

(6.7)

The transfer function of the *cascade* structure depicted in Fig. 6.4 is

$$G_{PID}(z; c) \triangleq \frac{K}{1 - z^{-1}} \left(1 + c_{z_1} z^{-1}\right) \left(1 + c_{z_2} z^{-1}\right), \qquad c \triangleq \left[K, c_{z_1}, c_{z_2}\right]^T,$$

(6.8)

corresponding, in the time domain, to

$$w_i[k] = e[k] + w_i[k-1],$$
$$w_1[k] = w_i[k] + c_{z_1} w_i[k-1],$$
$$w_2[k] = w_1[k] + c_{z_2} w_1[k-1],$$
$$u[k] = K w_2[k].$$

(6.9)

A *quantization/saturation block* is cascaded to each of the structures depicted in Figs. 6.2–6.4. As indicated in Fig. 6.5, such block first quantizes the control command $u_{PID}[k]$ into a low-resolution signal matching the DPWM resolution and then saturates the result against the maximum and minimum commands acceptable by the DPWM unit.

In Chapter 4, equations (4.15) are provided for converting the p-domain compensator design into a corresponding z-domain compensator expressed in the parallel form. Table 6.1 completes the picture by providing conversion formulas to convert the p-domain multiplicative form, defined by the

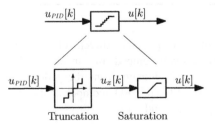

Truncation   Saturation   Figure 6.5   Quantization/saturation block.

TABLE 6.1 p-Domain to z-Domain Conversion Equations

Parallel Form $G_{PID}(z; \mathbf{K})$

$$K_p = G'_{PI\infty} G'_{PD0} \left(1 + \frac{\omega_{PI}}{\omega_{PD}} - \frac{2\omega_{PI}}{\omega_p} \right)$$

$$K_i = 2G'_{PI\infty} G'_{PD0} \frac{\omega_{PI}}{\omega_p} \tag{6.10}$$

$$K_d = \frac{G'_{PI\infty} G'_{PD0}}{2} \left(1 - \frac{\omega_{PI}}{\omega_p} \right) \left(\frac{\omega_p}{\omega_{PD}} - 1 \right)$$

Direct Form $G_{PID}(z; \mathbf{b})$

$$b_0 = \frac{G'_{PI\infty} G'_{PD0}}{2} \left(1 + \frac{\omega_{PI}}{\omega_{PD}} + \frac{\omega_p}{\omega_{PD}} + \frac{\omega_{PI}}{\omega_p} \right)$$

$$b_1 = G'_{PI\infty} G'_{PD0} \left(\frac{\omega_{PI}}{\omega_p} - \frac{\omega_p}{\omega_{PD}} \right) \tag{6.11}$$

$$b_2 = \frac{G'_{PI\infty} G'_{PD0}}{2} \left(1 - \frac{\omega_{PI}}{\omega_p} \right) \left(\frac{\omega_p}{\omega_{PD}} - 1 \right)$$

Cascade Form $G_{PID}(z; \mathbf{c})$

$$K = \frac{G'_{PI\infty} G'_{PD0}}{2} \left(1 + \frac{\omega_p}{\omega_{PD}} \right) \left(1 + \frac{\omega_{PI}}{\omega_p} \right)$$

$$c_{z_1} = \frac{\dfrac{\omega_{PI}}{\omega_p} - 1}{\dfrac{\omega_{PI}}{\omega_p} + 1} \tag{6.12}$$

$$c_{z_2} = \frac{\dfrac{\omega_{PD}}{\omega_p} - 1}{\dfrac{\omega_{PD}}{\omega_p} + 1}$$

---

parameters $\left(G'_{PI\infty}, G'_{PD0}, \omega_{PI}, \omega_{PD}, \omega_p \right)$, into the parallel, direct, or cascade z-domain form. Note that these formulas already embed the inverse bilinear transformation necessary to go back to the z-domain from the p-domain. One may verify that the special cases of PI and PD compensators are obtained, by letting $\left(\omega_{PD} \to \omega_p, G'_{PD0} \to 1 \right)$ and $\left(\omega_{PI} \to 0, G'_{PI\infty} \to 1 \right)$, respectively.

6.2   COEFFICIENT SCALING AND QUANTIZATION

Scaling and quantization are the two steps necessary to bring the compensator from its system-level formulation to a form suitable for implementation.

Coefficient scaling adapts the compensator gains to the A/D and DPWM gains present in the real system without altering the designed loop gain $T(z)$. The approach presented here, which is by no means the only option, is to scale the compensator coefficients in such a way that both the error signal and the control command assume the form of *integer* quantities. This choice, although somewhat arbitrary, is in line with the presentation made in Appendix B, which interprets each B2C signal as an integer.

Coefficient quantization, on the other hand, consists of a round-off of the compensator gains so that they can be stored in finite-length binary words, a process that invariably alters the compensator frequency response and therefore the loop gain. The number of bits used to hard code the coefficients needs to be related, in a way that has to be specified, to the quantization error on $G_{PID}(z)$.

Quantization effects associated with the compensator coefficients can be studied independently of the controller type—if hardwired or microprogrammed—and only depend on the compensator structure that is to be realized. Considering first the parallel form, scaling, and quantization amounts to successive manipulations on the coefficient vector K:

$$K = \begin{bmatrix} K_p \\ K_i \\ K_d \end{bmatrix} \xrightarrow{scaling} \check{K} = \begin{bmatrix} \check{K}_p \\ \check{K}_i \\ \check{K}_d \end{bmatrix} \xrightarrow{quantization} \tilde{K} = \begin{bmatrix} \tilde{K}_p \\ \tilde{K}_i \\ \tilde{K}_d \end{bmatrix}, \quad (6.13)$$

which induces corresponding changes in the compensator transfer function:

$$G_{PID}(z; K) \xrightarrow{scaling} G_{PID}(z; \check{K}) \xrightarrow{quantization} G_{PID}(z; \tilde{K}). \quad (6.14)$$

The same can be stated for the direct form,

$$b = \begin{bmatrix} b_0 \\ b_1 \\ b_2 \end{bmatrix} \xrightarrow{scaling} \check{b} = \begin{bmatrix} \check{b}_0 \\ \check{b}_1 \\ \check{b}_2 \end{bmatrix} \xrightarrow{quantization} \tilde{b} = \begin{bmatrix} \tilde{b}_0 \\ \tilde{b}_1 \\ \tilde{b}_2 \end{bmatrix}, \quad (6.15)$$

$$G_{PID}(z; b) \xrightarrow{scaling} G_{PID}(z; \check{b}) \xrightarrow{quantization} G_{PID}(z; \tilde{b}), \quad (6.16)$$

and for the cascade form,

$$c = \begin{bmatrix} K \\ c_{z_1} \\ c_{z_2} \end{bmatrix} \xrightarrow{scaling} \check{c} = \begin{bmatrix} \check{K} \\ \check{c}_{z_1} \\ \check{c}_{z_2} \end{bmatrix} \xrightarrow{quantization} \tilde{c} = \begin{bmatrix} \tilde{K} \\ \tilde{c}_{z_1} \\ \tilde{c}_{z_2} \end{bmatrix}, \quad (6.17)$$

$$G_{PID}(z; c) \xrightarrow{scaling} G_{PID}(z; \check{c}) \xrightarrow{quantization} G_{PID}(z; \tilde{c}). \quad (6.18)$$

6.2.1   Coefficients Scaling

In Chapters 4 and 5, the A/D converter is assumed to produce a quantized signal $v_s[k]$ having the same units as the sensed signal $v_s(t)$ and a quantization bin equal to $q_{v_s}^{(A/D)}$. In the controller implementation, the digital error is instead treated as an *integer* signal $\tilde{e}[k]$ having unity quantization bin $q_{\tilde{e}}^{(A/D)} = 1$.

Consider, for instance, the proportional action in the parallel structure

$$u_p[k] = K_p e[k], \tag{6.19}$$

in which the regulation error $e[k]$ has a granularity $q_e^{(A/D)} = q_{v_s}^{(A/D)}$. Scaling of the proportional gain is accomplished as

$$u_p[k] = \left(K_p q_{v_s}^{(A/D)} \right) \underbrace{\left(\frac{e[k]}{q_{v_s}^{(A/D)}} \right)}_{\tilde{e}[k]}, \tag{6.20}$$

where $\tilde{e}$ now has unity granularity $q_{\tilde{e}}^{(A/D)} = 1$. The above-mentioned operation formally attributes the A/D converter a gain equal to $1/q_{v_s}^{(A/D)}$ and simultaneously scales the proportional gain by $q_{v_s}^{(A/D)}$.

In a similar way, in the previous chapters, the compensator design examples assume a normalized DPWM carrier amplitude $N_r = 1$, which leads to an input control command $u[k]$ ranging from 0 to 1, with a quantization bin equal to q_u. A real DPWM, on the other hand, accepts an input binary word $\tilde{u}[k]$, which is more naturally interpreted as an integer value $\tilde{u}$ ranging from 0 to $N_r - 1$, N_r representing the word for which $D = 100\%$. For example, referring to the counter-based DPWM architecture described in Sections 2.4 and Section 5.1, N_r is the number of DPWM clock cycles per switching period. Rescaling of the compensator coefficients is accomplished by attributing the DPWM a gain equal to $1/N_r$ and redefining the control command as

$$\tilde{u}_p[k] = N_r u_p[k] = \underbrace{\left(K_p q_{v_s}^{(A/D)} N_r \right)}_{\check{K}_p} \tilde{e}[k], \tag{6.21}$$

which amounts to multiplying the proportional gain by N_r. The quantization bin of $\tilde{u}$ is $q_{\tilde{u}} = 1$.

Summarizing, the coefficient scaling for a $N_r = 1$ design is accomplished by the substitutions

$$K_p \rightarrow \quad \check{K}_p \triangleq \lambda K_p,$$

$$K_i \rightarrow \quad \check{K}_i \triangleq \lambda K_i, \tag{6.22}$$

$$K_d \rightarrow \quad \check{K}_d \triangleq \lambda K_d,$$

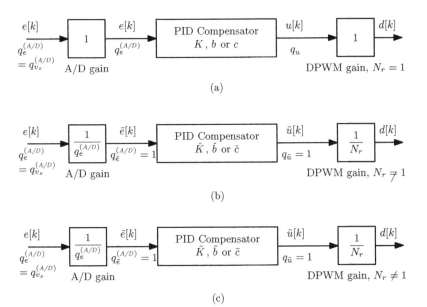

Figure 6.6   Coefficient scaling and quantization: (a) system-level view of the compensator as considered in the previous chapters and implementation-level view (b) before and (c) after coefficients quantization. Signals granularities are also indicated.

or, more compactly,

$$\boxed{K \xrightarrow{\text{scaling}} \check{K} \triangleq \lambda K},$$ (6.23)

where

$$\boxed{\lambda \triangleq q_{v_s}^{(A/D)} N_r}$$ (6.24)

is the *compensator scaling factor*, which depends on the A/D characteristic and the DPWM architecture.

As suggested by Fig. 6.6(a) and (b), result of the scaling process is the translation of the system-level compensator $G_{PID}(z; K)$ into a corresponding implementation-level compensator

$$\boxed{G_{PID}(z; \check{K}) = \lambda G_{PID}(z; K)}.$$ (6.25)

On the other hand, the uncompensated loop gain scales to $\check{T}_u(z) = T_u(z)/\lambda$ due to the equivalent gains attributed to the A/D converter and the DPWM. Therefore,

$$\check{T}(z) \triangleq G_{PID}(z; \check{K})\check{T}_u(z)$$

$$= (\lambda G_{PID}(z; K)) \left(\frac{T_u(z)}{\lambda} \right)$$

$$= T(z)$$

$$\Rightarrow \boxed{\check{T}(z) = T(z)}. \tag{6.26}$$

As anticipated, the *coefficient scaling does not alter the designed system loop gain* $T(z)$.

Coefficient scaling for the direct and cascade realizations can be derived based on similar reasoning,

$$b \xrightarrow{scaling} \check{b} \triangleq \lambda b = \begin{bmatrix} \lambda b_0 \\ \lambda b_1 \\ \lambda b_2 \end{bmatrix}, \tag{6.27}$$

$$c \xrightarrow{scaling} \check{c} \triangleq \begin{bmatrix} \lambda K \\ c_{z_1} \\ c_{z_2} \end{bmatrix}. \tag{6.28}$$

Observe that in the cascade form, only the component K of the coefficient vector c is scaled.

6.2.2 Coefficients Quantization

Contrary to the scaling step, coefficients rounding *does* alter the system loop gain $T(z)$. As presented in Appendix B, $\mathcal{Q}_n[\cdot]$ represents the n-bit round-off map for B2C quantities and $d_n c$ represents the absolute quantization error on c due to an n-bit round-off. With these notations, the quantization of scaled coefficients (6.22) can be written as

$$\tilde{K} = \begin{bmatrix} \mathcal{Q}_{n_p}\left[\check{K}_p\right] \\ \mathcal{Q}_{n_i}\left[\check{K}_i\right] \\ \mathcal{Q}_{n_d}\left[\check{K}_d\right] \end{bmatrix} = \begin{bmatrix} \check{K}_p + d_{n_p}\check{K}_p \\ \check{K}_i + d_{n_i}\check{K}_i \\ \check{K}_d + d_{n_d}\check{K}_d \end{bmatrix} = \check{K} + d\check{K}, \tag{6.29}$$

with

$$d\tilde{K} \triangleq \begin{bmatrix} d_{n_p}\check{K}_p \\ d_{n_i}\check{K}_i \\ d_{n_d}\check{K}_d \end{bmatrix}. \tag{6.30}$$

Word lengths n_p, n_i, and n_d in (6.29) represent the unknowns in the coefficient quantization problem. The objective is to determine these word lengths so as to satisfy the design constraints expressed by (6.3) and (6.4).

After coefficient quantization, referring to Fig. 6.6(c), the PID transfer function and the system loop gain become

$$G_{PID}(z; \tilde{K}) = \tilde{K}_p + \frac{\tilde{K}_i}{1 - z^{-1}} + \tilde{K}_d \left(1 - z^{-1}\right), \tag{6.31}$$

and

$$\tilde{T}(z) = G_{PID}(z; \tilde{K})\tilde{T}_u(z). \tag{6.32}$$

A goal of this section is to formulate the quantization problem from a general perspective by establishing a link between the constraints (6.3) and (6.4) and the quantities

$$\boxed{dG_{PID}(z; \tilde{K}) \triangleq G_{PID}(z; \tilde{K}) - G_{PID}(z; \check{K})} \tag{6.33}$$

and

$$\boxed{\delta G_{PID}(z; \check{K}) \triangleq \frac{dG_{PID}(z; \check{K})}{G_{PID}(z; \check{K})}}, \tag{6.34}$$

which are the *absolute and relative compensator sensitivity functions*, respectively. They quantify, in absolute or relative terms, the variation $G_{PID}(z; \check{K})$ undergoes due to the coefficient quantization process.

Combining (6.32) with (6.33) and (6.34), the expression for the quantized loop gain becomes

$$\tilde{T}(z) = \check{T}_u(z) \left(G_{PID}(z; \check{K}) + dG_{PID}(z; \check{K})\right), \tag{6.35}$$

and the relationship between $\tilde{T}(z)$ and $T(z)$ can then be formulated as

$$\boxed{\tilde{T}(z) = T(z) \left(1 + \delta G_{PID}(z; \check{K})\right)}. \tag{6.36}$$

Consider now Constraint I in (6.3), and quantize the coefficient vector $\check{K}$ so as to limit $\left|\delta G_{PID}(z; \check{K})\right|$ to within a specified amount $\varepsilon < 1$ at the target crossover frequency,

$$\left.\left|\delta G_{PID}(z; \check{K})\right|\right|_{\omega=\omega_c} < \varepsilon < 1. \tag{6.37}$$

With the aid of Fig. 6.7, one concludes that

$$1 - \varepsilon \leq \left.\left|1 + \delta G_{PID}(z; \check{K})\right|\right|_{\omega=\omega_c} \leq 1 + \varepsilon \tag{6.38}$$

and

$$-\arcsin(\varepsilon) \leq \left.\angle\left(1 + \delta G_{PID}(z; \check{K})\right)\right|_{\omega=\omega_c} \leq \arcsin(\varepsilon). \tag{6.39}$$

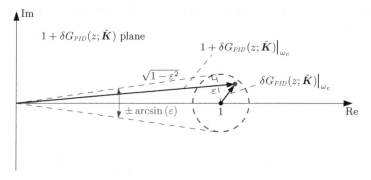

Figure 6.7 Maximum phase of $1 + \delta G_{PID}(z; \check{K})$ at $\omega = \omega_c$.

Therefore, from (6.36) and $|T(z)||_{\omega=\omega_c} = 1$, one has

$$1 - \varepsilon \leq |\tilde{T}(z)| \leq 1 + \varepsilon \tag{6.40}$$

and

$$-\arcsin(\varepsilon) \leq \angle\tilde{T}(z)|_{\omega=\omega_c} - \angle T(z)|_{\omega=\omega_c} \leq \arcsin(\varepsilon). \tag{6.41}$$

Note that a single constraint on $|\delta G_{PID}(z; \check{K})|$ at the target crossover frequency induces the corresponding constraints on both the gain and the phase of $\tilde{T}(z)$ at $\omega = \omega_c$. Comparing the above-mentioned relationships with (6.3) leads to

$$\boxed{\begin{aligned} \varepsilon_c &= \varepsilon, \\ \alpha_c &= \arcsin(\varepsilon). \end{aligned}} \tag{6.42}$$

For instance, if $\varepsilon = 1\%$, then the relative error of $|T(z)|$ at $\omega = \omega_c$ is *at most* 1%, while the phase error is *at most* equal to $\pm\alpha = \pm\arcsin(0.01) \approx \pm 0.57°$.

As for Constraint II in (6.4) on the low-frequency magnitude of $T(z)$, similar conclusions can be drawn. In particular, a constraint on $|\delta G_{PID}(z; \check{K})|$ at dc induces a corresponding constraint on $|\tilde{T}(z)|$,

$$\left|\delta G_{PID}(z; \check{K})\right|_{\omega=0} < \varepsilon_0 < 1 \Rightarrow \left|\frac{|\tilde{T}(z)| - |T(z)|}{|T(z)|}\right|_{\omega=0} < \varepsilon_0. \tag{6.43}$$

Both constraints (6.3) and (6.4) can be enforced by selecting word lengths (n_p, n_i, n_d) so as to limit $|\delta G_{PID}(z; \check{K})|$ at specific frequencies. This can be performed either analytically or, more rapidly, via Matlab® scripting as exemplified in the next section.

$\boxed{\textbf{Inset 6.1 – The Absolute Sensitivity Function}}$

The absolute sensitivity function (6.33) can be approximated as the differential of $G_{PID}(z; \check{K})$ with respect to the components of the coefficient vector $\check{K}$,

$$dG_{PID}(z; \check{K}) \approx \left. \frac{\partial G_{PID}(z; K)}{\partial K_p} \right|_{K=\check{K}} d_{n_p} \check{K}_p$$

$$+ \left. \frac{\partial G_{PID}(z; K)}{\partial K_i} \right|_{K=\check{K}} d_{n_i} \check{K}_i$$

$$+ \left. \frac{\partial G_{PID}(z; K)}{\partial K_d} \right|_{K=\check{K}} d_{n_d} \check{K}_d. \tag{6.44}$$

In the following, the above-mentioned approximation is indicated as

$$\boxed{ dG_{PID}(z; \check{K}) \approx \left. \frac{\partial G_{PID}(z; K)}{\partial K} \right|_{K=\check{K}} d\check{K} }. \tag{6.45}$$

Observe that whenever $G_{PID}(z; \check{K})$ is *linear* in the coefficients $\check{K}$, then the above-mentioned equation becomes *exact*. This is the case for the parallel and the direct structures, in which the vector coefficients $\check{K}$ and $\check{b}$ act linearly on the compensator transfer function. For the cascade realization, on the other hand, the dependence of $G_{PID}(z; \check{c})$ on $\check{c}$ is not linear, and the above-mentioned calculation only approximates $dG_{PID}(z; \check{c})$ when $d\check{c}$ is small.

6.3 VOLTAGE-MODE CONTROL EXAMPLE: COEFFICIENTS QUANTIZATION

The above-mentioned concepts are now employed to realize the PID compensator for the synchronous Buck converter example examined in the previous chapters. All three basic structures—parallel, direct, and cascade—are exemplified and compared. In all cases, the constraints (6.3) and (6.4) are specified as

$$\varepsilon = 1\% \Rightarrow \begin{cases} \varepsilon_c = 1\% \\ \alpha_c = \arcsin(\varepsilon) \approx 0.57° \end{cases} \quad \text{(Constraint I)},$$

$$\varepsilon_0 = 10\% \quad \text{(Constraint II)}. \tag{6.46}$$

Equation (4.39) reports the PID compensator coefficients for the synchronous Buck voltage-mode control designed in Chapter 2. For the same example, an 8-bit A/D converter over a quantization range of 2 V and a 10-bit DPWM are found to satisfy the basic no-limit-cycling conditions in Section 5.3. The A/D converter quantization step is therefore

$$q_{v_s}^{(A/D)} \approx 7.8 \text{ mV}. \tag{6.47}$$

As for the DPWM, regardless of its implementation, one can view its operation as based on an equivalent carrier ranging from 0 to $N_r - 1$, with

$$N_r = 2^{10} = 1024. \tag{6.48}$$

Hence, the compensator scaling factor is

$$\lambda = q_{v_s}^{(A/D)} N_r = 8 \text{ V}. \tag{6.49}$$

6.3.1   Parallel Structure

The scaled coefficients become

$$\check{K}_p \triangleq \lambda K_p \approx 24.76,$$
$$\check{K}_i \triangleq \lambda K_i \approx 0.5961, \tag{6.50}$$
$$\check{K}_d \triangleq \lambda K_d \approx 190.5.$$

An expression of the absolute sensitivity function of the parallel structure is

$$dG_{PID}(z; \check{K}) = \left. \frac{\partial G_{PID}(z; K)}{\partial K} \right|_{K = \check{K}} d\check{K}$$
$$= d_{n_p} \check{K}_p + \frac{d_{n_i} \check{K}_i}{1 - z^{-1}} + d_{n_d} \check{K}_d \left(1 - z^{-1}\right), \tag{6.51}$$

which resembles a PID-like transfer function having gains equal to the round-off errors on $\check{K}_p$, $\check{K}_i$, and $\check{K}_d$.

Consider first the low-frequency constraint. From (6.51), the expression of the relative sensitivity function at $\omega = 0$ depends solely on the integral gain quantization,

$$\left. \left| \delta G_{PID}(z; \check{K}) \right| \right|_{\omega=0} = \left| \delta_{n_i} \check{K}_i \right| = \frac{\left| d_{n_i} \check{K}_i \right|}{\check{K}_i}. \tag{6.52}$$

Therefore, n_i can be immediately determined from Constraint II. A minimum of 4 bits are necessary in order to satisfy $\left| \delta_{n_i} \check{K}_i \right| < \varepsilon_0 = 10\%$,

$$\check{K}_i = \mathcal{Q}_4 \left[\check{K}_i \right] = 0101_{\bar{2}} \times 2^{-3} = 0.625_{10}. \tag{6.53}$$

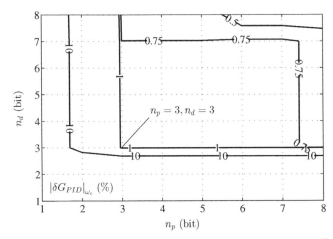

Figure 6.8 Compensator relative sensitivity function $\left|\delta G_{PID}(z;\check{\boldsymbol{K}})\right|$ (in %) at $\omega = \omega_c$ (parallel form).

Such choice yields a low-frequency relative error

$$\left.\left|\frac{|\tilde{T}(z)| - |T(z)|}{|T(z)|}\right|\right|_{\omega=0} \approx 4.8\% \tag{6.54}$$

in the loop gain magnitude.

As for the crossover frequency Constraint I, note that the PID-like structure of (6.51) suggests that $d_{n_i}\check{K}_i$ has negligible impact at the crossover frequency compared with $d_{n_p}\check{K}_p$ and $d_{n_d}\check{K}_d$. With this in mind, maintain the selected quantization for $\check{K}_i$ and investigate how $\delta G_{PID}(z;\check{\boldsymbol{K}})$ depends on n_p and n_d. The relative compensator sensitivity function, evaluated at $\omega = \omega_c$, is plotted in Fig. 6.8. In particular, the figure highlights constant error contours corresponding to $0.5\%, 0.75\%, 1\%$, and 10%. From this plot, one quickly recognizes that 3 bits are sufficient to represent both $\check{K}_p$ and $\check{K}_d$,

$$\tilde{K}_p \triangleq \mathcal{Q}_3\left[\check{K}_p\right] = 011_{\bar{2}} \times 2^3 = 24_{10},$$

$$\tilde{K}_d \triangleq \mathcal{Q}_3\left[\check{K}_d\right] = 011_{\bar{2}} \times 2^6 = 192_{10}. \tag{6.55}$$

With the above-mentioned choices, the magnitude and phase errors of $T(z)$ at $\omega = \omega_c$ are

$$\left||\tilde{T}(z)|_{\omega=\omega_c} - |T(z)|_{\omega=\omega_c}\right| \approx 0.75\%,$$

$$\angle\tilde{T}(z)|_{\omega=\omega_c} - \angle T(z)|_{\omega=\omega_c} \approx 0.36°. \tag{6.56}$$

and the corresponding compensator magnitude and phase responses are shown in Fig. 6.9. In the signal notation presented in Appendix B, the above choices for

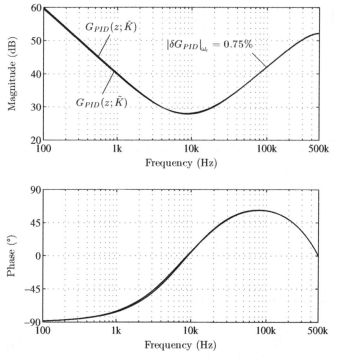

Figure 6.9   Bode plots of $G_{PID}(z; \tilde{K})$ and $G_{PID}(z; \check{K})$, $n_p = n_d = 3$, $n_i = 4$.

$(\tilde{K}_p, \tilde{K}_i, \tilde{K}_d)$ can be expressed as

$$\left[\tilde{K}_p\right]_3^3 = 011_{\tilde{2}},$$

$$\left[\tilde{K}_i\right]_{-3}^4 = 0101_{\tilde{2}}, \tag{6.57}$$

$$\left[\tilde{K}_d\right]_6^3 = 011_{\tilde{2}}.$$

6.3.2   Direct Structure

From Table 6.1, the coefficients b_0, b_1, and b_2 of the direct form can be derived from the p-domain parameters using (6.11). The result is

$$b_0 \approx 26.982,$$

$$b_1 \approx -50.72, \tag{6.58}$$

$$b_2 \approx 23.8.$$

In the direct form, scaling of the compensator amounts to scaling all the components of the coefficient vector by λ,

$$b_0 \rightarrow \breve{b}_0 \triangleq \lambda b_0 \approx 215.85,$$

$$b_1 \rightarrow \breve{b}_1 \triangleq \lambda b_1 \approx -405.76, \qquad (6.59)$$

$$b_2 \rightarrow \breve{b}_2 \triangleq \lambda b_2 \approx 190.5.$$

The scaled form of the compensator transfer function is therefore

$$G_{PID}(z; \breve{b}) = \frac{\breve{b}_0 + \breve{b}_1 z^{-1} + \breve{b}_2 z^{-2}}{1 - z^{-1}}, \qquad (6.60)$$

and the absolute sensitivity function of $G_{PID}(z; \breve{b})$ is

$$dG_{PID}(z; \breve{b}) = \frac{\partial G_{PID}(z; b)}{\partial b}\bigg|_{b=\breve{b}} d\breve{b}$$

$$= \frac{d_{n_0}\breve{b}_0 + d_{n_1}\breve{b}_1 z^{-1} + d_{n_2}\breve{b}_2 z^{-2}}{1 - z^{-1}}, \qquad (6.61)$$

where n_0, n_1, and n_2 are the word lengths used to express the quantized coefficients $\breve{b}_0$, $\breve{b}_1$, and $\breve{b}_2$, respectively.

Assume now $n_0 = n_1 = n_2 = n_b$ and evaluate the compensator relative sensitivity function at both $\omega = 0$ and $\omega = \omega_c$. Figure 6.10 illustrates the dependence of $\left|\delta G_{PID}(z; \breve{b})\right|_{\omega=0}$ and $\left|\delta G_{PID}(z; \breve{b})\right|_{\omega=\omega_c}$ on n_b. Observe that a minimum of $n_b = 9$ bits are required to satisfy the $\varepsilon = 1\%$ constraint, whereas the $\varepsilon_0 = 10\%$ constraint on the low-frequency loop gain magnitude requires $n_b = 12$ bits. $n_b = 12$ therefore represents the required resolution for the compensator coefficients, which evaluate to

$$\tilde{b}_0 = \mathcal{Q}_{12}\left[\breve{b}_0\right] = 011010111111_{\bar{2}} \times 2^{-3} = 215.875_{10},$$

$$\tilde{b}_1 = \mathcal{Q}_{12}\left[\breve{b}_1\right] = 100110101001_{\bar{2}} \times 2^{-2} = -405.75_{10}, \qquad (6.62)$$

$$\tilde{b}_2 = \mathcal{Q}_{12}\left[\breve{b}_2\right] = 010111110100_{\bar{2}} \times 2^{-3} = 190.5_{10}.$$

A more careful examination of this result shows that coefficient $\tilde{b}_2$ can actually be stored in a 10-bit word,

$$\tilde{b}_2 = 010111110100_{\bar{2}} \times 2^{-3}$$

$$= 0101111101_{\bar{2}} \times 2^{-1}$$

$$= 190.5_{10}. \qquad (6.63)$$

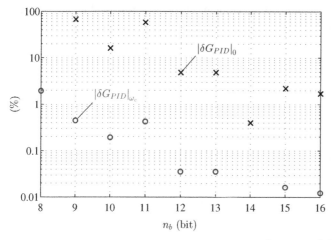

Figure 6.10 Compensator relative sensitivity function $\left|\delta G_{PID}(z;\breve{b})\right|$ (in %) at $\omega = 0$ and $\omega = \omega_c$ (direct form).

The above-mentioned results can be summarized as follows:

$$\left[\tilde{b}_0\right]_{-3}^{12} = 011010111111_{\bar{2}},$$

$$\left[\tilde{b}_1\right]_{-2}^{12} = 100110101001_{\bar{2}}, \qquad (6.64)$$

$$\left[\tilde{b}_2\right]_{-1}^{10} = 0101111101_{\bar{2}}.$$

A comparison between Bode plots of $G_{PID}(z;\breve{b})$ and $G_{PID}(z;\tilde{b})$ with $n_b = 12$ is illustrated in Fig. 6.11. In the case of direct realization, the error in the frequency response away from the crossover frequency ω_c tends to be more severe than what is observed for the parallel realization in Fig. 6.9, and the required coefficient resolution is ultimately dominated by the low-frequency constraint. Such higher sensitivity of direct forms to coefficient round-off is well known from the literature, which is why the direct realization of a PID compensator should in general be avoided.

6.3.3 Cascade Structure

As a last example, consider the cascade realization of the compensator. The coefficient vector $c = \left[K, c_{z_1}, c_{z_2}\right]^T$ can be derived from the p-domain design using (6.12),

$$K \approx 26.98,$$

$$c_{z_1} \approx -0.9691, \qquad (6.65)$$

$$c_{z_2} \approx -0.9107.$$

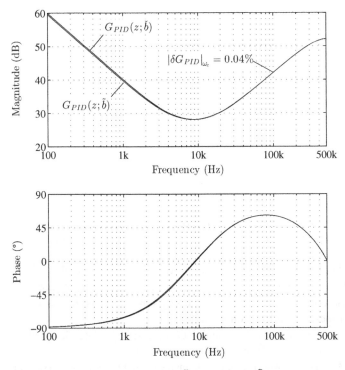

Figure 6.11   Bode plots of $G_{PID}(z; \check{\boldsymbol{b}})$ and $G_{PID}(z; \tilde{\boldsymbol{b}})$, $n_b = 12$.

In the cascade realization, scaling only concerns the coefficient K, as the zeros of the unscaled and scaled compensators must coincide. Therefore

$$K \to \check{K} \triangleq \lambda K \approx 215.85,$$

$$c_{z_1} \to \check{c}_{z_1} \triangleq c_{z_1} \approx -0.9691, \qquad (6.66)$$

$$c_{z_2} \to \check{c}_{z_2} \triangleq c_{z_2} \approx -0.9107.$$

The scaled version of the cascade form is

$$G_{PID}(z; \check{\boldsymbol{c}}) = \frac{\check{K}}{1 - z^{-1}} \left(1 + \check{c}_{z_1} z^{-1}\right) \left(1 + \check{c}_{z_2} z^{-1}\right), \qquad (6.67)$$

and its absolute sensitivity function is

$$
dG_{PID}(z; \check{c}) \approx \left. \frac{\partial G_{PID}(z; c)}{\partial c} \right|_{c=\check{c}} d\check{c}
$$

$$
= \frac{\left(1 + \check{c}_{z_1} z^{-1}\right)\left(1 + \check{c}_{z_2} z^{-1}\right)}{1 - z^{-1}} d_{n_k} \check{K}
$$

$$
+ \check{K} \frac{1 + \check{c}_{z_2} z^{-1}}{1 - z^{-1}} z^{-1} d_{n_{z_1}} \check{c}_{z_1}
$$

$$
+ \check{K} \frac{1 + \check{c}_{z_1} z^{-1}}{1 - z^{-1}} z^{-1} d_{n_{z_2}} \check{c}_{z_2}. \tag{6.68}
$$

As anticipated in Inset 6.1, the cascade structure is *not* linear with respect to $\check{c}$, and the above-mentioned approximation provides a correct estimate of the absolute error on $G_{PID}(z; \check{c})$ only when $d\check{c}$ is small.

Assume now that both $\check{c}_{z_1}$ and $\check{c}_{z_2}$ are quantized with the same number of bits $n_{z_1} = n_{z_2} = n_z$, whereas n_k bits are employed to round-off the compensator gain $\check{K}$. Next, evaluate the compensator relative sensitivity function at both $\omega = 0$ and $\omega = \omega_c$ and consider their dependence on n_k and n_z.

Contour maps of $|\delta G_{PID}(z; \check{c})|_{\omega=0}$ and $|\delta G_{PID}(z; \check{c})|_{\omega=\omega_c}$ are depicted in Fig. 6.12(a) and (b), respectively. From the plots, one concludes that $n_k = 3$ and $n_z = 6$ bits would satisfy the low-frequency Constraint II, but not the crossover frequency Constraint I, which would require 6 bits for all the coefficients. Therefore, choose $n_k = n_z = 6$ bits and evaluate the quantized compensator coefficients as

$$
\tilde{K} = \mathcal{Q}_6\left[\check{K}\right] = 011011_{\bar{2}} \times 2^3 = 216_{10},
$$

$$
\tilde{c}_{z_1} = \mathcal{Q}_6\left[\check{c}_{z_1}\right] = 100001_{\bar{2}} \times 2^{-5} = -0.96875_{10}, \tag{6.69}
$$

$$
\tilde{c}_{z_2} = \mathcal{Q}_6\left[\check{c}_{z_2}\right] = 100011_{\bar{2}} \times 2^{-5} = -0.90625_{10}.
$$

In signal notation, we have

$$
\left[\tilde{K}\right]_3^6 = 011011_{\bar{2}},
$$

$$
\left[\tilde{c}_{z_1}\right]_{-5}^6 = 100001_{\bar{2}}, \tag{6.70}
$$

$$
\left[\tilde{c}_{z_2}\right]_{-5}^6 = 100011_{\bar{2}}.
$$

Figure 6.13 compares the Bode plots of $G_{PID}(z; \check{c})$ and $G_{PID}(z; \tilde{c})$ with such quantization.

In general, the cascade form provides a more uniform round-off error throughout the frequency span and a slightly larger sensitivity to coefficient quantization compared to the parallel form. This usually leads to longer required word lengths for coefficient representation. However, the cascade form offers direct access to both the

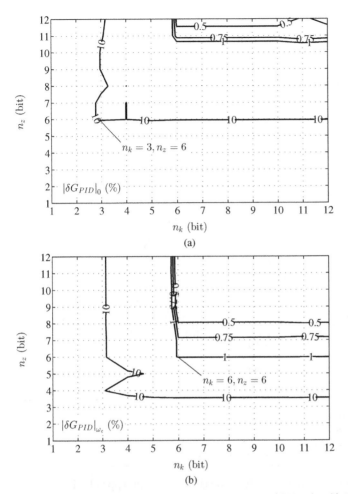

Figure 6.12   Compensator relative sensitivity function $|\delta G_{PID}(z; \check{c})|$ (in %) at (a) $\omega = 0$ and (b) $\omega = \omega_c$ (cascade form).

compensator zeros and its overall gain. This makes the cascade structure very convenient in implementing an on-line-programmable compensator frequency response.

Inset 6.2 – Evaluating the Sensitivity Function Using Matlab®

Contour plots such as those depicted in Figs 6.8 and 6.12 are generated using Matlab®in the following manner.

Consider, for definiteness, the parallel realization. Start by evaluating scaled coefficients $\check{K}_p$, $\check{K}_i$, and $\check{K}_d$ from K_p, K_i, and K_d and by evaluating the unquantized compensator transfer function $G_{PID}(z; \check{K})$:

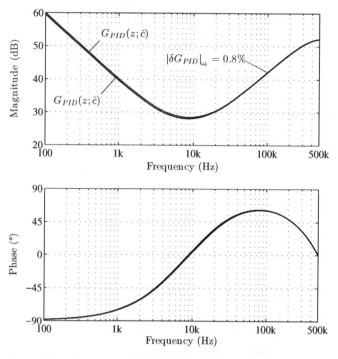

Figure 6.13   Bode plots of $G_{PID}(z; \check{c})$ and $G_{PID}(z; \tilde{c})$, $n_k = n_z = 6$.

```
Kps   =   Kp*(qAD*N_r);
Kis   =   Ki*(qAD*N_r);
Kds   =   Kd*(qAD*N_r);
Gczs  =   Kps+Kis/(1-z^-1)+Kds*(1-z^-1);
```

Next, quantize the integral gain using function Qn reported in Appendix B:

```
wki   =   Qn(Kis,4);
```

Calculation of the compensator relative sensitivity function can then be accomplished by evaluating (6.51) on a discrete grid of (n_p, n_d) pairs defined by two vectors np_v and nd_v.

```
np_v   =   [1:1:8];
nd_v   =   [1:1:8];

err =   zeros(length(nd_v),length(np_v));

for ip=1:length(np_v)
    wkp =   Qn(Kps,np_v(ip));
    for id=1:length(nd_v)
        wkd =   Qn(Kds,nd_v(id));
        Gcz =   wkp.xq+wki.xq/(1-z^-1)+wkd.xq*(1-z^-1);
```

```
        dGcz       =   wkp.dx+wki.dx/(1-z^-1)+wkd.dx*(1-z^-1);
        err(id,ip) =   100*abs(freqresp(dGcz,wc)/freqresp(Gczs,wc));
    end;
end;
```

where wc represents the angular crossover frequency ω_c. Values of $\delta G_{PID}(z; \tilde{K})$ evaluated at $\omega = \omega_c$ are now stored into matrix err. Contour plots can then be generated using Matlab® contour command.

6.4 FIXED-POINT CONTROLLER IMPLEMENTATION

A starting point for the fixed-point implementation step is the block diagram reported in Fig. 6.14, obtained from Fig. 6.2 after the original compensator coefficients have been quantized. In the following, quantization of $\tilde{u}_{PID}$ into $\tilde{u}$ is modeled as *truncation*,

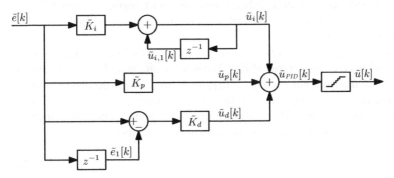

Figure 6.14   Parallel PID realization after coefficient scaling and quantization.

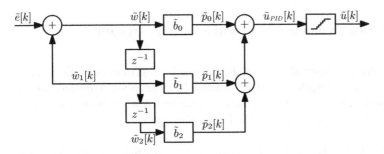

Figure 6.15   Direct PID realization after coefficient scaling and quantization.

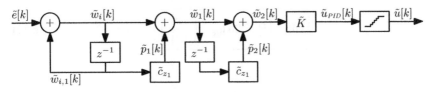

Figure 6.16   Cascade PID realization after coefficient scaling and quantization.

that is, as removal of an appropriate number of least significant digits from the control command.

The fixed-point implementation step is strongly dependent on the controller target platform. As anticipated in the introduction of this chapter, emphasis is given here on *custom* controller design for hardware-based compensators. In this context, the designer has the capability to fully tailor every aspect of the controller implementation according to the application. Resolution, clocking scheme, type, and complexity of the controller arithmetic blocks are all degrees of freedom available to the designer. On the other hand, in a software-based (microcontrolled) design, the hardware infrastructure is fixed, and the implementation step becomes the task of making the best use of an already existing digital processor.

Perhaps the most distinctive difference between software-based and hardware-based designs is the existence, in the former, of a well-specified *word length* for manipulating digital signals. In a 16-bit microcontroller architecture, for instance, the arithmetic logic unit (ALU) handles 16-bit operands and produces 16-bit results. The designer of a software program must normalize signals in the controller structure in such a way that truncation effects are negligible, and signal saturations do not occur.

On the other hand, in a hardware-based design, every arithmetic operation can be designed with a custom number of bits. The word length is no longer constrained by the hardware platform but becomes a degree of freedom the designer can tailor to the specific application requirements. In this case, an objective of the fixed-point implementation step is to establish, for every signal $\tilde{x}[k]$ in the controller realization, the number of bits n and the scale q required for its B2C representation,

$$[\tilde{x}]_q^n , \tag{6.71}$$

an operation often referred to as *word length determination* or *word length estimation*.

6.4.1 Effective Dynamic Range and Hardware Dynamic Range

Crucial to the determination of both n and q is the estimation of the signal's *dynamic range*. In short, word length n must be sufficiently large to accommodate for the typical excursion of the signal during a "severe" disturbance, whereas the scale q must be sufficiently fine not to cause abnormal controller operation due to quantization effects.

The *effective dynamic range* of $\tilde{x}$, or simply *dynamic range*, is here defined as the ratio between the signal's *upper bound* $\lceil \tilde{x} \rceil$ and its *lower bound* $\lfloor \tilde{x} \rfloor$

$$\boxed{\mathrm{DR}_{\mathit{eff}} [\tilde{x}] \triangleq 20 \log_{10} \left(\frac{\lceil \tilde{x} \rceil}{\lfloor \tilde{x} \rfloor} \right).} \tag{6.72}$$

- The *upper bound* $\lceil \tilde{x} \rceil$ of $\tilde{x}$ represents the largest value reached by $|\tilde{x}|$. An estimate of $\lceil \tilde{x} \rceil$ can be drawn from the most severe transient the control system is expected to handle. Guidelines for a first-cut estimation of a signal's upper bound are mentioned in Section 6.4.2.

- The *lower bound* $\lfloor \tilde{x} \rfloor$ of $\tilde{x}$ represents the smallest nonzero value of $|\tilde{x}|$ that needs to be resolved for proper control operation. Estimation of $\lfloor \tilde{x} \rfloor$ is often crucial to the hardware optimization of the controller and is best performed via a simulation-based investigation.

Suppose now that $\tilde{x}$ is to be represented by a B2C quantity $[\tilde{x}]_q^n$, where both n and q are to be determined. By *hardware dynamic range* of $[\tilde{x}]_q^n$, one refers to the ratio between a signal's upper bound and its scale,

$$
\boxed{\mathrm{DR}_{hw}[\tilde{x}]_q^n \triangleq 20 \log_{10}\left(\frac{\lceil \tilde{x} \rceil}{2^q}\right).}
\tag{6.73}
$$

Observe that the *maximum* hardware dynamic range of $\tilde{x}$ is uniquely limited by n:

$$
\begin{aligned}
\mathrm{DR}_{hw}[\tilde{x}]_q^n &\leq 20 \log_{10}\left(\frac{2^{n-1}2^q}{2^q}\right) \\
&= (20 \log_{10} 2)(n-1) \\
&\approx 6.02(n-1).
\end{aligned}
\tag{6.74}
$$

The effective dynamic range is a property of a signal, and it measures the distance between the smallest and largest values the signals assumes during its evolution. On the other hand, the hardware dynamic range is a property of a B2C quantity and is limited by the hardware number of bits. The idea is to design the hardware dynamic range so as to entirely host the signal, that is, accurately represent its lower bound $\lfloor \tilde{x} \rfloor$ without appreciable quantization effects, and simultaneously represent its largest value without saturations.

Establishing a link between the effective and hardware dynamic ranges is relatively simple. First, choose a scale 2^q so as to represent the signal's lower bound with the required accuracy,

$$
\lfloor \tilde{x} \rfloor = [\lfloor \tilde{x} \rfloor]_q^l \times 2^q.
\tag{6.75}
$$

The hardware dynamic range is then equal to the effective dynamic range of $\tilde{x}$, plus the hardware dynamic range required to represent $\lfloor \tilde{x} \rfloor$,

$$\mathrm{DR}_{hw}[\tilde{x}]_q^n = 20\log_{10}\left(\frac{\lceil\tilde{x}\rceil}{2^q}\right)$$

$$= 20\log_{10}\left(\frac{\lceil\tilde{x}\rceil}{\lfloor\tilde{x}\rfloor}\frac{\lfloor\tilde{x}\rfloor}{2^q}\right)$$

$$= \mathrm{DR}_{\textit{eff}}\,[x] + \mathrm{DR}_{hw}[\lfloor\tilde{x}\rfloor]_q^l. \tag{6.76}$$

From (6.74), the minimum number of bits required to completely host the hardware dynamic range $\mathrm{DR}_{hw}[\tilde{x}]_q^n$ is

$$\boxed{n = 1 + \mathrm{ceil}\left(\frac{\mathrm{DR}_{hw}[\tilde{x}]_q^n}{20\log_{10}2}\right),} \tag{6.77}$$

where ceil(c) denotes the smallest integer greater than or equal to c.

In practice, the above-mentioned concepts are applied as follows:

- Determine the signal's upper and lower bounds $\lceil\tilde{x}\rceil$ and $\lfloor\tilde{x}\rfloor$ either analytically or with the aid of computer simulations.
- Select the scale 2^q in order to represent $\lfloor\tilde{x}\rfloor$ with sufficient accuracy.
- Determine the required hardware dynamic range from (6.73).
- Determine the required word length n from (6.77).

6.4.2 Upper Bound of a Signal and the L^1-Norm

Let $\tilde{x}[k]$ be a generic signal in the controller implementation. Initially, $(k < 0)\,e[k] = 0$, and $\tilde{x}[k]$ is constant at some $\tilde{X}_0$. Evolution of $\tilde{x}[k]$ in response to $\tilde{e}[k]$ can be written in terms of a certain impulse response $h_x[k]$,

$$\tilde{x}[k] = \tilde{X}_0 + \sum_{i=0}^{k} h_x[k-i]\tilde{e}[i]. \tag{6.78}$$

Examine two cases:

- *Unbiased signals* are such that $\tilde{X}_0 = 0$, examples being the proportional and derivative terms $\tilde{u}_p$ and $\tilde{u}_d$.
- *Biased signals* are such that $\tilde{X}_0 \neq 0$. In the standard PID structure examined in this book, a controller signal is biased if and only if $h_x[k]$ contains the integral action, that is, if signal $\tilde{x}[k]$ is cascaded to the integrator.

An upper bound of an unbiased signal can be derived analytically under the assumption that the digitized error signal $\tilde{e}[k]$ remains bounded, a condition expressed as

$$|\tilde{e}[k]| \leq \tilde{e}_{max} \Rightarrow \lceil\tilde{e}\rceil = \tilde{e}_{max}. \tag{6.79}$$

From (6.79), an upper bound for $\tilde{x}[k]$ is then

$$
\begin{aligned}
|\tilde{x}[k]| &\leq \left| \sum_{i=0}^{k} h_x[k-i]\tilde{e}[i] \right| \\
&\leq \sum_{i=0}^{k} |h_x[k-i]|\,|\tilde{e}[i]| \\
&\leq \left(\sum_{i=0}^{k} |h_x[i]| \right) \lceil \tilde{e} \rceil \\
&\leq \left(\sum_{i=0}^{+\infty} |h_x[i]| \right) \lceil \tilde{e} \rceil,
\end{aligned}
\tag{6.80}
$$

or

$$
|\tilde{x}[k]| \leq \|h_x\|_1 \lceil \tilde{e} \rceil,
\tag{6.81}
$$

where $\|h_x\|_1$ denotes the L^1-norm of h_x,

$$
\|h_x\|_1 \triangleq \sum_{i=0}^{+\infty} |h_x[i]|.
\tag{6.82}
$$

Notice that the right-hand side sum always converges, as h_x does not contain the integral action.

In conclusion, the upper bound estimation of unbiased signals according to the L^1-norm criterion implies

$$
\boxed{\lceil \tilde{x} \rceil = \|h_x\|_1 \lceil \tilde{e} \rceil}.
\tag{6.83}
$$

The L^1-norm criterion for establishing a signal's upper bound represents, in general, a very conservative approach to word length estimation. The reason is that the L^1-norm upper bound is evaluated from the worst-case input sequence that maximizes $|\tilde{x}|$. Such sequence is hardly, if ever, encountered during the system operation. Nonetheless, the L^1 criterion can represent the starting point for a more refined simulation-based analysis.

For biased signals, the situation is usually more complex because the steady-state value $\tilde{X}_0$ of $\tilde{x}[k]$ depends on the converter operating point and, in general, changes with it. A first-cut estimation of the signal's upper bound considers the range of variation of $\tilde{X}_0$ as the system duty cycle varies between 0 and 1, that is, as the steady-state controller output $\tilde{U}$ varies between 0 and $N_r - 1$,

$$
\boxed{\lceil \tilde{x} \rceil = \max_{0 \leq \tilde{U} \leq N_r - 1} \tilde{X}_0(\tilde{U})}.
\tag{6.84}
$$

Note, however, that additional overhead must sometimes be allowed with respect to the steady-state estimation due to the signal's dynamics causing potential saturations during transient events.

6.5 VOLTAGE-MODE CONVERTER EXAMPLE: FIXED-POINT IMPLEMENTATION

In this section, fixed-point realizations of the three basic PID structures considered in this chapter are discussed, continuing the study of the synchronous Buck converter voltage-mode control example.

It is first necessary to clarify how the digitized error signal $\tilde{e}$ is computed with the aid of Fig. 6.17, which refers, for ease of representation, to a hypothetical 3-bit A/D converter operating over a full-scale range of 2 V. It is assumed here that the A/D converter outputs the digitized version of the sampled voltage v_s as an *unsigned* binary word and that the digitized output ranges from 000_2 to 111_2 as the analog A/D input v_s ranges from 0 to 2 V.

The A/D output is first converted into a positive B2C word $[\tilde{v}_s]_0^4$ by padding the MSB with a 0. Then, the error is calculated by subtracting the B2C-converted A/D output from the digital set point, the latter also expressed as a 4-bit B2C word $\left[\tilde{V}_{ref}\right]_0^4$,

$$[\tilde{e}]_0^4 \Leftarrow \left[\tilde{V}_{ref}\right]_0^4 - [\tilde{v}_s]_0^4. \tag{6.85}$$

In the synchronous Buck converter example, an 8-bit A/D converter with a 2-V full-scale range is used. Therefore,

$$\boxed{[\tilde{e}]_0^9 \Leftarrow \left[\tilde{V}_{ref}\right]_0^9 - [\tilde{v}_s]_0^9.} \tag{6.86}$$

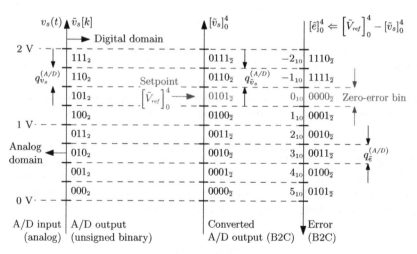

Figure 6.17   Digital error calculation from the A/D output for a 3-bit A/D converter.

A 10-bit DPWM is employed in the considered example. The control command $\tilde{u}$ is a biased signal ranging from 0 to $N_r - 1 = 1023_{10}$ with a granularity $q_{\tilde{u}} = 1$. Hence, $\tilde{u}$ can be represented over a scale 2^0, while the required hardware dynamic range and word length are

$$\mathrm{DR}_{hw}[\tilde{u}]_0^n = 20 \log_{10} \frac{1023_{10}}{2^0} \approx 60.2 \text{ dB}$$

$$\Rightarrow n = 1 + \mathrm{ceil}\left(\frac{\mathrm{DR}_{hw}[\tilde{u}]_0^n}{20 \log_{10} 2}\right) = 11.$$

$$(6.87)$$

Controller output $\tilde{u}$ is then represented by an 11-bit word over a scale 2^0,

$$\boxed{[\tilde{u}]_0^{11}}.$$

$$(6.88)$$

Observe that an extra bit appears because the represented range of a B2C quantity is always bipolar. In practice, $\tilde{u}$ is guaranteed to remain in the positive half range by the output limiter block, and the extra bit is dropped before the signal is latched by the 10-bit DPWM.

Next, identify the worst-case transient condition for this considered application example. Figure 6.18 depicts the digitized error signal during a 0 A $\rightarrow$ 5 A $\rightarrow$ 0 A sequence of step-load transients. The simulation is carried out with Matlab® including both A/D and DPWM quantizations and employing quantized coefficients $\tilde{K}_p$, $\tilde{K}_i$, and $\tilde{K}_d$ determined in Section 6.3.1. On the other hand, control calculations are simulated with the full Matlab® floating-point precision. Treating this transient as a worst-case reference, an upper bound

$$\boxed{\lceil \tilde{e} \rceil = 7 \text{ LSB}}$$

$$(6.89)$$

for the digitized error signal is assumed. In general, startup behavior or other types of transients may impact the worst-case error signal.

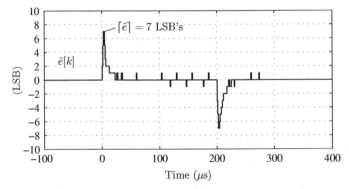

Figure 6.18   Digitized error signal during a 0 A $\rightarrow$ 5 A $\rightarrow$ 0 A step-up/step-down load transient.

6.5.1 Parallel Realization

Figure 6.14 reports the parallel PID realization structure after the coefficient quantization process. Proportional, integral, and derivative terms during the 0 A → 5 A → 0 A step load transient treated as the worst-case scenario are shown in Fig. 6.19. Observe that both $\tilde{u}_p$ and $\tilde{u}_d$ are unbiased signals, whereas $\tilde{u}_i$ is biased.

Consider first the proportional term $\tilde{u}_p$,

$$\tilde{u}_p[k] = \tilde{K}_p \tilde{e}[k]. \tag{6.90}$$

Owing to the granularity of $\tilde{e}$ being $q_{\tilde{e}} = 1$, $\tilde{u}_p$ always remains a multiple of $\tilde{K}_p = 24_{10}$. Lower bound of $\tilde{u}_p$ is then

$$\lfloor \tilde{u}_p \rfloor = \tilde{K}_p = \underbrace{\left[\tilde{K}_p\right]_3}_{011_{\bar{2}}} \times 2^3, \tag{6.91}$$

as per (6.57). The scale of $\lfloor \tilde{u}_p \rfloor$ is therefore equal to that of $\tilde{K}_p$, or 2^3. Representing $\tilde{u}_p$ on a finer scale would not improve accuracy of the product, whereas a coarser scale would inevitably lead to an equivalent truncation of $\tilde{K}_p$, which should be avoided. Hence, 2^3 is the best choice for the scale of $\tilde{u}_p$.

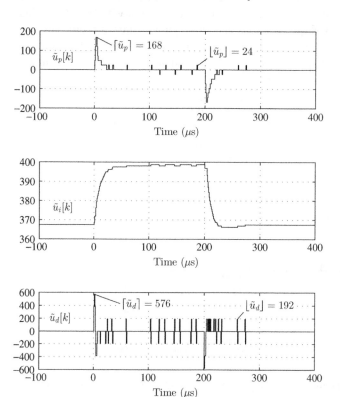

Figure 6.19 Proportional, integral, and derivative terms during a 0 A → 5 A → 0 A step-up/step-down load transient.

The upper bound of $\tilde{u}_p$ is simply

$$\lceil \tilde{u}_p \rceil = \tilde{K}_p \lceil \tilde{e} \rceil = 168_{10}, \qquad (6.92)$$

as also highlighted in Fig. 6.19. It can be verified that such estimate coincides with the L^1-norm of $\tilde{u}$. In fact, for a proportional relationship, the L^1 upper bound always coincides with the true peak value of the signal.

The required dynamic range and word length for hosting $\tilde{u}_p$ in a B2C word are

$$\mathrm{DR}_{hw} \lceil \tilde{u}_p \rceil_3^n = 20 \log_{10} \frac{168_{10}}{2^3} \approx 26.44 \text{ dB}$$

$$\Rightarrow n = 1 + \mathrm{ceil}\left(\frac{\mathrm{DR}_{hw} \lceil \tilde{u}_p \rceil_3^n}{20 \log_{10} 2} \right) = 6. \qquad (6.93)$$

From the above-mentioned derivation, $\tilde{u}_p$ can be stored in a 6-bit-long B2C word and represented over a scale equal to 2^3,

$$\left[\tilde{u}_p \right]_3^6 \Leftarrow \left[\tilde{K}_p \right]_3^3 \times [\tilde{e}]_0^9, \qquad (6.94)$$

where a *saturated multiplication* is employed.

Consider next the derivative term

$$\tilde{u}_d[k] = \tilde{K}_d \left(\tilde{e}[k] - \tilde{e}[k-1] \right). \qquad (6.95)$$

Once again, $\tilde{u}_d$ is always an integer multiple of $\tilde{K}_d = 192_{10}$, which from (6.57) is exactly represented over a scale of 2^6,

$$\lfloor \tilde{u}_d \rfloor = \tilde{K}_d = \underbrace{\left[\tilde{K}_d \right]_6^3}_{011_{\bar{2}}} \times 2^6. \qquad (6.96)$$

As for $\tilde{u}_d$'s upper bound, from Fig. 6.19, one has

$$\lceil \tilde{u}_d \rceil = 576_{10}. \qquad (6.97)$$

On the other hand, the L^1-norm upper bound predicted by (6.81) is

$$\lceil \tilde{u}_d \rceil = 2\tilde{K}_d \tilde{e}_{max} = 2688_{10}, \qquad (6.98)$$

much larger than the simulated value, the reason being the worst-case nature of the L^1-norm criterion. Nonetheless, the word length of $\tilde{u}_d$ is calculated according to the latter estimate of $\lceil \tilde{u}_d \rceil$.

Proceeding as done with the proportional term, the hardware dynamic range and the word length required to host $\tilde{u}_d$ are

$$\text{DR}_{hw}[\tilde{u}_d]_6^n = 20\log_{10}\frac{2688_{10}}{2^6} \approx 32.5 \text{ dB}$$

$$\Rightarrow n = 1 + \text{ceil}\left(\frac{\text{DR}_{hw}[\tilde{u}_d]_6^n}{20\log_{10}2}\right) = 7. \tag{6.99}$$

According to the foregoing arguments, $\tilde{u}_d$ can be stored as a 7-bit-long B2C word with a scale equal to 2^6,

$$[\tilde{u}_d]_6^7 \Leftarrow \left[\tilde{K}_d\right]_6^3 \times [\tilde{e}]_0^9. \tag{6.100}$$

Finally, examine the integral term $\tilde{u}_i$. In every instant in time, the integrator output is always a multiple of $\tilde{K}_i = 0.625_{10}$, as already discussed in Section 5.3.2 in the context of the role the integral gain may have in triggering limit cycle oscillations. Therefore,

$$\lfloor\tilde{u}_i\rfloor = \tilde{K}_i = \underbrace{\left[\tilde{K}_i\right]_{-3}^4}_{0101_{\overline{2}}} \times 2^{-3}, \tag{6.101}$$

and $\tilde{u}_i$ is represented on the same scale as $\tilde{K}_i$, that is, 2^{-3}.

As $\tilde{u}_i$ is a biased signal, its upper bound is determined according to criterion (6.84),

$$\lceil\tilde{u}_i\rceil = N_r - 1 = 1023_{10}. \tag{6.102}$$

Proceeding as before for the word length determination, we obtain

$$\text{DR}_{hw}[\tilde{u}_i]_{-3}^n = 20\log_{10}\frac{1023_{10}}{2^{-3}} \approx 78.26 \text{ dB}$$

$$\Rightarrow n = 1 + \text{ceil}\left(\frac{\text{DR}_{hw}[\tilde{u}_i]_{-3}^n}{20\log_{10}2}\right) = 14. \tag{6.103}$$

The integral term can therefore be hosted in a 14-bit B2C word with a scale equal to 2^{-3}. Observe that the integral term requires the largest hardware dynamic range among the three control terms. In fact, $\tilde{u}_i$ must simultaneously fine-tune the control command *and* cover the entire operating range of the digital pulse width modulator.

The word length of $\tilde{w}_i = \tilde{K}_i\tilde{e}$ can be determined as done for the proportional term. Its upper bound is

$$\lceil\tilde{w}_i\rceil = \tilde{K}_i\tilde{e}_{max} = 4.375_{10}, \tag{6.104}$$

and therefore

$$\mathrm{DR}_{hw}[\tilde{w}_i]_{-3}^n = 20 \log_{10} \frac{[\tilde{w}_i]}{2^{-3}} \approx 30.9 \text{ dB}$$

$$\Rightarrow n = 1 + \mathrm{ceil}\left(\frac{\mathrm{DR}_{hw}[\tilde{w}_i]_{-3}^n}{20 \log_{10} 2}\right) = 7. \qquad (6.105)$$

Signal $\tilde{w}_i$ is then obtained from $\tilde{e}$ via a saturated multiplication

$$[\tilde{w}_i]_{-3}^7 \Leftarrow \left[\tilde{K}_i\right]_{-3}^4 \times [\tilde{e}]_0^9. \qquad (6.106)$$

The overall high-resolution PID control signal $\tilde{u}_{PID}$ is

$$\tilde{u}_{PID} = \tilde{u}_p + \tilde{u}_i + \tilde{u}_d. \qquad (6.107)$$

In the B2C implementation, the above-mentioned addition requires a preliminary alignment of both $\tilde{u}_p$ and $\tilde{u}_d$ to $\tilde{u}_i$,

$$[\tilde{u}_p]_{-3}^{12} \leftarrow [\tilde{u}_p]_3^6,$$
$$[\tilde{u}_d]_{-3}^{16} \leftarrow [\tilde{u}_d]_6^7, \qquad (6.108)$$

$$[\tilde{u}_{PID}]_{-3}^{14} \Leftarrow [\tilde{u}_p]_{-3}^{12} + [\tilde{u}_i]_{-3}^{14} + [\tilde{u}_d]_{-3}^{16}. \qquad (6.109)$$

The last step is the quantization/saturation of $\tilde{u}_{PID}$ necessary to have the compensator output $\tilde{u}$ matching the DPWM word length and resolution. Following the block diagram shown in Fig. 6.5, $\tilde{u}_{PID}$ is first truncated into the intermediate signal $\tilde{u}_x$,

$$[\tilde{u}_x]_0^{11} \leftarrow [\tilde{u}_{PID}]_{-3}^{14}, \qquad (6.110)$$

and the result is then limited between 0 and $N_r - 1 = 1023_{10}$,

$$[\tilde{u}]_0^{11} \leftarrow \begin{cases} 1023_{10} & \text{if } [\tilde{u}_x]_0^{11} > 1023_{10}, \\ 0 & \text{if } [\tilde{u}_x]_0^{11} < 0, \\ [\tilde{u}_x]_0^{11} & \text{otherwise.} \end{cases} \qquad (6.111)$$

Notice that, in practice, comparison against 1023_{10} is unnecessary, as $[\tilde{u}_x]_0^{11}$ is an 11-bit B2C word.

Figure 6.20 illustrates the block diagram of the fixed-point implementation of the parallel structure determined earlier.

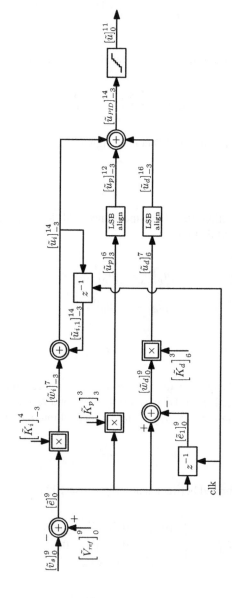

Figure 6.20 Fixed-point implementation of the parallel structure of the PID compensator.

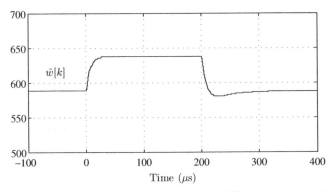

Figure 6.21   Direct implementation: signal $\tilde{w}[k]$ during a 0 A $\rightarrow$ 5 A $\rightarrow$ 0 A step-up/step-down load transient.

6.5.2   Direct Realization

Consider now the direct realization of the PID compensator illustrated in Fig. 6.15. In this structure, signal $\tilde{w}$ is the accumulation of all the past error samples,

$$\tilde{w}[k] = \tilde{w}[k-1] + \tilde{e}[k], \tag{6.112}$$

whereas $\tilde{w}_1$ and $\tilde{w}_2$ are one-step and two-step delayed versions of $\tilde{w}$, respectively. As for signals $\tilde{p}_0$, $\tilde{p}_1$, and $\tilde{p}_2$, they are, respectively, proportional to $\tilde{w}$, $\tilde{w}_1$, and $\tilde{w}_2$. Note that all the signals in the realization are biased.

Signal $\tilde{w}$ inherits the granularity $q_{\tilde{e}} = 1$ of $\tilde{e}$ and can therefore be represented on the same scale 2^0 as $\tilde{e}$,

$$\lfloor \tilde{w} \rfloor = \lfloor \tilde{e} \rfloor = 1_{10}. \tag{6.113}$$

As for the upper bound of $\tilde{w}$, its steady-state value is related to the compensator output command $\tilde{u}$ by

$$\tilde{W} = \frac{\tilde{U}}{\tilde{b}_0 + \tilde{b}_1 + \tilde{b}_2}. \tag{6.114}$$

Upper bound of $\tilde{w}$ is, following (6.84),

$$\lceil \tilde{w} \rceil = \frac{N_r - 1}{\tilde{b}_0 + \tilde{b}_1 + \tilde{b}_2} \approx 1637_{10}, \tag{6.115}$$

The hardware dynamic range and the word length required for $\tilde{w}$ are

$$\mathrm{DR}_{hw}[\tilde{w}]_0^n = 20 \log_{10} \frac{\lceil \tilde{w} \rceil}{2^0} \approx 64.3 \text{ dB}$$

$$\Rightarrow n = 1 + \mathrm{ceil}\left(\frac{\mathrm{DR}_{hw}[\tilde{w}]_0^n}{20 \log_{10} 2} \right) = 12, \tag{6.116}$$

which also represents the required word length for $\tilde{w}_{acc}$.

Examine next the signals

$$\tilde{p}_0[k] = \tilde{b}_0 \tilde{w}[k],$$

$$\tilde{p}_1[k] = \tilde{b}_1 \tilde{w}_1[k], \qquad (6.117)$$

$$\tilde{p}_2[k] = \tilde{b}_2 \tilde{w}_2[k],$$

whose simulated waveforms are shown in Fig. 6.22. Given the unity granularity of $\tilde{w}$, the scales of $\tilde{p}_0$, $\tilde{p}_1$, and $\tilde{p}_2$ are given by the scales of $\tilde{b}_0$, $\tilde{b}_1$, and $\tilde{b}_2$, respectively,

$$\lfloor \tilde{p}_0 \rfloor = \tilde{b}_0 = \underbrace{\left[\tilde{b}_0\right]_{-3}^{12}}_{011010111111_{\bar{2}}} \times 2^{-3},$$

$$\lfloor \tilde{p}_1 \rfloor = \tilde{b}_1 = \underbrace{\left[\tilde{b}_1\right]_{-2}^{12}}_{100110101001_{\bar{2}}} \times 2^{-2}, \qquad (6.118)$$

$$\lfloor \tilde{p}_2 \rfloor = \tilde{b}_2 = \underbrace{\left[\tilde{b}_2\right]_{-1}^{10}}_{0101111101_{\bar{2}}} \times 2^{-1},$$

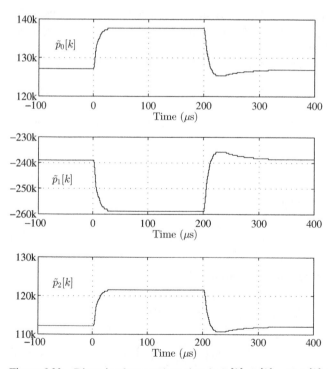

Figure 6.22 Direct implementation: signals $\tilde{p}_0[k]$, $\tilde{p}_1[k]$, and $\tilde{p}_2[k]$ during a $0\,A \rightarrow 5\,A \rightarrow 0\,A$ step-up/step-down load transient.

as per (6.64). Estimation of upper bounds for $\tilde{p}_0$, $\tilde{p}_1$, and $\tilde{p}_2$ is accomplished by scaling the upper bound of $\tilde{w}$ by $\tilde{b}_0$, $\tilde{b}_1$, and $\tilde{b}_2$, respectively,

$$\lceil \tilde{p}_0 \rceil = \lceil \tilde{w} \rceil \tilde{b}_0 = \frac{\tilde{b}_0 \left(N_r - 1 \right)}{\tilde{b}_0 + \tilde{b}_1 + \tilde{b}_2} = 353344.2_{10},$$

$$\lceil \tilde{p}_1 \rceil = \lceil \tilde{w} \rceil \tilde{b}_1 = \frac{\tilde{b}_1 \left(N_r - 1 \right)}{\tilde{b}_0 + \tilde{b}_1 + \tilde{b}_2} = 664131.6_{10}, \tag{6.119}$$

$$\lceil \tilde{p}_2 \rceil = \lceil \tilde{w} \rceil \tilde{b}_2 = \frac{\tilde{b}_2 \left(N_r - 1 \right)}{\tilde{b}_0 + \tilde{b}_1 + \tilde{b}_2} = 311810.4_{10}.$$

According to the foregoing calculations, dynamic ranges and the required word lengths for $\tilde{p}_0$, $\tilde{p}_1$, and $\tilde{p}_2$ are

$$\mathrm{DR}_{hw}[\tilde{p}_0]^n_{-3} \triangleq 20 \log_{10} \frac{[\tilde{p}_0]}{2^{-3}} \approx 129 \text{ dB} \Rightarrow n = 1 + \mathrm{ceil} \left(\frac{\mathrm{DR}_{hw}[\tilde{p}_0]^n_{-3}}{20 \log_{10} 2} \right) = 23,$$

$$\mathrm{DR}_{hw}[\tilde{p}_1]^n_{-2} \triangleq 20 \log_{10} \frac{[\tilde{p}_1]}{2^{-2}} \approx 128 \text{ dB} \Rightarrow n = 1 + \mathrm{ceil} \left(\frac{\mathrm{DR}_{hw}[\tilde{p}_1]^n_{-2}}{20 \log_{10} 2} \right) = 23,$$

$$\mathrm{DR}_{hw}[\tilde{p}_2]^n_{-1} \triangleq 20 \log_{10} \frac{[\tilde{p}_2]}{2^{-1}} \approx 116 \text{ dB} \Rightarrow n = 1 + \mathrm{ceil} \left(\frac{\mathrm{DR}_{hw}[\tilde{p}_2]^n_{-1}}{20 \log_{10} 2} \right) = 21.$$
$$\tag{6.120}$$

The high-resolution PID command $\tilde{u}_{PID}$ can be evaluated as a saturated sum of $\tilde{p}_0$, $\tilde{p}_1$, and $\tilde{p}_2$. Before the summation, an alignment of $\tilde{p}_1$ and $\tilde{p}_2$ to $\tilde{p}_0$ is necessary,

$$[\tilde{p}_1]^{24}_{-3} \leftarrow [\tilde{p}_1]^{23}_{-2},$$
$$[\tilde{p}_2]^{23}_{-3} \leftarrow [\tilde{p}_2]^{21}_{-1}, \tag{6.121}$$

$$[\tilde{u}_{PID}]^{14}_{-3} \Leftarrow [\tilde{p}_0]^{23}_{-3} + [\tilde{p}_1]^{24}_{-3} + [\tilde{p}_2]^{23}_{-3}. \tag{6.122}$$

Finally, as with the parallel implementation, the output control command $\tilde{u}$ is a truncated and limited version of $\tilde{u}_{PID}$,

$$[\tilde{u}_x]^{11}_0 \leftarrow [\tilde{u}_{PID}]^{14}_{-3}, \tag{6.123}$$

$$[\tilde{u}]^{11}_0 \leftarrow \begin{cases} 0 & \text{if } [\tilde{u}_x]^{11}_0 < 0, \\ [\tilde{u}_x]^{11}_0 & \text{otherwise.} \end{cases} \tag{6.124}$$

Figure 6.23 illustrates the fixed-point implementation of the direct structure discussed in this section.

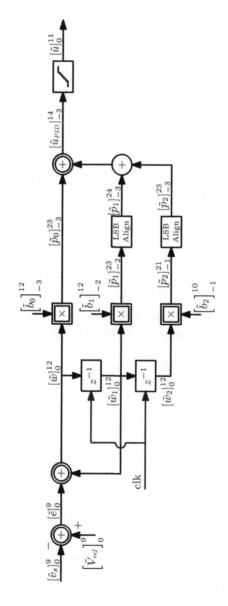

Figure 6.23   Fixed-point implementation of the direct structure.

6.5.3   Cascade Realization

In the cascade realization—Fig. 6.16—the first step toward word length determination is the dynamic range analysis of the three signals $\tilde{w}_i$, $\tilde{w}_1$, and $\tilde{w}_2$. Their transient responses during the $0\,\mathrm{A} \rightarrow 5\,\mathrm{A} \rightarrow 0\,\mathrm{A}$ sequence taken as the worst case are illustrated in Fig. 6.24.

Signal $\tilde{w}_i$ is obtained by integration of $\tilde{e}[k]$ exactly as signal $\tilde{w}$ of the direct structure. Hence,

$$\lfloor \tilde{w}_i \rfloor = \lfloor \tilde{e} \rfloor = 1. \tag{6.125}$$

Furthermore, $\tilde{w}_i$ is a biased signal whose steady-state value is

$$\tilde{W}_i = \frac{\tilde{U}}{\tilde{K} \left(1 + \tilde{c}_{z_1} \right) \left(1 + \tilde{c}_{z_2} \right)}. \tag{6.126}$$

As shown in Fig. 6.24, dynamics of $\tilde{w}_i$ are sufficiently slow and essentially overshoot-free. It follows that one can safely take (6.84) as the design criterion and define

$$\lceil \tilde{w}_i \rceil = \frac{N_r - 1}{\tilde{K} \left(1 + \tilde{c}_{z_1} \right) \left(1 + \tilde{c}_{z_2} \right)} \approx 1617_{10} \tag{6.127}$$

as the signal's upper bound. It follows that

$$\mathrm{DR}_{hw}[\tilde{w}_i]_0^n \triangleq 20 \log_{10} \frac{\lceil \tilde{w}_i \rceil}{2^0} \approx 64.2 \text{ dB} \Rightarrow n = 1 + \mathrm{ceil}\left(\frac{\mathrm{DR}_{hw}[\tilde{w}_1]_{-5}^n}{20 \log_{10} 2} \right) = 12. \tag{6.128}$$

Examine now signals $\tilde{w}_1$ and $\tilde{w}_2$. Their lower bounds are

$$\lfloor \tilde{w}_1 \rfloor = \left(1 + \tilde{c}_{z_1} \right) \lfloor \tilde{w}_i \rfloor = 0.03125_{10} = 01_{\bar{2}} \times 2^{-5} \tag{6.129}$$

and

$$\begin{aligned} \lfloor \tilde{w}_2 \rfloor &= \left(1 + \tilde{c}_{z_2} \right) \lfloor \tilde{w}_1 \rfloor \\ &= \left(1 + \tilde{c}_{z_2} \right) \left(1 + \tilde{c}_{z_1} \right) \lfloor \tilde{w}_i \rfloor \\ &= 0.0029296875_{10} \\ &= 011_{\bar{2}} \times 2^{-10}. \end{aligned} \tag{6.130}$$

Therefore, these signals can be exactly represented over a scale of 2^{-5} and 2^{-10}, respectively. As for their upper bounds, $\tilde{w}_1$ and $\tilde{w}_2$ are biased signals as well. Assess their steady-state range of operation first. As for $\tilde{w}_1$,

$$0 \leq \tilde{W}_1 \leq \frac{N_r - 1}{\tilde{K} \left(1 + \tilde{c}_{z_2} \right)} \approx 50.2, \tag{6.131}$$

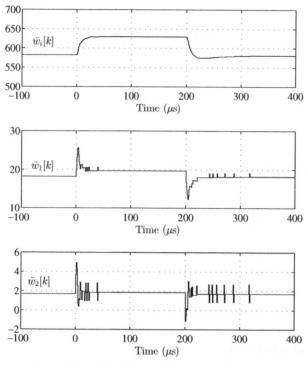

Figure 6.24 Cascade implementation: signals $\tilde{w}_1[k]$, $\tilde{w}_2[k]$, and $\tilde{w}_3[k]$ during a 0 A → 5 A → 0 A step-up/step-down load transient.

whereas for $\tilde{w}_2$

$$0 \leq \tilde{W}_2 \leq \frac{N_r - 1}{\tilde{K}} \approx 4.74. \tag{6.132}$$

Both of the above-mentioned inequalities express steady-state boundaries and do not hold, in principle, during transient events. Indeed, from Fig. 6.24, it is possible to see that the large overshoots occurring during the transient are likely to drive these signals into saturation if no dynamic overhead is included. In the first-cut design presented here, steady-state upper bounds are extended by a factor of two,

$$\lceil \tilde{w}_1 \rceil \approx 101_{10},$$
$$\lceil \tilde{w}_2 \rceil \approx 9.47_{10}. \tag{6.133}$$

It follows that

$$\mathrm{DR}_{hw}[\tilde{w}_1]_{-5}^n \triangleq 20 \log_{10} \frac{\lceil \tilde{w}_1 \rceil}{2^{-5}} \approx 70.2 \ \mathrm{dB} \Rightarrow n = 1 + \mathrm{ceil}\left(\frac{\mathrm{DR}_{hw}[\tilde{w}_2]_{-5}^n}{20 \log_{10} 2}\right) = 13$$

$$\mathrm{DR}_{hw}[\tilde{w}_2]_{-10}^n \triangleq 20 \log_{10} \frac{\lceil \tilde{w}_2 \rceil}{2^{-10}} \approx 79.7 \ \mathrm{dB} \Rightarrow n = 1 + \mathrm{ceil}\left(\frac{\mathrm{DR}_{hw}[\tilde{w}_3]_{-10}^n}{20 \log_{10} 2}\right) = 15.$$
$$\tag{6.134}$$

High-resolution control signal $\tilde{u}_{PID}$ is equal to $\tilde{w}_2$ multiplied by $\tilde{K}$. The required scale and the upper bound for $\tilde{u}_{PID}$ are then straightforwardly determined,

$$\lfloor \tilde{u}_{PID} \rfloor = \tilde{K} \lfloor \tilde{w}_2 \rfloor = 0.6328125_{10} = 01010001_{\tilde{2}} \times 2^{-7},$$

$$\lceil \tilde{u}_{PID} \rceil = \tilde{K} \lceil \tilde{w}_2 \rceil \approx 1070_{10}.$$

(6.135)

It follows that

$$\mathrm{DR}_{hw}[\tilde{u}_{PID}]^n_{-7} = 20 \log_{10} \frac{\lceil \tilde{u}_{PID} \rceil}{2^{-7}} \approx 102.7 \text{ dB}$$

$$\Rightarrow n = 1 + \text{ceil}\left(\frac{\mathrm{DR}_{hw}[\tilde{u}_{PID}]^n_{-7}}{20 \log_{10} 2} \right) = 19.$$

(6.136)

The compensator output $\tilde{u}$ is obtained, as in the parallel structure, by truncation and saturation,

$$[\tilde{u}_x]^{12}_0 \leftarrow [\tilde{u}_{PID}]^{19}_{-7},$$

(6.137)

$$[\tilde{u}]^{11}_0 \leftarrow \begin{cases} 1023_{10} & \text{if } [\tilde{u}_x]^{12}_0 > 1023_{10}, \\ 0 & \text{if } [\tilde{u}_x]^{12}_0 < 0, \\ [\tilde{u}_x]^{12}_0 & \text{otherwise.} \end{cases}$$

(6.138)

Note that, in this case, saturation against 1023_{10} must be explicitly implemented.

The above-mentioned analysis is not complete yet though, as one still needs to determine the word lengths of the partial products

$$\tilde{p}_1 \triangleq \tilde{w}_i[k-1]\tilde{c}_{z_1}$$

(6.139)

and

$$\tilde{p}_2 \triangleq \tilde{w}_1[k-1]\tilde{c}_{z_2}.$$

(6.140)

Application of the above-mentioned procedure yields

$$[\tilde{p}_1]^{17}_{-5}$$

(6.141)

and

$$[\tilde{p}_2]^{18}_{-10}.$$

(6.142)

The complete fixed-point realization of the cascade structure is illustrated in Fig. 6.25.

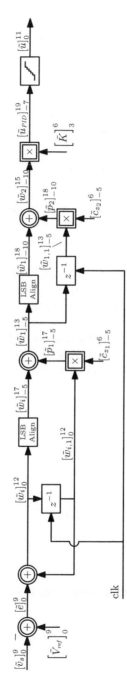

Figure 6.25   Fixed-point implementation of the cascade structure.

6.5.4   Linear versus Quantized System Response

As a conclusion of this section dedicated to fixed-point implementation examples, it is instructive to compare the actual closed-loop response of the system—including coefficient quantization effects and fixed-point arithmetic—against the "unquantized" response designed based on linearized small-signal models.

Figure 6.26 reports the 0 to 5 A step-load transients of the output voltage, inductor current, and digital control command. Two simulations are carried out to generate this plot:

- In the "unquantized" case, no A/D nor DPWM quantizations are performed. Furthermore, control calculations are executed using the full Matlab® double precision floating-point arithmetic.

- The second simulation, on the other hand, accounts for A/D and DPWM finite resolutions, and PID calculations are executed in fixed-point precision.

The simulated PID structure is, in both cases, the parallel realization.

The comparison highlights a number of features that are commonly observed in an actual digital control system. First, observe that steady-state output voltage waveforms before the transient are slightly different, that is, the two systems regulate at

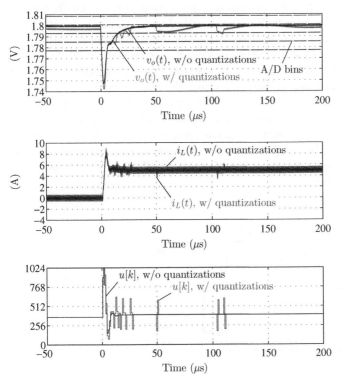

Figure 6.26   Comparison between linear and quantized closed-loop response to a 0 A → 5 A step load.

different steady-state operating points. Once A/D quantization is considered, in fact, the digital controller can only regulate the sampled waveform to within the A/D quantization bin. To better visualize this effect, A/D quantization levels are superimposed to the voltage waveforms. Notice that the steady-state error between the linear and quantized controllers lies within the zero-error bin.

Consider next the dynamics of the transient. Over the timescale of the transient—approximately 100 μs—the two responses essentially achieve the same performance, meaning that quantization effects do not disrupt the designed closed-loop dynamics. This has been the main objective of this chapter—to design quantizations so that they have limited impact on the system behavior. Differences in the two responses are nonetheless visible as the voltage waveform crosses the various A/D levels. One can verify that the large, impulsive control command variations occurring at every A/D level crossing is essentially due to the derivative action.

6.6   HDL IMPLEMENTATION OF THE CONTROLLER

Considering again the synchronous Buck converter case study, the complete HDL coding of the PID compensator is presented in this section. In particular, the parallel PID structure is exemplified in both VHDL and Verilog languages [171–179].

In HDL-coding the PID structure, it is of utmost importance to separate the *combinational* portion of the controller from the *sequential* portion. While the former groups all the arithmetic, logic, and bitwise manipulations required for control operation, the latter serves the purpose of updating the compensator state variables in a clocked manner. For instance, Fig. 6.27 sketches the subdivision of the parallel PID structure into its combinational part, which calculates $\tilde{u}[k]$ from the input $\tilde{e}[k]$ and states $\tilde{e}[k-1]$ and $\tilde{u}_i[k-1]$, and the sequential part that updates the state vector on a sampling cycle basis. Such separation between combinational and sequential portion is the first step toward the generation of robust, portable, and, most importantly, *synthesizable* HDL code.

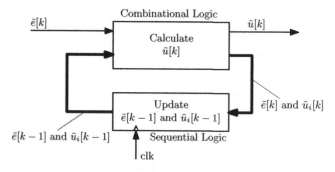

Figure 6.27   Separation between combinational and sequential portions of the controller.

6.6.1   VHDL Example

In the following example, a digital signal `clk` acts as a clock for the positive edge-triggered PID state machine. The digitized error signal, represented by a B2C word $[\tilde{e}]_0^9$, is VHDL coded as a 9-bit `signed` input signal named e. Similarly, the overall PID control command $[\tilde{u}]_0^{11}$ is represented by an 11-bit `signed` output signal u. VHDL signals `Kp`, `Ki`, and `Kd` represent the proportional, integral, and derivative gains, respectively.

Saturated arithmetic is systematically employed, assuming the presence, in the design, of the two entities `saturated_adder` and `saturated_multiplier` defined in Inset B.12 of Appendix B. Observe that, although the implementation—or *architecture*, to employ VHDL terminology—of such arithmetic blocks may differ from what is exemplified in Inset B.12, it is implicit that these blocks remain purely combinational.

Entity and architecture declarations for the parallel PID realization are as follows:

```
1   library IEEE;
2   use IEEE.STD_LOGIC_1164.ALL;
3   use IEEE.NUMERIC_STD.ALL;
4
5   entity PID is
6
7       port (
8           clk :   in std_logic;
9
10          Kp      :   in signed(2 downto 0);   --   [3,3]
11          Ki      :   in signed(3 downto 0);   --   [4,-3]
12          Kd      :   in signed(2 downto 0);   --   [3,6]
13
14          e   :   in signed(8 downto 0);          --   [9,0]
15          u   :   out signed(10 downto 0)         --   [11,0]
16      );
17
18  end PID;
19
20  architecture PID_parallel of PID is
21
22      component saturated_adder is
23          generic (
24              n, p, m :   integer     --   m <= max(n,p)+1
25              );
26
27          port (
28              x       :   in signed(n-1 downto 0);
29              y       :   in signed(p-1 downto 0);
30              z       :   out signed(m-1 downto 0);
31              OV      :   out std_logic;
32              op      :   in std_logic
33              );
34      end component;
```

```
35
36      component saturated_multiplier is
37         generic (
38            n, p, m :   integer     --  m<=n+p
39            );
40
41         port (
42            x        :   in signed(n-1 downto 0);
43            y        :   in signed(p-1 downto 0);
44            z        :   out signed(m-1 downto 0);
45            OV       :   out std_logic
46            );
47      end component;
48
49      signal e1       :   signed(8 downto 0); --  [9,0]
50
51      signal up       :   signed(5 downto 0); --  [6,3]
52
53      signal ui, ui1  :   signed(13 downto 0);   --  [14,-3]
54      signal wi   :   signed(6 downto 0);     --  [7,-3]
55
56      signal ud       :   signed(6 downto 0); --  [7,6]
57      signal wd   :   signed(8 downto 0);     --  [9,0]
58
59      signal upd      :   signed(16 downto 0);   --  [17,-3]
60      signal upid     :   signed(13 downto 0);   --  [14,-3]
61      signal ux       :   signed(10 downto 0);   --  [11,0]
62
63  begin
64
65      --  **************************************************
66      --  Combinational part
67      --  **************************************************
68      Kp_mult :   saturated_multiplier
69         generic map (n=>9, p=>3, m=>6)
70         port map (x=>e, y=>Kp, z=>up, OV=>open);
71
72      Ki_mult :   saturated_multiplier
73         generic map (n=>9, p=>4, m=>7)
74         port map (x=>e, y=>Ki, z=>wi, OV=>open);
75
76      Ki_add  :   saturated_adder
77         generic map (n=>14, p=>7, m=>14)
78         port map (x=>ui1, y=> wi, z=> ui, OV=>open, op=>'1');
79
80      Kd_sub  :   saturated_adder
81         generic map (n=>9, p=>9, m=>9)
82         port map (x=>e, y=>e1, z=>wd, OV=>open, op=>'0');
83
84      Kd_mult :   saturated_multiplier
85         generic map (n=>9, p=>3, m=>7)
86         port map (x=>wd, y=>Kd, z=>ud, OV=>open);
87
88      upd_add :   saturated_adder
89         generic map (n=>12, p=>16, m=>17)
```

```
90          port map (x=>(up&"000000"), y=>(ud&"000000000"),
91             z=> upd, OV=>open, op=>'1');
92
93   upid_add    :   saturated_adder
94          generic map (n=>14, p=>17, m=>14)
95          port map (x=>ui, y=>upd, z=>upid, OV=>open, op=>'1');
96
97   ux  <=  upid(13 downto 3);
98   u   <=  (others=>'0') when ux(10)='1' else ux;
99
100  --   **************************************************
101  --   Sequential part
102  --   **************************************************
103  pid_proc   :   process(clk)
104          begin
105              if (clk'event AND clk='1') then
106                  ui1 <=  ui;
107                  e1  <=  e;
108              end if;
109          end process;
110
111  end PID_parallel;
```

The above-mentioned architecture definition closely follows the implementation-level block diagram of Fig. 6.20. All arithmetic operations implementing control calculations pertain to the combinational part of the PID machine, and so does the quantization/saturation step implemented in lines 97 and 98. It is to be noted that explicit checking for $[\tilde{u}_x]_0^{11} > 1023_{10}$ is, as anticipated, unnecessary, as it is automatically ensured by the saturated addition used for computing $[\tilde{u}_{PID}]_{-3}^{13}$, followed by the $[\tilde{u}_x]_0^{11} \leftarrow [\tilde{u}_{PID}]_{-3}^{13}$ truncation. Hence, only the check for $[\tilde{u}_x]_0^{11} < 0$ is implemented in line 98 by examination of the sign bit of ux.

As for the sequential part of the PID, it is implemented in line 103 as a process statement labeled pid_proc, clocked by clk, which updates registers ui1 and e1 representing signals $\tilde{e}[k-1]$ and $\tilde{u}_i[k-1]$.

6.6.2   Verilog Example

A Verilog version of the PID module discussed earlier is given by the following code:

```
1    module PID(clk,Kp,Ki,Kd,e,u);
2
3        input clk;
4
5        input signed [2:0]  Kp;      // [3,3]
6        input signed [3:0]  Ki;      // [4,-3]
7        input signed [2:0]  Kd;      // [3,6]
8
9        input signed     [8:0] e;            // [9,0]
```

```
10      output signed   [10:0] u;          //  [11,0]
11
12      reg signed  [8:0] e1;              //  [9,0]
13
14      wire signed [5:0] up;              //  [6,3]
15
16      wire signed [13:0] ui;             //  [14,-3]
17      reg signed  [13:0] ui1;            //  [14,-3]
18      wire signed [6:0] wi;              //  [7,-3]
19
20      wire signed [6:0] ud;              //  [7,6]
21      wire signed [8:0] wd;              //  [9,0]
22
23      wire signed [16:0] upd;            //  [17,-3]
24      wire signed [13:0] upid;           //  [14,-3]
25      wire signed [10:0] ux;             //  [11,0]
26
27      //  ****************************************
28      //  Combinational part
29      //  ****************************************
30      saturated_multiplier   Kp_mult (e,Kp,up,,);
31          defparam Kp_mult.n=9, Kp_mult.p=3, Kp_mult.m=6;
32
33      saturated_multiplier   Ki_mult (e,Ki,wi,,);
34          defparam   Ki_mult.n=9, Ki_mult.p=4, Ki_mult.m=7;
35
36      saturated_adder Ki_add  (ui1,wi,ui,,1'b1);
37          defparam Ki_add.n=14, Ki_add.p=7, Ki_add.m=14;
38
39      saturated_adder Kd_sub  (e,e1,wd,,1'b0);
40          defparam   Kd_sub.n=9, Kd_sub.p=9, Kd_sub.m=9;
41
42      saturated_multiplier   Kd_mult (wd,Kd,ud,,);
43          defparam Kd_mult.n=9, Kd_mult.p=3, Kd_mult.m=7;
44
45      saturated_adder upd_add ({up,6'b0},{ud,9'b0},upd,1'b1);
46          defparam upd_add.n=12, upd_add.p=16, upd_add.m=17;
47
48      saturated_adder upid_add   (upd,ui,upid,,1'b1);
49          defparam   upid_add.n=17, upid_add.p=14, upid_add.m=14;
50
51      assign ux   =   upid[13:3];
52      assign u    =   (ux[10]==1'b1) ?   11'b0   :   ux;
53
54      //  ****************************************
55      //  Sequential part
56      //  ****************************************
57      always  @(posedge clk)
58          begin
59              ui1 <=  ui;
60              e1  <=  e;
61          end
62
63  endmodule
```

As in the VHDL example, the module makes use of two saturated arithmetic blocks, namely `saturated_adder` and `saturated_multiplier`, which have been reported in Inset B.13. The module is clearly separated into its combinational and sequential part, the latter coded in line 57 as a positive edge-triggered `always` statement.

6.7   SUMMARY OF KEY POINTS

- The implementation step consists of (i) scaling and quantization of the compensator coefficients and (ii) realization of the control law in a finite-precision fixed-point arithmetic environment. Both steps irreversibly alter the ideally designed loop gain, and selection of the quantization resolution must be performed with the objective of constraining the error to a negligible amount.

- The implementation step is closely related to the controller structure. In this chapter, three PID realizations are considered, namely the parallel, the direct, and the cascade structures.

- Quantization of the compensator coefficient must be guided by the compensator relative sensitivity function. This quantity can be related to the relative error on the system loop gain. Typical constraints concern the loop gain error at the desired crossover frequency, as well as loop gain errors in the low-frequency region. Once appropriate constraints to the compensator sensitivity function are set, the required resolution for the compensator coefficients can be determined.

- Implementation of the control law in a fixed-point arithmetic environment consists in determining the resolution and the scale of every binary quantity in the controller structure. To this end, the main tool is represented by the dynamic range of each signal. The hardware dynamic range—that is, the required wordlength—should be large enough to (i) host the entire signal excursion without saturations and (ii) represent each signal with a sufficiently fine resolution not to trigger quantization-related effects. In assessing these limits, an estimation of the severity of the transients the controller is expected to handle is of great importance.

CHAPTER 7

DIGITAL AUTOTUNING

In the standard design flow, a control loop is designed based on power converter models assuming known system and controller parameters. With this approach, the closed-loop performance is necessarily sensitive to variations in system dynamics due to changes in operating conditions, as well as tolerances and drifts in parameter values. Furthermore, the converter parasitics are notoriously difficult to model, and an accurate load model may not be available at all at the time the control loop is designed. With analog compensators, usually implemented using op-amp circuits with fixed on-chip or off-chip RC networks, a worst-case approach is usually adopted, and the resulting control loop is rarely optimal for the particular application, with no possibility of adaptation to specific operating conditions. A robust design capable of handling wide variations in dynamics is overly conservative over most of the expected range, resulting in degraded performance. Furthermore, the design flow must be reiterated for each new application.

With the much increased flexibility and programmability of digital controllers, opportunities are open to incorporate intelligent control algorithms to improve dynamic responses and robustness over a wider range of possible operating points and over wider ranges of system parameters. Digital autotuning, that is, the ability to automatically tune the controller parameters in response to the actual system dynamics, represents a significant departure from the conventional design flow. A digital controller ideally becomes a "plug and play" unit capable of identifying the key characteristics of the power converter and the load and of self-calibrating accordingly in order to meet predefined or user-defined design constraints. In addition to one-time tuning controllers, *adaptive tuning* techniques are capable of maintaining the control loop performance over time by adjusting the compensation to variations in system parameters. Furthermore, the capability of a single controller integrated circuit to handle a variety of converter parameters or even different converter topologies can enhance versatility and greatly simplify the design process, eliminating the need for differentiated production lines. Numerous advances have been made in the area of practical autotuning digital control algorithms and implementation techniques [81–101], and the area is subject to ongoing research and development efforts.

In this chapter, after a brief introduction to digital autotuning in Section 7.1 and a summary of programmable PID structures in Section 7.2, two autotuning techniques are presented in more detail: an injection-based approach in Section 7.3 and a relay

Digital Control of High-Frequency Switched-Mode Power Converters, First Edition.
Luca Corradini, Dragan Maksimović, Paolo Mattavelli, and Regan Zane.
© 2015 The Institute of Electrical and Electronics Engineers, Inc. Published 2015 by John Wiley & Sons, Inc.

feedback-based approach in Section 7.4. Practical implementation issues are briefly reviewed in Section 7.5. Key points are summarized in Section 7.6.

7.1 INTRODUCTION TO DIGITAL AUTOTUNING

The term "autotuning" can refer to different specific functions:

- *One-step tuning*, which has the goal of calibrating the control system from an initially "safe" condition, which guarantees stability but does not guarantee the necessary dynamic performances. The tuning step can be performed only once or at specific points, for example, upon system turn-on.

- *Performance tracking*, in which an autotuning step is repeated in time in order to keep track of the process parametric variations due to changes in operating conditions or drifts in component parameters.

- *Adaptive tuning*, in which autotuning is performed *continuously* throughout the system operation.

A number of autotuning techniques proposed in the literature adopt a philosophy similar to the well-known techniques of manual PID tuning, such as the Ziegler–Nichols methods. These techniques are based on field measurements of the plant. A limited number, for example, two or three, of key plant parameters are first identified from measurements and are subsequently employed for calibrating the controller gains according to predefined tuning formulas. If an increase in complexity can be tolerated, a more extensive nonparametric identification of the converter frequency response can performed by embedding frequency analysis capabilities into the digital controller [79, 80, 180–184]. Once the converter frequency response is determined, automatic PID tuning can be accomplished against any design constraints formulated in the frequency domain (such as crossover frequency, phase and gain margin etc.) [96].

Regardless of the specific approach, *identification* and *tuning* are the two basic operations performed by any digital autotuner. The identification/tuning steps can be either performed as two distinct phases of the tuning process or, in other cases, nested into a main tuning loop that operates so as to iteratively minimize a given *tuning error*, the definition of which depends on the autotuner implementation.

The identification step invariably involves perturbing the power converter operation around its steady-state operating point. During the tuning step, the operating point is usually maintained, either in open-loop or in closed-loop using a "safe," low-bandwidth controller. In the techniques discussed in this chapter, the converter is feedback-controlled at all times. A major advantage of such *closed-loop* autotuning is that regulation of the output voltage is not affected by the tuning process. On the other hand, precautions must be taken not to perturb the system heavily, so as to compromise regulation accuracy or load operation.

Two main approaches to perturbing the power converter have been explored in the literature of closed-loop autotuning:

- *Limit cycling*, which has in Chapter 5 been discussed as a generally undesirable steady-state disturbance, can be purposely induced as a way to create a small-signal perturbation around the converter operating point [81, 83, 84, 86–90]. Frequency and amplitude of the limit cycle can carry relevant information for identification purposes. The limit cycle is typically induced by purposely inserting a strong nonlinearity in the digital loop. *Relay feedback* autotuners discussed in Section 7.4 are a class of approaches following this philosophy.

- A *digital perturbation* can be explicitly generated by the autotuning system and superimposed to the control command [81, 91–93, 95–97, 101]. As frequency, amplitude, and waveshape of the digital perturbation are under the autotuner's control, this method is overall more reliable and offers a wider set of possibilities with respect to the limit cycle-based approach. On the other hand, a more complex implementation is usually required. One of these techniques is discussed in Section 7.3.

Tuning invariably relies on a programmable compensator structure allowing for on-the-fly adjustments of the compensator parameters—PID gains in the case of a PID compensator. A brief overview of programmable PID structures is provided in Section 7.2.

7.2   PROGRAMMABLE PID STRUCTURES

Programmable versions of the basic PID structures examined in the previous chapters can easily be obtained by replacing the constant gain blocks with full digital multipliers. The realizations most commonly employed for digital autotuning are the parallel structure and the cascade structure described in detail in Chapter 6. In these structures, adjustments of the PID gains have clear and easily understandable effects on the PID frequency response. The direct realization, on the contrary, is less attractive for digital autotuning because a modification of either of the coefficients b_0, b_1, or b_2 affects the PID gains and zeros in a complex manner.

Figure 7.1 illustrates a programmable version of the parallel PID realization. The low-frequency, mid-frequency, and high-frequency portions of the PID response can be accessed by acting on K_i, K_p, and K_d, respectively. For example, Fig. 7.2 exemplifies the change in frequency response that a digital PID undergoes when the proportional gain $\check{K}_p$ is adjusted from 50% to 200% of a nominal value. Changes in $\check{K}_p$ mostly affect the mid-frequency magnitude response. The phase response is strongly affected by $\check{K}_p$ due to the induced change in the PID zeros locations. The effect of a change in the derivative gain on the same PID frequency response is illustrated in Fig. 7.3. The magnitude response at high frequencies changes proportionally to $\check{K}_d$. Observe, however, that the phase response is affected as well in the high-frequency range. This observation leads to a general conclusion that affecting one of the PID parameters impacts *both* magnitude and phase response of the compensator. The designer of an autotuning system must therefore be aware of the interacting nature of the parameters of *any* PID realization.

The cascade structure is useful when the autotuner requires direct access to the compensator zeros and to the overall gain. In the cascade form seen in Chapter 6,

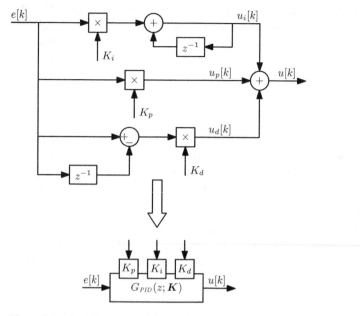

Figure 7.1   (top) Programmable PID in parallel form and (bottom) block diagram symbol.

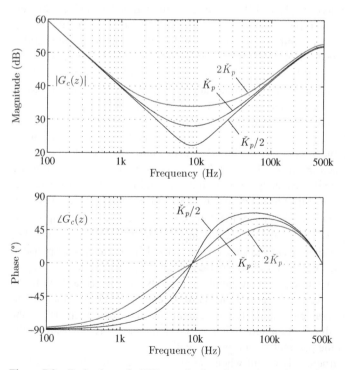

Figure 7.2   Bode plots of a PID transfer function under a change in the proportional gain.

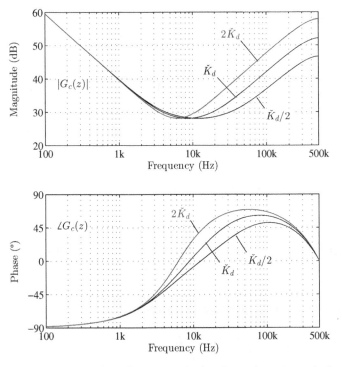

Figure 7.3   Bode plots of a PID transfer function under a change in the derivative gain.

however, acting on c_{z_1} or c_{z_2} has the side effect of affecting the low-frequency portion of the PID frequency response as well. Given the PID transfer function in the cascaded form,

$$G_{PID}(z; c) = K \frac{\left(1 + c_{z_1} z^{-1}\right)\left(1 + c_{z_2} z^{-1}\right)}{1 - z^{-1}}, \tag{7.1}$$

the low-frequency asymptotic form becomes

$$G_{PID}(z; c) \overset{low\,freq.}{\approx} K \frac{\left(1 + c_{z_1}\right)\left(1 + c_{z_2}\right)}{1 - z^{-1}}. \tag{7.2}$$

It can be seen that modifications in c_{z_1} or c_{z_2} affect the low-frequency asymptote. This is generally undesirable, as it may impact the closed-loop stability during the tuning process that performs adjustments in c_{z_1} and c_{z_2}. One way to avoid this interaction is to use an alternative realization of the cascade structure,

$$G_{PID}(z; \boldsymbol{\kappa}) \triangleq \frac{K_i}{1 - z^{-1}}\left(1 - \kappa_1 + \kappa_1 z^{-1}\right)\left(1 - \kappa_2 + \kappa_2 z^{-1}\right), \qquad \boldsymbol{\kappa} \triangleq [K_i, \kappa_1, \kappa_2]^T, \tag{7.3}$$

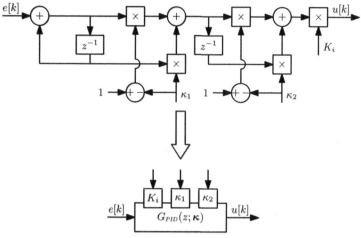

Figure 7.4 (top) Programmable PID in the cascade form and (bottom) block diagram symbol.

which leads to the following low-frequency asymptotic form:

$$G_{PID}(z; \boldsymbol{\kappa}) \overset{low\,freq.}{\approx} \frac{K_i}{1 - z^{-1}}. \tag{7.4}$$

In this form, the overall gain and the integral gain coincide, while the PID zeros in the z-plane are located at

$$z_{1,2} = -\frac{\kappa_{1,2}}{1 - \kappa_{1,2}}. \tag{7.5}$$

This programmable cascade structure is depicted in Fig. 7.4.

Another useful programmable structure is illustrated in Fig. 7.5, a hybrid realization where the integral term acts in parallel to the proportional-derivative action, which is in turn implemented in a cascade form,

$$G_{PID}(z; \boldsymbol{h}) \triangleq \frac{K_i}{1 - z^{-1}} + K_{PD}\left(1 - \kappa + \kappa z^{-1}\right), \qquad \boldsymbol{h} \triangleq [K_i, K_{PD}, \kappa]^T. \tag{7.6}$$

This structure is easily reduced to a parallel PI architecture by imposing $\kappa = 0$ or to a full PID form with direct access to the overall PD gain K_{PD} and zero location.

Programmability inevitably entails an increase in hardware complexity of the digital compensator. While multiplication by a constant is highly optimized by HDL synthesis tools and hardwired as a combination of binary shifts and additions, the complexity of a programmable structure depends on the number of bits required by the true digital multipliers. This, in turn, is related to how robust the autotuning system must be and, therefore, how wide the set of representable PID transfer functions needs to be.

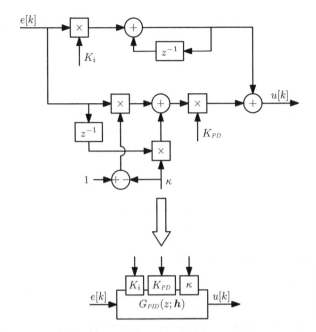

Figure 7.5   (top)
Programmable PID in
hybrid form and (bottom)
block diagram symbol.

7.3   AUTOTUNING VIA INJECTION OF A DIGITAL PERTURBATION

A general block diagram of an injection-based autotuning approach is illustrated in
Fig. 7.6. The autotuning system injects a digital perturbation signal $u_{pert}[k]$ into the

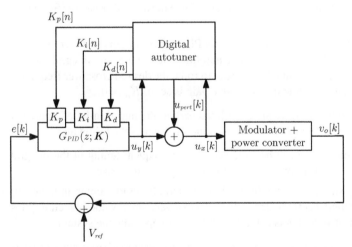

Figure 7.6   A block diagram of an autotuning approach based on injection of a digital
perturbation.

feedback loop, superimposed to the PID output $u_y[k]$. The overall control command that modulates the converter is therefore

$$u_x[k] \triangleq u_y[k] + u_{pert}[k].$$ (7.7)

Simultaneously, signals $u_y[k]$ and $u_x[k]$ before and after the injection point are monitored. One may note that the injection of $u_{pert}[k]$ and monitoring of $u_y[k]$ and $u_x[k]$ resembles the injection performed in the Middlebrook's loop gain measurement approach [78], where the system loop gain is found by evaluating the magnitude and phase relationships between $u_y[k]$ and $u_x[k]$ at the injection frequency.

The tuning process adjusts the compensator gains until the correct amplitude and phase relationships are established between the ac components of $u_x[k]$ and $u_y[k]$. As these depend on the system loop gain *at the injection frequency*, this method can be used to tune for a predefined bandwidth ω_c and phase margin φ_m. If the system loop gain is $T(z)$, the *tuning target* can be expressed as

$$\boxed{T(e^{j\omega_c T_s}) = -e^{j\varphi_m}}.$$ (7.8)

The complex constraint (7.8) translates into two real constraints on the PID compensator. Therefore, the basic version of the injection-based approach performs a *two-parameters tuning*, that is, it is only capable of determining two out of the three PID parameters, whereas the remaining one must be kept fixed during the tuning process. This identifies two versions of the approach:

- *PD tuning*. In this scenario, which is the one considered in more detail in the remaining part of this section, an integral gain K_i is set to a safe value and kept fixed during the tuning process. The autotuner determines the proportional and the derivative gains K_p and K_d in order to force the desired crossover frequency and phase margin.

- *PI tuning*. This scenario assumes that a PI compensation is sufficient in order to safely control the converter. The autotuner determines the integral and the proportional gains K_i and K_p in order to force the desired crossover frequency and phase margin, while the derivative gain is set to $K_d = 0$.

In addition to the basic versions outlined earlier, the injection-based approach can be employed to handle more complex scenarios, such as the following:

- The PD tuning phase can be followed by a subsequent tuning of the integral gain K_i aimed at improving the low-frequency gain response.

- A PI tuning step can be conditionally executed after a PD tuning in the event that the PD tuning sets the derivative gain to a very small value (which implies that no derivative action is actually needed to compensate the system).

In the following, the theory of operation of the injection-based autotuning is first outlined in its most general formulation, followed by a detailed implementation of a PD autotuner.

7.3.1 Theory of Operation

Let $u_{pert}[k]$ be a digital sinewave of small amplitude $\hat{u}_m$ oscillating at frequency ω_p,

$$u_{pert}[k] \triangleq \hat{u}_m \sin(\omega_p kT_s + \varphi_p). \tag{7.9}$$

Assuming that the closed-loop system reacts linearly in response to the small perturbation $u_{pert}[k]$, small-signal ac components $\hat{u}_x[k]$ and $\hat{u}_y[k]$ arise at the perturbation frequency ω_p on top of u_x and u_y,

$$
\begin{aligned}
u_x[k] &= U + \hat{u}_x[k] = U + \hat{u}_{x,m} \sin\left(\omega_p kT_s + \varphi_x\right), \\
u_y[k] &= U + \hat{u}_y[k] = U + \hat{u}_{y,m} \sin\left(\omega_p kT_s + \varphi_y\right),
\end{aligned}
\tag{7.10}
$$

where U is the steady-state value of the control command as determined by the unperturbed feedback loop.

As the system is excited at a single frequency and responds linearly given the small-signal assumption, the autotuner's operation is more conveniently described in terms of *phasor analysis*. In the following, $\vec{a}$ denotes the phasor corresponding to an ac oscillation $a[k]$,

$$
\begin{aligned}
a[k] &= a_m \sin\left(\omega kT_s + \varphi_a\right), \\
&\downarrow \\
\vec{a} &\triangleq a_m e^{j\varphi_a}.
\end{aligned}
\tag{7.11}
$$

Phasor $\vec{a}$ is a complex number representable by a vector in the complex plane.

Small-signal-wise, (7.7) becomes

$$\hat{u}_x[k] = \hat{u}_y[k] + u_{pert}[k], \tag{7.12}$$

and therefore, in phasor notation,

$$
\underset{\triangleq \hat{u}_{x,m} e^{j\varphi_y}}{\vec{u}_x} = \underset{\triangleq \hat{u}_{y,m} e^{j\varphi_x}}{\vec{u}_y} + \underset{\triangleq \hat{u}_m e^{j\varphi_p}}{\vec{u}_{pert}}.
\tag{7.13}
$$

The phasor relationship between $\vec{u}_y$ and $\vec{u}_x$ is, then,

$$\boxed{\frac{\vec{u}_y}{\vec{u}_x} = -T(e^{j\omega_p T_s})}. \tag{7.14}$$

This relationship is the theoretical foundation of the method. If the PID gains are adjusted in such a way that (i) phasors $\vec{u}_x$ and $\vec{u}_y$ have equal magnitude and (ii) $\vec{u}_x$ lags $\vec{u}_y$ by φ_m degrees, then

$$\boxed{\vec{u}_x = \vec{u}_y e^{-j\varphi_m}}. \tag{7.15}$$

From (7.14), it follows that

$$\frac{\vec{u}_y}{\vec{u}_x} = e^{j\varphi_m} = -T(e^{j\omega_p T_s}) \Rightarrow \boxed{T(e^{j\omega_p T_s}) = -e^{j\varphi_m}}, \qquad (7.16)$$

which corresponds to (7.8) for $\omega_p = \omega_c$. In other words, condition (7.15) is an equivalent expression of the tuning target.

During the tuning process, on the other hand, (7.15) is not met, and amplitude and phase errors between $\hat{u}_x[k]$ and $\hat{u}_y[k]$ can be processed to decide how to adjust the PID parameters. In particular:

- The control bandwidth must be increased or decreased depending on whether $|\vec{u}_y| < |\vec{u}_x|$ or $|\vec{u}_y| > |\vec{u}_x|$, respectively.

- The loop gain phase at ω_p must be increased or decreased depending on whether $\vec{u}_x$ lags $\vec{u}_y$ by less or more than φ_m degrees.

Given the tuning target (7.15), define the *tuning error* $\vec{\varepsilon}$ as the phasor difference between $\vec{u}_x$ and the φ_m-delayed version of $\vec{u}_y$,

$$\boxed{\vec{\varepsilon} \triangleq \vec{u}_x - \vec{u}_y e^{-j\varphi_m}}. \qquad (7.17)$$

As a phasor, the tuning error $\vec{\varepsilon}$ summarizes the errors in both the phase margin and the crossover frequency. To better understand how, consider first the phasor diagram exemplified in Fig. 7.7(a), which corresponds to a generic situation *during*

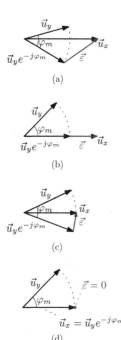

(a)

(b)

(c)

(d)

Figure 7.7   Phasor diagram (a) during the tuning process, (b) with zero error for the phase margin, (c) with zero error for the crossover frequency, and (d) after the tuning target has been reached.

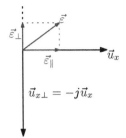

Figure 7.8   Decomposition of the tuning error $\vec{\varepsilon}$ into a component $\vec{\varepsilon}_\parallel$ parallel to $\vec{u}_x$ and an orthogonal component $\vec{\varepsilon}_\perp$.

the tuning process. Assume now that the phase margin error is nulled, resulting in the phasor diagram of Fig. 7.7(b). The φ_m-delayed phasor $\vec{u}_y e^{-j\varphi_m}$ is now parallel to $\vec{u}_x$ but has a different magnitude as a result of the crossover frequency not yet being equal to ω_p. Notice also how $\vec{\varepsilon}$ is parallel to $\vec{u}_x$. Next, Fig. 7.7(c) illustrates the case when the crossover frequency error is zero, whereas a nonzero phase margin error still keeps $\vec{u}_y e^{-j\varphi_m}$ misaligned with respect to $\vec{u}_x$. The tuning error $\vec{\varepsilon}$ is essentially orthogonal to $\vec{u}_x$. Once both the phase margin and crossover frequency errors are corrected, the phasor diagram becomes as shown in Fig. 7.7(d).

From the above-mentioned discussion, it can be seen how the tuning error can be decomposed into a component $\vec{\varepsilon}_\parallel$ parallel to $\vec{u}_x$, and the corresponding orthogonal component $\vec{\varepsilon}_\perp$, parallel to $\vec{u}_{x\perp} \triangleq -j\vec{u}_x$,

$$\vec{\varepsilon} = \vec{\varepsilon}_\parallel + \vec{\varepsilon}_\perp. \tag{7.18}$$

Such decomposition is illustrated in Fig. 7.8, the significance of which can be described as follows:

- $\vec{\varepsilon}_\parallel$ can be thought of originating mostly from the error in the crossover frequency. Tuning for $\vec{\varepsilon}_\parallel = 0$ means tuning the loop crossover frequency.

- $\vec{\varepsilon}_\perp$, on the other hand, can be mostly associated with the error in-phase margin. Tuning for $\vec{\varepsilon}_\perp = 0$ therefore means tuning the loop phase margin.

These conclusions allow construction of an autotuner using two feedback loops, a *crossover frequency tuning loop* correcting $\vec{\varepsilon}_\parallel$ and a *phase margin tuning loop* correcting $\vec{\varepsilon}_\perp$.

The remaining item to be discussed before showing the autotuner block diagram is how to construct signals proportional to $\vec{\varepsilon}_\parallel$ and $\vec{\varepsilon}_\perp$ in the time domain. Inset 7.1 deals with a general way to accomplish this, and how the projection operation relates to the phasor representations of signals.

Inset 7.1 – Signal projection and its phasor representation

Consider two phasors $\vec{a}$ and $\vec{b}$ associated with two time-domain oscillations $a[k]$ and $b[k]$, respectively,

$$\begin{aligned} \vec{a} &\leftrightarrow a[k] = a_m \sin(\omega k T_s + \varphi_a), \\ \vec{b} &\leftrightarrow b[k] = b_m \sin(\omega k T_s + \varphi_b). \end{aligned} \tag{7.19}$$

Multiplication of the time-domain signals yields

$$p_\parallel[k] \triangleq a[k] \times b[k] = a_m \sin(\omega k T_s + \varphi_a) \times b_m \sin(\omega k T_s + \varphi_b)$$

$$= \frac{a_m b_m}{2} \left(\cos(\varphi_b - \varphi_a) - \cos(2\omega k T_s + \varphi_a + \varphi_b) \right). \tag{7.20}$$

The dc term

$$\bar{p}_\parallel \triangleq \frac{a_m b_m}{2} \cos(\varphi_b - \varphi_a) = \frac{b_m}{2} \underbrace{a_m \cos(\varphi_b - \varphi_a)}_{\text{Component of } \vec{a} \text{ along } \vec{b}} \tag{7.21}$$

contains a factor proportional to the projection of $\vec{a}$ along $\vec{b}$. Similarly, multiplication of $a[k]$ by the 90°-delayed version $b_\perp[k]$ of $b[k]$ yields

$$p_\perp[k] \triangleq a[k] \times b_\perp[k] = a_m \sin(\omega k T_s + \varphi_a) \times b_m \sin\left(\omega k T_s + \varphi_b - \frac{\pi}{2}\right)$$

$$= -a_m \sin(\omega k T_s + \varphi_a) \times b_m \cos(\omega k T_s + \varphi_b)$$

$$= \frac{a_m b_m}{2} \left(\sin(\varphi_b - \varphi_a) - \sin(2\omega k T_s + \varphi_a + \varphi_b) \right), \tag{7.22}$$

and the dc term

$$\bar{p}_\perp \triangleq \frac{a_m b_m}{2} \sin(\varphi_b - \varphi_a) = \frac{b_m}{2} \underbrace{a_m \sin(\varphi_b - \varphi_a)}_{\text{Component of } \vec{a} \text{ along } \vec{b}_\perp \triangleq -j\vec{b}} \tag{7.23}$$

contains a factor proportional to the component of $\vec{a}$ orthogonal to $\vec{b}$.

In summary, the above-mentioned time-domain multiplications allow extraction of the in-phase and quadrature components of $a[k]$ with respect to $b[k]$ as the dc components of time-domain products. The second terms in (7.20) and (7.22), on the other hand, represent oscillations at *twice* the frequency ω and act as disturbances for the intended decomposition.

Lastly, note that quantities

$$\bar{p}_\parallel \triangleq \frac{a_m b_m}{2} \cos(\varphi_b - \varphi_a) \quad \text{and} \quad \bar{p}_\perp \triangleq \frac{a_m b_m}{2} \sin(\varphi_b - \varphi_a) \tag{7.24}$$

can be thought of as the real and imaginary components of the complex quantity

$$\frac{\vec{a}^* \times \vec{b}}{2} = \frac{a_m b_m}{2} e^{j(\varphi_b - \varphi_a)}$$

$$= \frac{a_m b_m}{2} \left(\cos(\varphi_b - \varphi_a) + j \sin(\varphi_b - \varphi_a) \right)$$

$$= \bar{p}_\parallel + j\bar{p}_\perp, \tag{7.25}$$

$\vec{a}^*$ indicating the phasor complex conjugate of $\vec{a}$.

7.3.2   Implementation of a PD Autotuner

A block diagram of the PD autotuning system is reported in Fig. 7.9, with a programmable PID compensator in the parallel form. First, dc components are eliminated from $u_x[k]$ and $u_y[k]$ by subtraction of the steady-state value U of the control action. In practice, the integral term $u_i[k]$ of the control action can be employed as U. This is possible, thanks to the inherently filtered nature of this signal, which makes it insensitive to the perturbation $u_{pert}[k]$. Next, two delay blocks are employed to generate a φ_m-delayed version $\hat{u}_{y,\varphi_m}[k]$ of $\hat{u}_y[k]$ and a 90°-delayed version $\hat{u}_{x\perp}[k]$ of $\hat{u}_x[k]$. The time-domain tuning error $\varepsilon[k]$ is then calculated as

$$\varepsilon[k] \triangleq \hat{u}_x[k] - \hat{u}_{y,\varphi_m}[k], \tag{7.26}$$

which is the time-domain counterpart of (7.15). With $\varepsilon[k]$ now available, the time-domain products $p_\|[k]$ and $p_\perp[k]$ are calculated as

$$p_\|[k] \triangleq \varepsilon[k] \times \hat{u}_x[k],$$
$$p_\perp[k] \triangleq \varepsilon[k] \times \hat{u}_{x\perp}[k]. \tag{7.27}$$

From Inset 7.1, the dc component $\bar{p}_\|$ of $p_\|[k]$ is related to $\bar{\varepsilon}_\|$, whereas dc component $\bar{p}_\perp$ of $p_\perp[k]$ is related to $\bar{\varepsilon}_\perp$. On the basis of the foregoing discussions, forcing $\bar{p}_\|$ to zero accomplishes tuning of the crossover frequency. Similarly, forcing $\bar{p}_\perp$ to zero amounts to tuning for the desired phase margin φ_m.

In the PD tuning discussed here, the integral gain K_i is kept fixed, while the proportional and the derivative gains K_p and K_d are adjusted in order to force $\bar{\varepsilon} = 0$. This is accomplished by two integral compensators inserted in the tuning loop, which

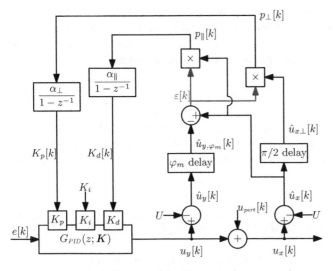

Figure 7.9   A block diagram of the PD autotuning approach.

generate K_d and K_p as the result of the accumulation of $p_\parallel[k]$ and $p_\perp[k]$,

$$K_d[k] = \alpha_\parallel p_\parallel[k] + K_d[k-1],$$
$$K_p[k] = \alpha_\perp p_\perp[k] + K_p[k-1], \tag{7.28}$$

where coefficients $\alpha_\parallel > 0$ and $\alpha_\perp > 0$ determine the speed of the tuning loop as well as its stability.

The reason why K_d is employed to tune for $\vec{\varepsilon}_\parallel = 0$ is that K_d has a major influence on the high-frequency PID gain and directly affects the loop gain crossover frequency. On the other hand, K_p mostly affects the high-frequency phase response by displacing the PD zero. For this reason, the phase margin tuning $\vec{\varepsilon}_\perp = 0$ is accomplished by acting on K_p.

The tuning control law (7.28) correctly implements a negative—that is, corrective—adjustment:

- A negative error in the crossover frequency implies a positive $\vec{\varepsilon}_\parallel$ (i.e., oriented as $\vec{u}_x$). In turn, a positive $\vec{\varepsilon}_\parallel$ determines an increase in K_d and, therefore, an increase in crossover frequency.

- A negative error in the phase margin implies a negative $\vec{\varepsilon}_\perp$ (oriented opposite to $\vec{u}_{x\perp}$). In turn, a negative $\vec{\varepsilon}_\perp$ determines a decrease in K_p, which moves the PD zero to lower frequencies, thus increasing the phase boost.

Negative feedback allows the tuning integrators to null the dc components of $p_\parallel[k]$ and $p_\perp[k]$ and force the system to $\vec{\varepsilon} = 0$. Observe that the oscillating terms at twice ω_p present in the products $p_\parallel[k]$ and $p_\perp[k]$ have the sole effect of introducing a small oscillation on top of $K_p[k]$ and $K_d[k]$. As the tuning loop is normally much slower than the main converter regulation loop, an oscillation at $2\omega_p$ is very well filtered by the tuning integrators.

Implementation of the φ_m delay, as well as of the 90° delay, can be performed by simple digital filtering. Recall first that a phase delay corresponding to one switching step is described by the operator z^{-1}. A fractional delay comprised between 0 and one switching period, on the other hand, can be implemented as

$$F_a(z) \triangleq 1 - a + az^{-1}, \qquad 0 \le a \le 1. \tag{7.29}$$

Bode plots of $F_a(z)$ for various values of a are reported in Fig. 7.10. The phase delay increases with a, is zero for $a = 0$, and corresponds to a one-step delay for $a = 1$. Parameter a is related to the intended phase rotation $\varphi = \angle F_a(e^{j\omega T_s})$ at the perturbation frequency ω_p by

$$a = \frac{\tan \varphi}{\left(1 - \cos(\omega_p T_s)\right)\tan \varphi - \sin(\omega_p T_s)} \approx \frac{\tan|\varphi|}{\omega_p T_s}, \tag{7.30}$$

where the approximation holds well for frequencies up to one-tenth of the switching rate.

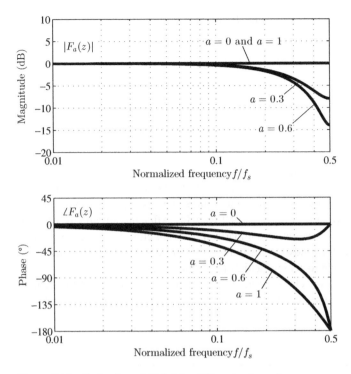

Figure 7.10   Bode plots of $F_a(z)$ for different values of a.

When a phase delay comprised between N and $N + 1$ switching steps needs to be realized, one can simply cascade an N-step delay z^{-N} to a suitably designed fractional delay $F_a(z)$,

$$F(z) \triangleq z^{-N} F_a(z) = z^{-N} \left(1 - a + a z^{-1} \right). \tag{7.31}$$

This solution allows delaying the signal by the specified amount with negligible attenuation. Furthermore, whenever the required delay is exactly a multiple NT_s of the switching period, the above-mentioned filtering reduces to a pure N-steps delay by setting $a = 0$.

7.3.3   Simulation Example

The injection-based PD autotuning approach is now evaluated in simulation on the synchronous Buck voltage-mode control example examined throughout the previous chapters. As a first simulation test, assume nominal values of the power converter parameters, but the initial values of the proportional and the derivative compensator gains are only 20% of their target value. From (6.50),

$$\check{K}_p = \frac{24.76}{5} \approx 4.95,$$

$$\check{K}_i = 0.5961,$$

$$\check{K}_d = \frac{190.5}{5} \approx 38.1.$$

(7.32)

Recall from Chapter 4 that target values have been designed to achieve a crossover frequency of 100 kHz and a 45° phase margin. With the initial values so modified, both the control bandwidth and the phase margin become severely compromised, as illustrated by the loop gain Bode plots of Fig. 7.11. After the autotuning process, they are restored to specification. A comparison between closed-loop responses to a 0 A→5 A step load before and after the tuning is reported in Fig. 7.12 demonstrating how the autotuner is able to achieve a high-performance control loop.

The tuning process itself is reported in Fig. 7.13 in terms of the time-domain evolution of the PID gains. The autotuner is enabled at $t = 0$ and quickly adjusts the

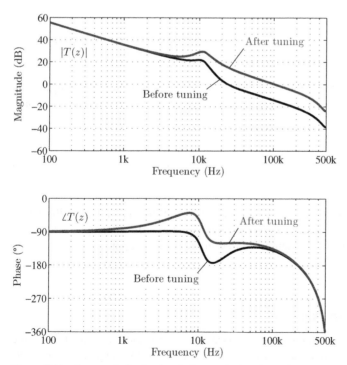

Figure 7.11 Loop gain Bode plots before and after the tuning process, first simulation scenario.

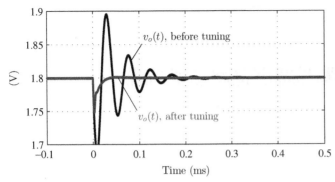

Figure 7.12   Closed-loop response to a 0 A→5 A step load before and after the tuning process, first simulation scenario.

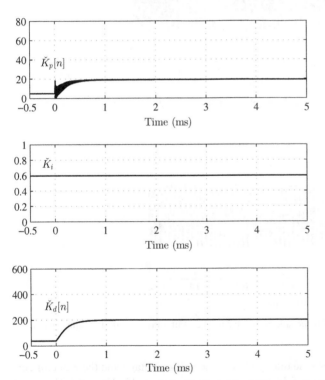

Figure 7.13   Proportional, integral, and derivative PID gains during the tuning process, first simulation scenario.

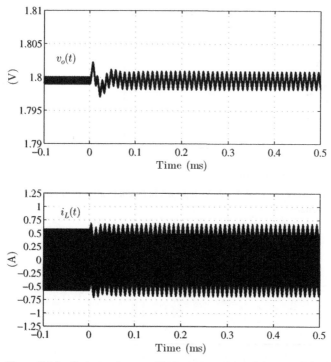

Figure 7.14 Output voltage and inductor current at the onset of the tuning process, first simulation scenario.

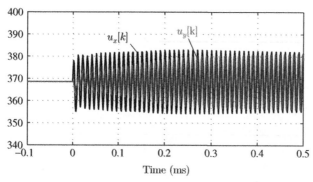

Figure 7.15 Evolution of $u_x[k]$ and $u_y[k]$ during the tuning process, first simulation scenario.

compensator gains toward the tuning target. The output voltage and the inductor current waveforms at the onset of the tuning process are reported in Fig. 7.14. Evolution of the signals $u_x[k]$ and $u_y[k]$ during the tuning process is depicted in Fig. 7.15.

A second simulation scenario assumes, as an initial condition for the tuning process, that the nominal compensator values are as determined in Chapter 6 and that the output capacitance in the power converter is two times larger than the value assumed in the design process,

$$C' = 2 \times C = 2 \times 200 \ \mu\text{F} = 400 \ \mu\text{F}. \qquad (7.33)$$

The larger output filter capacitance implies a lower resonant frequency with respect to the design assumptions and, given the ≈ -20 dB/decade slope of the system loop gain above such frequency, approximately half the crossover frequency with respect to the design target $f_c = 100$ kHz. This scenario is commonly encountered in practice, where the load presents an uncertain capacitive load for the point-of-load converter.

The tuning process is illustrated in Fig. 7.16. The derivative gain is tuned to roughly twice the initial value to compensate for the corresponding lower gain of the plant. Simultaneously, the proportional gain is increased, decreasing the PID phase boost around 100 kHz. The reason for this behavior is easily understood from Fig. 7.3, which clearly shows that an increase in K_d brings, as a side effect, an increase in the PID phase boost. The autotuner counteracts this effect by increasing K_p, which decreases the PID leading action as illustrated in Fig. 7.2. This is an example of the interacting nature of the PID gains on the compensator magnitude and phase responses.

The Bode plots of the system loop gains before and after the tuning step are reported in Fig. 7.17, while Fig. 7.18 illustrates the closed-loop response of the

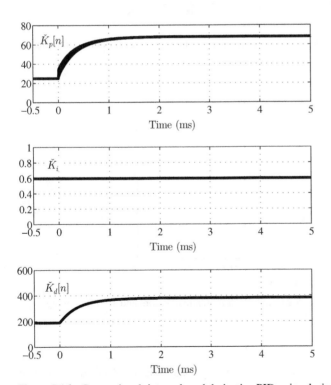

Figure 7.16   Proportional, integral, and derivative PID gains during the tuning process, second simulation scenario.

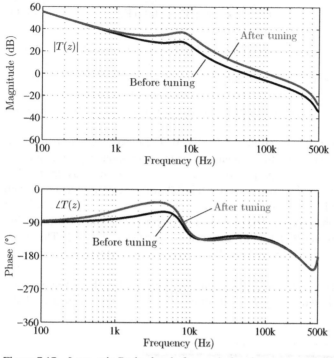

Figure 7.17 Loop gain Bode plots before and after the tuning process, second simulation scenario.

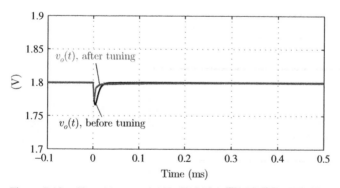

Figure 7.18 Closed-loop response to a 0 A→5 A step load before and after the tuning process, second simulation scenario.

pre- and posttuned converter to a 0- to 5-A step load variation. One may note that the autotuner is able to improve the dynamic response in the presence of substantial uncertainty in the value of the output filter capacitance.

7.3.4 Small-Signal Analysis of the PD Autotuning Loop

As discussed in Section 7.3.2, parameters $\alpha_\parallel$ and $\alpha_\perp$ in the PD autotuner shown in Fig. 7.9 affect the dynamic response of the tuning process. It is understood that larger $\alpha_\parallel$ and $\alpha_\perp$ result in faster tuning, but quantitative results are needed in order to design the tuning loops properly. This section is focused on a small-signal analysis of the dynamics of the PD autotuning loops.

Referring to Fig. 7.19, the autotuning system described in Section 7.3.2 is a multiloop feedback with two feedback paths, one acting to tune K_d and another acting to tune K_p.

It is important to clarify first what the "small-signal analysis" means for the autotuning system of Fig. 7.9. As suggested in Fig. 7.19, the autotuning feedback must ideally be cut at both K_d and K_p, an operation that defines signal pairs $(K_{d,x}, K_{p,x})$ and $(K_{d,y}, K_{p,y})$. The end goal of the small-signal analysis is to understand how small perturbations $(\hat{K}_{d,x}, \hat{K}_{p,x})$ on $K_{d,x}$ and $K_{p,x}$ propagate throughout the multiloop feedback to produce corresponding perturbations $(\hat{K}_{d,y}, \hat{K}_{p,y})$ on $K_{d,y}$ and $K_{p,y}$.

In the determination of this dependence, two complications arise:

- As already seen, products $p_\parallel$ and $p_\perp$ contain not only a dc component but also an oscillating component at twice the perturbation frequency ω_p. For small-signal analysis purposes, only the small variations in the low-frequency components $\overline{p}_\parallel[n]$ and $\overline{p}_\perp[n]$ are relevant, and any information related to the oscillating component is disregarded. This is analogous to an averaging approximation applied to the tuning loop dynamics.

- A small perturbation $(\hat{K}_{d,x}, \hat{K}_{p,x})$ alters the amplitude/phase relationships between $\hat{u}_x[k]$ and $\hat{u}_y[k]$, which in turn oscillate at ω_p due to the input perturbation $u_{pert}[k]$. It is precisely these slow changes in the *phasor* relationships

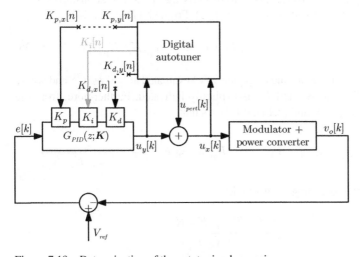

Figure 7.19   Determination of the autotuning loop gains.

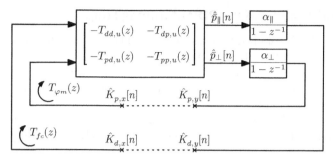

Figure 7.20   Small-signal block diagram of the autotuning loop.

between $\vec{u}_x$ and $\vec{u}_y$ that must be described by the small-signal analysis, in a way similar to dynamic phasor analysis of resonant converters.

With the above-mentioned remarks in mind, a small-signal representation of the autotuning loop is reported in Fig. 7.20. Similarly to what has been done in the previous chapters, the loop is subdivided into its uncompensated dynamics and its compensation. The uncompensated loop gains $T_{dd,u}(z)$, $T_{dp,u}(z)$, $T_{pd,u}(z)$ and $T_{pp,u}(z)$ relate the small-signal perturbations $(\hat{K}_{d,x}, \hat{K}_{p,x})$ on $K_{d,x}$ and $K_{p,x}$ to the small-signal perturbations $(\hat{\bar{p}}_{\parallel}, \hat{\bar{p}}_{\perp})$ on $\bar{p}_{\parallel}$ and $\bar{p}_{\perp}$,

$$
\begin{bmatrix} \hat{\bar{p}}_{\parallel}(z) \\ \hat{\bar{p}}_{\perp}(z) \end{bmatrix} = \begin{bmatrix} -T_{dd,u}(z) & -T_{dp,u}(z) \\ -T_{pd,u}(z) & -T_{pp,u}(z) \end{bmatrix} \begin{bmatrix} \hat{K}_{d,x}(z) \\ \hat{K}_{p,x}(z) \end{bmatrix}. \tag{7.34}
$$

It will be shown below that the uncompensated loop gains are, in fact, frequency independent. Consequently, the entire dynamics of the tuning process is determined by the integral compensators. To see this, start by deriving a phasor relationship between the tuning error $\vec{\varepsilon}$ and the system loop gain $T(z)$. From (7.17) and (7.14),

$$
\vec{\varepsilon} = \vec{u}_x - \vec{u}_y e^{-j\varphi_m} = \vec{u}_x \left(1 + e^{-j\varphi_m} T(\omega_p; K_d, K_p)\right), \tag{7.35}
$$

where $T(\omega_p; K_d, K_p)$ denotes the loop gain $T(z)$ evaluated at $z = e^{j\omega_p T_s}$ and where dependence of $T(z)$ on K_d and K_p is explicitly indicated. Furthermore, from the injection equations (7.13) and (7.14), it follows that

$$
\vec{u}_x = \vec{u}_y + \vec{u}_{pert} = \vec{u}_{pert} - T(\omega_p; K_d, K_p)\vec{u}_x \Rightarrow \vec{u}_x = \frac{\vec{u}_{pert}}{1 + T(\omega_p; K_d, K_p)}. \tag{7.36}
$$

Combination of the foregoing results yields

$$
\vec{\varepsilon} = \frac{1 + e^{-j\varphi_m} T(\omega_p; K_d, K_p)}{1 + T(\omega_p; K_d, K_p)} \vec{u}_{pert}. \tag{7.37}
$$

This is the nonlinear phasor relationship between compensator gains K_d and K_p and the phasor representation $\vec{\varepsilon}$ of the tuning error.

Next, dependence of $\overline{p}_\parallel$ and $\overline{p}_\perp$ on $\vec{\varepsilon}$ must be determined. From (7.27), these signals are obtained by multiplication of $\varepsilon[k]$ with $\hat{u}_x[k]$ and $\hat{u}_{x\perp}[k]$, respectively. From (7.25), then, complex quantity

$$\frac{\vec{\varepsilon}^* \times \vec{u}_x}{2} = \frac{\vec{\varepsilon}^*}{2} \times \frac{\vec{u}_{pert}}{1 + T(\omega_p; K_d, K_p)} \tag{7.38}$$

has $\overline{p}_\parallel$ and $\overline{p}_\perp$ as its real and imaginary parts, respectively. Combining the above-mentioned equation with (7.37), one has

$$\frac{\vec{\varepsilon}^* \times \vec{u}_x}{2} = \frac{\left(1 + e^{j\varphi_m} T(\omega_p; K_d, K_p)\right) |\vec{u}_{pert}|^2}{2 \left|1 + T(\omega_p; K_d, K_p)\right|^2} \triangleq f(K_d, K_p). \tag{7.39}$$

Complex function f of the real quantities K_d and K_p summarizes the nonlinear relationship between the compensator gains and quantities $\overline{p}_\parallel$ and $\overline{p}_\perp$,

$$\begin{aligned} \overline{p}_\parallel &= \Re[f(K_d, K_p)], \\ \overline{p}_\perp &= \Im[f(K_d, K_p)], \end{aligned} \tag{7.40}$$

illustrated in Fig. 7.21 in block diagram form. Linearization of the foregoing equations with respect to K_d and K_p provides the four small-signal uncompensated loop gains of Fig. 7.20,

$$\begin{bmatrix} -T_{dd,u}(z) & -T_{dp,u}(z) \\ -T_{pd,u}(z) & -T_{pp,u}(z) \end{bmatrix} \triangleq \begin{bmatrix} \dfrac{\partial \Re[f]}{\partial K_d} & \dfrac{\partial \Re[f]}{\partial K_p} \\ \dfrac{\partial \Im[f]}{\partial K_d} & \dfrac{\partial \Im[f]}{\partial K_p} \end{bmatrix}, \tag{7.41}$$

where partial derivatives are intended to be evaluated at a given steady-state operating point for the autotuner. As anticipated, uncompensated loop gains turn out to be constants because the nonlinear model from which they are derived expresses static relationships between phasors, and there are no complex dynamics associated with it. Nonetheless, the approximations made to arrive at the above-mentioned result are entirely acceptable for an accurate small-signal analysis and design of the tuning loop.

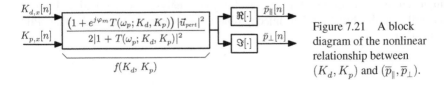

Figure 7.21   A block diagram of the nonlinear relationship between (K_d, K_p) and $(\overline{p}_\parallel, \overline{p}_\perp)$.

With the uncompensated loop gains so derived, consider now the problem of designing the autotuning loop compensation. One approach would be to treat the autotuning loop as a two-input, two-output system and as design compensations for K_d and K_p simultaneously. In the following, a simplified approach is undertaken, more similar to the design of the multiloop controller already encountered in Chapter 4 for the synchronous Buck converter. More precisely, a fast inner loop acting on K_p is designed for the phase margin tuning, whereas a slower, outer loop acts on K_d and tune the system crossover frequency. The rationale behind this design choice is to have the phase margin tuning sufficiently fast so that a stable phase margin is guaranteed at all times, while the crossover frequency can be adjusted more slowly.

With the K_d loop *open*, the compensated loop gain of the phase margin tuning loop is simply

$$\boxed{T_{\varphi_m}(z) \triangleq \frac{\alpha_\perp}{1 - z^{-1}} T_{pp,u}}. \tag{7.42}$$

Selecting $\alpha_\perp$ for a desired crossover frequency is then straightforward.

Once the integral compensator of the phase margin tuning loop is inserted and the loop closed, one has $\hat{K}_{p,y} = \hat{K}_{p,x}$, and equations governing the dynamics of K_d become

$$\hat{K}_{d,y}(z) = -T_{dd,u}\hat{K}_{d,x}(z) - T_{dp,u}\hat{K}_{p,x}(z),$$

$$\hat{K}_{p,x}(z) = \hat{K}_{p,y}(z) = -\frac{T_{pd}(z)}{1 + T_{\varphi_m}(z)}\hat{K}_{d,x}(z), \tag{7.43}$$

with

$$T_{pd}(z) \triangleq \frac{\alpha_\perp}{1 - z^{-1}} T_{pd,u}. \tag{7.44}$$

Therefore, the uncompensated dynamics of the crossover frequency tuning loop is

$$\boxed{T_{f_c,u}(z) = -\frac{\hat{K}_{d,y}(z)}{\hat{K}_{d,x}(z)} = T_{dd,u} - \frac{T_{dp,u}T_{pd}(z)}{1 + T_{\varphi_m}(z)}}, \tag{7.45}$$

and the compensated crossover frequency tuning loop gain is

$$\boxed{T_{f_c}(z) = \frac{\alpha_\parallel}{1 - z^{-1}} T_{f_c,u}(z)}. \tag{7.46}$$

Bode plots of $T_{\varphi_m}(z)$ and $T_{f_c}(z)$ for the PD autotuner tested by simulations in Section 7.3.3 are illustrated in Fig. 7.22.

Although developed for the specific case of PD tuning, the methodology employed in this section for the small-signal analysis of the tuning loop is generally applicable to other injection-based autotuning cases and other PID realizations.

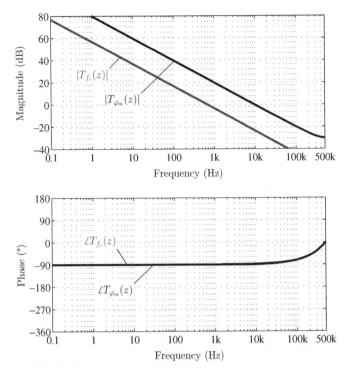

Figure 7.22   Bode diagrams of the crossover frequency and phase margin tuning loop gains.

7.4   DIGITAL AUTOTUNING BASED ON RELAY FEEDBACK

Relay feedback tuning of PID compensators has been long proposed in the literature as a mean for automatically identifying the so-called *ultimate period* of a plant, defined as the period of the oscillation that arises when the closed-loop system is brought to have zero phase margin by a proportional compensator of suitably large gain [85, 83]. With the ultimate period so identified, the frequency-domain version of the Ziegler–Nichols method provides a first-cut choice for the compensator parameters [82, 85, 127]. More recent formulations of the relay feedback autotuning propose extensions of the method for identifying the process response at an arbitrary frequency [84].

This section discusses the theory of digital relay feedback operation, as well as a method proposed in the literature to use it as a digital autotuning method for dc–dc switched-mode power converters. The implementation presented here is based on the approach described in [89]. A modification of the relay feedback method for a more robust crossover frequency tuning is discussed in [90].

7.4.1 Theory of Operation

A *relay* is an instantaneous nonlinear system implementing the function $e_r = f_r(e)$ defined as

$$e_r = f_r(e) = \begin{cases} +A_r & e > 0, \\ -A_r & e \leq 0, \end{cases} \tag{7.47}$$

with $A_r > 0$ defined as the relay amplitude. When inserted into an existing feedback loop, the strong nonlinearity of the relay triggers a limit-cycle oscillation whose frequency and amplitude carry information related to the plant, which can be used for tuning purposes.

To present the theory in a suitably general manner, consider the feedback system depicted in Fig. 7.23, where a relay block has been inserted in front of the compensator $G_c(z)$. In the following, $G_c(z)$ is always assumed to contain an integral action. Furthermore, a filter $F(z)$ is interposed between the compensator and the power converter.

Owing to the presence of the relay block, the system depicted in Fig. 7.23 oscillates at a certain frequency f_{osc}. An intuitive explanation for the existence of such oscillation is that, due to the integral action present in $G_c(z)$, signal $e_r[k]$ must average to zero over time. If this were not the case, the relay block would output a positive or negative signal and force $G_c(z)$ out of this condition. As the relay output $e_r[k]$ cannot be zero instantaneously by virtue of (7.47), the only way $e_r[k]$ can have zero average value is for the relay input $e[k]$ to periodically oscillate around zero. Assuming such oscillation to be essentially sinusoidal due to the low-pass nature of the power converter, the relay block output $e_r[k]$ is a square wave of amplitude $+A_r$ and fundamental frequency f_{osc}. In the language of nonlinear systems, the oscillation that arises in the system because of the relay nonlinearity is another example of *limit cycling*.

The frequency at which such limit cycle sustains itself must fulfill the requirement that

$$\underset{\text{Phase shift of } G_c(z) \text{ at } f_{osc}}{\angle G_c(f_{osc})} + \underset{\text{Phase shift of } T_u(z) \text{ at } f_{osc}}{\angle T_u(f_{osc})} + \underset{\text{Phase shift of } F(z) \text{ at } f_{osc}}{\angle F(f_{osc})} = -\pi, \tag{7.48}$$

where $T_u(z)$ represents, as usual, the uncompensated small-signal transfer function of the converter. Observe that the relay block does not contribute to the loop phase response, as the fundamental component of $e_r[k]$ is in phase with $e[k]$.

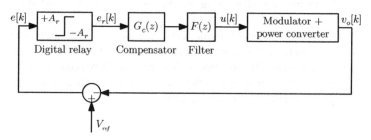

Figure 7.23   Block diagram of a feedback loop with a digital relay.

This qualitative explanation can be made more formal and insightful with the aid of the *describing function theory*. The describing function of a static nonlinearity such as the one implemented by the relay function (7.47) describes how the nonlinearity propagates, in amplitude and phase, an input sinusoidal signal when only the fundamental component at the output is observed. In the case of the relay, the describing function of (7.47) is

$$\Psi(a_{osc}) \triangleq \frac{4A_r}{\pi a_{osc}},$$ (7.49)

where a_{osc} represents the oscillation amplitude as measured on $e[k]$. Equation (7.49) simply expresses the ratio between the fundamental component amplitude $4A_r/\pi$ of $e_r[k]$ and the input amplitude a_{osc} of $e[k]$. Quantity $\Psi(a_{osc})$ is real, implying that zero phase shift is introduced by the relay block.

With the describing function so defined, a *nonlinear loop gain* $T_{NL}(z; a_{osc})$ can be introduced as

$$T_{NL}(z; a_{osc}) \triangleq \Psi(a_{osc})T(z)F(z),$$ (7.50)

where $T(z) = G_c(z)T_u(z)$ is the conventional small-signal loop gain considered in the previous chapters. The oscillation condition is then formally expressed as

$$1 + T_{NL}(z; a_{osc}) = 0,$$ (7.51)

known as the *first-order harmonic balance equation* [185], which can be interpreted as a generalization of the Barkhausen condition for oscillations in linear feedback systems. Equation (7.51) leads to the phase balance equation (7.48) and to a magnitude balance equation

$$\frac{4A_r}{\pi a_{osc}} T_{f_{osc}} F_{f_{osc}} = 1,$$ (7.52)

with $F_{f_{osc}}$ and $T_{f_{osc}}$ magnitudes of $F(z)$ and $T(z)$ at f_{osc}. The magnitude balance equation states that it is the relay gain—as expressed by $\Psi(a_{osc})$—that makes the nonlinear loop gain magnitude equal to one. If the relay were removed from the system, the oscillation would decay. Furthermore, the oscillation amplitude a_{osc} is determined by the magnitude response of $F(z)$ and $T(z)$ at f_{osc}.

7.4.2 Implementation of a Digital Relay Feedback Autotuner

The above-mentioned equations set the framework necessary for describing the basics of relay feedback autotuning. Assume that the compensator is implemented in the PID programmable cascade form (7.3),

$$G_c(z) = G_{PID}(z; \kappa) = \frac{K_i}{1 - z^{-1}} \left(1 - \kappa_1 + \kappa_1 z^{-1}\right)\left(1 - \kappa_2 + \kappa_2 z^{-1}\right).$$ (7.53)

The two compensator zeros $z_{1,2}$ are located at

$$z_{1,2} = -\frac{\kappa_{1,2}}{1 - \kappa_{1,2}}, \tag{7.54}$$

and it can be seen that $0 \leq z_{1,2} < 1$ as long as $\kappa_{1,2} \leq 0$.
    The tuning approach is a three-step procedure [89]:

1. Tuning of z_1 by placing it at the converter resonant frequency
2. Tuning of z_2 to achieve a desired phase margin
3. Tuning of the overall PID gain K_i to achieve a desired crossover frequency

The general block diagram of the feedback configuration during the tuning phases is reported in Fig. 7.24.

Phase 1: Tuning of z_1   Initially, $\kappa_1 = \kappa_2 = 0$, so that

$$G_{PID}(z; \kappa) = \frac{K_i}{1 - z^{-1}}. \tag{7.55}$$

Furthermore, let K_i be sufficiently small to realize a safe, low-bandwidth compensation for the converter. If the filtering block is removed from the system ($F(z) = 1$), the phase balance equation becomes

$$\angle G_{PID}(f_{osc}) + \angle T_u(f_{osc}) = -\pi$$
$$\Rightarrow \angle T_u(f_{osc}) = -\pi - \angle G_{PID}(f_{osc}) = -\frac{\pi}{2}, \tag{7.56}$$

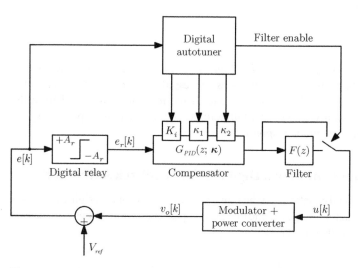

Figure 7.24   Feedback loop configuration during the tuning phases.

as $\angle G_{PID}(f_{osc}) \approx -\pi/2$. For a Buck converter, the frequency at which its control-to-output phase response is equal to $-\pi/2$ is very close to the resonant frequency f_0 of the converter LC filter. In other words,

$$f_{osc} \approx f_0, \tag{7.57}$$

and the limit cycle that arises in the system can be used to identify the converter resonant frequency.

Digitally, measurement of the oscillation frequency can be implemented with a simple counter, clocked at the switching rate f_s and reset at the beginning of the measuring process, and kept in free-running state for a predetermined number N of oscillation periods. An oscillation period can be detected either from the zero crossings of the error signal $e[k]$ or from the sign changes of $e_r[k]$, which is more robust when a small hysteresis window is implemented in the relay function to eliminate chattering noise incoming from the sensing path. If N_s is the number of counter ticks at the end of the measurement, we have

$$N_s T_s = N T_{osc} \Rightarrow \frac{f_s}{f_{osc}} = \frac{N_s}{N}. \tag{7.58}$$

To see how κ_1 is to be selected in order to place the first PID zero in correspondence with the converter resonant frequency, consider the approximation

$$\begin{aligned}
\left(1 - \kappa_1 + \kappa_1 z^{-1}\right)\big|_{z=e^{j\omega T_s}} &= 1 - \kappa_1 + \kappa_1 e^{-j\omega T_s} \\
&\approx 1 - \kappa_1 + \kappa_1 \left(1 - j\omega T_s\right) \\
&= 1 - j\kappa_1\omega T_s, \tag{7.59}
\end{aligned}$$

which suggests

$$\omega_{z1} \approx -\frac{f_s}{\kappa_1} \tag{7.60}$$

as an estimate for the frequency break point associated with z_1. Combining the above-mentioned result with (7.58), choosing

$$\boxed{\kappa_1 \leftarrow -\frac{1}{2\pi}\frac{f_s}{f_{osc}} = -\frac{1}{2\pi}\frac{N_s}{N}} \tag{7.61}$$

forces $\omega_{z1} \approx 2\pi f_{osc}$, as intended. Observe that quantity $2\pi N$ is known to the autotuner and that the above-mentioned operation reduces to storing into κ_1 a scaled version of the counter ticks N_s acquired during the Phase 1 measurement.

Phase 2: Tuning of z_2 The second PID zero is placed in order to achieve a desired phase margin φ_m at the intended crossover frequency. More precisely, this means that z_2 must be positioned so that the *linear* part of the loop gain has a phase response equal to $-\pi + \varphi_m$ at the target f_c, or

$$\angle T(f_c) = \angle G_{PID}(f_c) + \angle T_u(f_c) = -\pi + \varphi_m. \tag{7.62}$$

Plugging the above-mentioned target relationship into the phase balance equation (7.48) *and assuming $f_c = f_{osc}$* yields

$$-\pi + \varphi_m + \angle F(f_c) = -\pi \Rightarrow \angle F(f_c) = -\varphi_m, \tag{7.63}$$

which can be interpreted as follows: if a filter $F(z)$ lagging φ_m degrees at the target crossover frequency f_c is introduced in the relay feedback loop, and if z_2 is tuned so that $f_{osc} = f_c$, then the phase margin constraint (7.62) for the linear loop gain is satisfied.

The above-mentioned idea is implemented as follows. An initial value for κ_2 is first selected, keeping κ_1 as already determined in the previous tuning phase via (7.61). Next, a search algorithm adjusts κ_2 until $f_{osc} = f_c$. One way to do this is by using the flowchart of Fig. 7.25, which implements a form of binary search. As oscillation occurs at the frequency where the phase boost induced by z_2 exactly compensates the filter phase lag, decreasing the frequency break point ω_{z2} results in an increase in f_{osc}. Conversely, increasing ω_{z2} moves f_{osc} to lower values. The flowchart of Fig. 7.25 employs this principle, accounting for the fact that any change of κ_2 in the positive direction moves the frequency break point ω_{z2} of the PID zero to higher

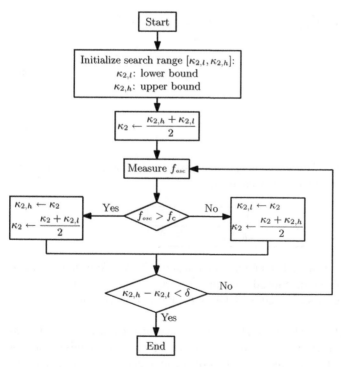

Figure 7.25   Flowchart for the binary search of κ_2.

frequencies, while decreasing κ_2 has the opposite effect. This tuning phase ends when the upper and lower search bounds for κ_2, respectively, indicated as $\kappa_{2,h}$ and $\kappa_{2,l}$ in Fig. 7.25, differ by less than a predetermined amount δ.

An important point to note here is that the above-mentioned frequency search relies on the monotonicity of the phase response of $T(z)F(z)$. If the phase response of such plant does not monotonically decrease with increasing frequency, frequency intervals appear in which the relay-induced limit cycle is unstable. Corresponding frequencies then become unreachable by the relay autotuner, compromising the effectiveness of the method.

Phase 3: Tuning of K_i   With κ_1 and κ_2 now determined, the PID gain K_i must be adjusted for the desired crossover frequency. Measurement of the oscillation amplitude a_{osc}, if performed with adequate accuracy, can provide the scale factor necessary to adjust K_i. To see this, consider the magnitude balance equation (7.52) after the tuning of z_2, that is, when the system oscillates at $f_{osc} = f_c$,

$$\frac{4A_r}{\pi a_{osc}} T_{f_{osc}} F_{f_{osc}} = 1 \Rightarrow T_{f_{osc}} = \frac{\pi a_{osc}}{4A_r F_{f_{osc}}}. \tag{7.64}$$

Hence, if a_{osc} can be measured, the magnitude balance equation provides the value of $T_{f_{osc}}$ at the target crossover frequency. Scaling K_i by the reciprocal of this value

$$\boxed{K_i \leftarrow \frac{4A_r F_{f_{osc}}}{\pi a_{osc}} \times K_i} \tag{7.65}$$

performs the intended tuning. After the removal of the relay block and the filter, one has $T_{f_{osc}} = 1$ as desired.

7.4.3   Simulation Example

Simulation results of relay feedback autotuning of the synchronous Buck converter example are reported in Figs. 7.26–7.29. Tuning starts at $t = 0$ from an initial closed-loop condition in which a low-bandwidth integral compensator is regulating the output voltage. Once enabled, the autotuner enters phase 1 and the system starts oscillating close to the converter resonant frequency. In this condition, the oscillation period is measured over $N = 10$ oscillation intervals, allowing to tune z_1 as explained previously. Next, phase 2 begins and κ_2 undergoes a number of consecutive adjustments until the system oscillates at $f_{osc} \approx f_c \approx 100$ kHz. At the end of this phase, the linear loop gain has a phase lag equal to $-\pi + \varphi_m = -180° + 45° = -135°$ at f_c. This is shown in Fig. 7.29, in which the linear loop gain $T(z)$ is illustrated at the various tuning steps. The oscillation amplitude a_{osc} is also measured during phase 2 and used in phase 3 to calculate the overall PID gain K_i and position f_c at the desired value of 100 kHz.

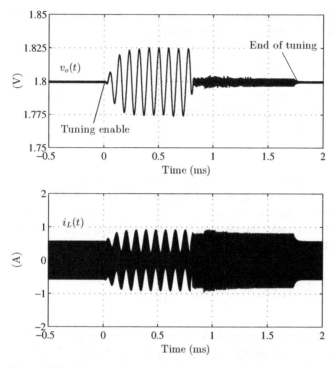

Figure 7.26 Output voltage and inductor current during the tuning process.

7.5 IMPLEMENTATION ISSUES

In the previous sections, operating principles are presented for two different digital autotuning techniques. In practical digital realizations of these autotuners, a number of items have to be carefully considered, some of which are briefly summarized here:

- *Output voltage perturbation.* Any identification approach inevitably perturbs the plant under investigation. In general, such perturbation must be limited to acceptable values not only to prevent potentially destructive situations for the load but also to guarantee normal operation of the converter. In the injection-based approach, the amplitude of output voltage oscillation is clearly related to the amplitude of $u_{pert}[k]$ and can therefore be controlled by acting on $\hat{u}_m$. An injection amplitude controller is included, for instance, in the stability margin monitor presented in [183], which is based on the similar principles as the injection-based autotuner presented in this chapter.

  In the relay feedback autotuner, the oscillation amplitude can be reduced by acting on the relay amplitude A_r. Observe, however, that reducing a_{osc} impacts the accuracy of the crossover frequency tuning during phase 3. A modified relay feedback autotuner for more robust and precise crossover frequency tuning is described in [90].

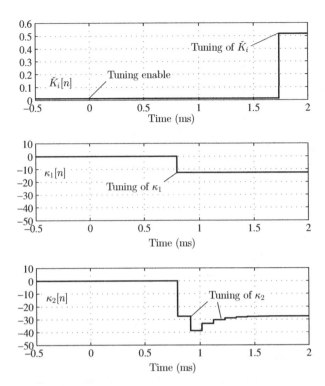

Figure 7.27   PID gains during the tuning process.

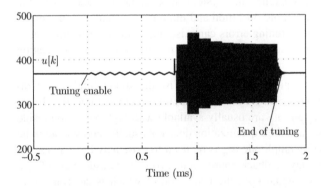

Figure 7.28   Control command during the tuning process.

- *Perturbation signal.* The injection-based approach presented in Section 7.3 makes use of a sinusoidal perturbation signal $u_{pert}[k]$. In a low-complexity digital implementation, however, generation of a sinusoidal signal of prescribed amplitude and frequency may present a problem. For this reason, practical implementations of this method employ a square-wave waveshape for $u_{pert}[k]$, which is much easier to generate [92, 93, 95, 97]. However, a square-wave

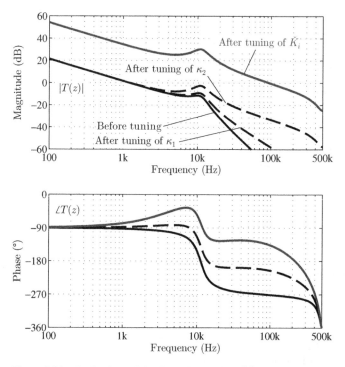

Figure 7.29 Bode plots of the linear loop gain $T(z)$ at various steps of the tuning process.

perturbation no longer excites the system at a single frequency, and the phasor analysis of the tuning operation and the tuning dynamics is no longer applicable. In particular, tuning errors can arise due to the intermodulation of perturbation harmonics occurring as a result of the time-domain multiplication operation.

- *Quantization effects.* Tuning errors can also arise due to amplitude quantization of the signals processed by the digital autotuner. In the injection-based approach, signals u_x and u_y are usually available with high resolution inside the digital controller, and their quantization does not usually represent an issue unless the perturbation amplitude $\hat{u}_m$ is reduced to extremely low values. On the other hand, the relay feedback approach involves the measurement of a_{osc} from the digital error signal $e[k]$, the resolution of which is determined by the A/D converter. For this reason, accuracy of the relay feedback autotuner strongly degrades when operating on extremely small output voltage perturbations.

7.6   SUMMARY OF KEY POINTS

- Autotuning refers to the digital controller ability to adjust its parameters in response to on-line identification of the loop dynamics. Various forms of autotuning include one-step tuning, performance tracking, and adaptive tuning.

- Identification and tuning are the two basic steps performed by any digital autotuner. They can be performed in sequence or nested into a tuning loop that adjusts the compensator parameters iteratively. Depending on the autotuning technique, identification can involve just a few relevant parameters or the entire converter frequency response.

- Tuning can be open-loop or closed-loop according to the system configuration during the tuning step. In this chapter, only closed-loop autotuning techniques are discussed, in which the power converter is feedback-controlled at all times.

- Various methods for perturbing the system to be identified exist. A digital perturbation signal can be generated inside the digital control system and superimposed to the control command. Alternatively, several autotuning techniques rely on intentionally triggering limit cycle oscillations in the system.

- A programmable compensator structure is required for on-line adjustments of the compensation gains. The parallel and cascade structures are the most common because the mid-frequency and high-frequency portions of the compensator response can be easily adjusted without altering the low-frequency behavior. This is a fundamental requirement in order not to compromise system stability during the tuning process.

- In the injection-based autotuning approach, a digital perturbation is superimposed to the control command, while the autotuner monitors the signals before and after the injection point. By ensuring a proper amplitude and phase relationship between the two, the feedback loop can be tuned for a desired crossover frequency and phase margin. In its basic form, this is a two-parameter tuning and can be used for PD or PI structures.

- In the relay feedback autotuner, the perturbation is a limit cycle triggered by the insertion of a strong nonlinearity inside the feedback loop. Frequency and amplitude of such limit cycle carry information on the plant under control and are used by the digital autotuner to achieve a desired crossover frequency and phase margin.

DISCRETE-TIME LINEAR SYSTEMS AND THE Z-TRANSFORM

This appendix presents a brief introduction to discrete-time systems and the Z-transform. Properties of discrete-time systems are first presented in the time domain, based on the theory of constant coefficients difference equations. The Z-transform is then introduced as a tool for analyzing discrete-time systems in the frequency domain.

A.1 DIFFERENCE EQUATIONS

In the following text, *linear*, *causal*, and *time-invariant* discrete-time systems having a single input $u[k]$ and a single output $y[k]$ are considered. Such systems are described by linear, constant coefficients difference equations of the type

$$y[k] = \sum_{i=1}^{N} a_i y[k-i] + \sum_{i=0}^{M} b_i u[k-i] \,, \tag{A.1}$$

with the further assumption that coefficients a_i and b_i are real and that $a_N \neq 0$ and $b_M \neq 0$.

System output $y[k]$ is uniquely determined, for every $k \geq 0$, from the input signal $u[k]$, $k \geq 0$, and from the *initial conditions* on $y[k]$, $-N \leq k \leq -1$,

$$
\begin{aligned}
y[-1] &= y_{-1}, \\
y[-2] &= y_{-2}, \\
&\cdots \\
y[-N] &= y_{-N},
\end{aligned}
\tag{A.2}
$$

Digital Control of High-Frequency Switched-Mode Power Converters, First Edition.
Luca Corradini, Dragan Maksimović, Paolo Mattavelli, and Regan Zane.
© 2015 The Institute of Electrical and Electronics Engineers, Inc. Published 2015 by John Wiley & Sons, Inc.

and on $u[k]$, $-M \leq k \leq -1$,

$$
\begin{aligned}
u[-1] &= u_{-1}, \\
u[-2] &= u_{-2}, \\
&\cdots \\
u[-M] &= u_{-M}.
\end{aligned}
\tag{A.3}
$$

Example A.1.1:   Consider a first-order system ($N = 1$, $M = 0$)

$$
y[k] = ay[k-1] + bu[k],
\tag{A.4}
$$

with initial condition $y[-1] = y_{-1}$ and a generic input $u[k]$. The first few samples of the output signal $y[k]$ can be derived by inspection from (A.4):

$$
\begin{array}{c|c|c}
k & u[k] & y[k] \\
\hline
-1 & 0 & y_{-1} \\
0 & u[0] & ay_{-1} + bu[0] \\
1 & u[1] & a^2 y_{-1} + abu[0] + bu[1] \\
2 & u[2] & a^3 y_{-1} + a^2 bu[0] + abu[1] + bu[2] \\
\cdots & \cdots & \cdots
\end{array}
\tag{A.5}
$$

In general,

$$
y[k] = a^{k+1} y_{-1} + \sum_{i=0}^{k} u[i] b a^{k-i}.
\tag{A.6}
$$

Owing to system linearity, the overall response $y[k]$ can always be expressed as a linear superposition of the *forced response* $y_f[k]$ and the *free response* $y_o[k]$,

$$
y[k] = y_f[k] + y_o[k],
\tag{A.7}
$$

where

- The forced response $y_f[k]$ is the system response to the input $u[k]$, $k \geq 0$, when initial conditions are all zero: $y_{-1} = \ldots y_{-N} = 0$ and $u_{-1} = \ldots u_{-M} = 0$.
- The free response $y_o[k]$ is the system's evolution due to the initial conditions only, with zero input: $u[k] = 0$ for all $k \geq 0$.

A.1.1   Forced Response

Any causal input $u[k]$ can be represented as a linear superposition of discrete pulses,

$$
u[k] = \sum_{i=0}^{+\infty} u[i]\delta[k-i], \qquad k \geq 0,
\tag{A.8}
$$

where

$$
\delta[k] = \begin{cases} 1, & k = 0 \\ 0, & k \neq 0 \end{cases}
\tag{A.9}
$$

represents the unit discrete pulse, sometimes referred to as *Kronecker delta*. From system linearity, it follows that the forced response to a generic input $u[k]$ can be expressed in terms of the *impulse response* $h[k]$, that is, the response to the unit discrete pulse. Expression of the forced response of the system to a generic $u[k]$ is the superposition of impulse responses,

$$\boxed{y_f[k] = \sum_{i=0}^{k} u[i]h[k-i] = \sum_{i=0}^{k} h[i]u[k-i]}. \tag{A.10}$$

Note that system causality implies $h[k] = 0$ for all $k < 0$. Operation expressed by (A.10) is called the *discrete convolution* between $h[k]$ and $u[k]$.

Example A.1.2: The impulse response of the first-order system examined in example A.1.1 can be determined by inspection as

$$h[k] = ba^k, \quad k \geq 0. \tag{A.11}$$

The system forced output is, therefore,

$$y_f[k] = \sum_{i=0}^{k} u[i]h[k-i] = \sum_{i=0}^{k} u[i]ba^{k-i}, \tag{A.12}$$

which is the second term of the overall response (A.6).

A.1.2   Free Response

Consider the system's *characteristic equation*

$$\boxed{z^N - a_1 z^{N-1} - a_2 z^{N-2} - \ldots a_{N-1}z - a_N = 0}, \quad z \in \mathbb{C}, \tag{A.13}$$

and suppose that it has N_r real roots p_i and N_c pairs of complex conjugate roots $r_i e^{\pm j\theta_i}$,

$$p_i, \quad i = 1 \ldots N_r, \tag{A.14}$$

$$r_i e^{\pm j\theta_i}, \quad r_i > 0, \quad i = 1 \ldots N_c. \tag{A.15}$$

For simplicity, assume that all the roots are *simple*, that is, their multiplicity as roots of the characteristic equation is equal to one. By Gauss' fundamental theorem of algebra, $N = N_r + 2N_c$.

When $M \leq N$, the system free response is a linear combination of system *modes*,

$$\boxed{y_o[k] = \sum_{i=1}^{N_r} A_i p_i^k + \sum_{i=1}^{N_c} \left(B_i r_i^k \cos k\theta_i + \tilde{B}_i r_i^k \sin k\theta_i \right)}, \tag{A.16}$$

where the N arbitrary real coefficients A_i, B_i, and $\tilde{B}_i$ are determined from the initial conditions by imposing the value of $y_o[k]$ in its first N samples,

$$y_o[0] = \sum_{i=1}^{N} a_i y_o[0-i] + \sum_{i=0}^{M} b_i u[0-i],$$

$$y_o[1] = \sum_{i=1}^{N} a_i y_o[1-i] + \sum_{i=0}^{M} b_i u[1-i],$$

$$\ldots \quad = \quad \ldots$$

$$y_o[N-1] = \sum_{i=1}^{N} a_i y_o[N-1-i] + \sum_{i=0}^{M} b_i u[N-1-i].$$

Observe that the right-hand sides of the foregoing equations are all known from the initial conditions.

When $M > N$, the free response is a linear combination of the system modes, plus an initial sequence of finite length,

$$y_o[k] = \sum_{i=0}^{M-N-1} q_i \delta[k-i] + \sum_{i=1}^{N_r} A_i p_i^k + \sum_{i=1}^{N_c} \left(B_i r_i^k \cos k\theta_i + \tilde{B}_i r_i^k \sin k\theta_i \right),$$

$$(\text{A}.17)$$

where coefficients q_i are functions—here not shown—of the initial conditions. In general, there are in this case M arbitrary constants uniquely determined from the initial conditions,

$$y_o[0] = \sum_{i=1}^{N} a_i y_o[0-i] + \sum_{i=0}^{M} b_i u[0-i],$$

$$y_o[1] = \sum_{i=1}^{N} a_i y_o[1-i] + \sum_{i=0}^{M} b_i u[1-i],$$

$$\ldots \quad = \quad \ldots$$

$$y_o[M-1] = \sum_{i=1}^{N} a_i y_o[M-1-i] + \sum_{i=0}^{M} b_i u[M-1-i].$$

Example A.1.3: The characteristic equation of the first-order system considered in Example A.1.1 is

$$z - a = 0, \qquad (\text{A}.18)$$

with the corresponding mode a^k. Free evolution of the system is therefore of the form

$$y_o[k] = Aa^k. \qquad (\text{A}.19)$$

By imposing $y_o[0] = ay_o[-1] = ay_{-1}$, one has $A = ay_{-1}$ and therefore

$$y_o[k] = a^{k+1}y_{-1}, \tag{A.20}$$

which represents the first term in (A.6).

Example A.1.4:   Consider the system

$$y[k] = a_1 y[k-1] + b_0 u[k] + b_1 u[k-1] + b_2 u[k-2], \tag{A.21}$$

where $M = 2$ and $N = 1$. The free response therefore consists of an initial sequence of finite length (one sample long, since $M - N - 1 = 0$), plus a term proportional to the only system mode $a_1{}^k$. To check this, consider the first few samples of the free response,

k	$u[k]$	$y[k]$
-2	u_{-2}	$-$
-1	u_{-1}	y_{-1}
0	0	$a_1 y_{-1} + b_1 u_{-1} + b_2 u_{-2}$
1	0	$a_1 \left(a_1 y_{-1} + b_1 u_{-1} + b_2 u_{-2}\right) + b_2 u_{-1}$
2	0	$a_1{}^2 \left(a_1 y_{-1} + b_1 u_{-1} + b_2 u_{-2}\right) + b_2 a_1 u_{-1}$
$\ldots$	$\ldots$	$\ldots$

(A.22)

Except for $k = 0$, a generic output $y[k]$ can be written as

$$y[k] = \left(a_1 y_{-1} + b_1 u_{-1} + b_2 u_{-2} + \frac{b_2}{a_1}u_{-1}\right)a_1{}^k. \tag{A.23}$$

For $k = 0$, on the other hand, the term $-\dfrac{b_2}{a_1}u_{-1}$ must be subtracted from the foregoing expression. In general, for $k \geq 0$, we have

$$y[k] = -\frac{b_2}{a_1}u_{-1}\delta[k] + \left(a_1 y_{-1} + b_1 u_{-1} + b_2 u_{-2} + \frac{b_2}{a_1}u_{-1}\right)a_1{}^k, \tag{A.24}$$

which has the general form (A.17).

A.1.3   Impulse Response and System Modes

It can be shown that, when $M < N$, the impulse response itself is a linear superposition of the system modes,

$$h[k] = \sum_{i=1}^{N_r} A_i p_i^k + \sum_{i=1}^{N_c}\left(B_i r_i^k \cos k\theta_i + \tilde{B}_i r_i^k \sin k\theta_i\right), \tag{A.25}$$

where coefficients A_i, B_i, and $\tilde{B}_i$ are determined from the a_i's and b_i's.

When $M \geq N$, on the other hand, an initial sequence of length $M - N + 1$ appears,

$$h[k] = \sum_{i=0}^{M-N} c_i \delta[k-i] + \sum_{i=1}^{N_r} A_i p_i^k + \sum_{i=1}^{N_c} \left(B_i r_i^k \cos k\theta_i + \tilde{B}_i r_i^k \sin k\theta_i \right), \quad (A.26)$$

where coefficients c_i are, again, functions of the a_i's and b_i's.

If $N = 0$, the system response is a function of the input signal only, and the impulse response consists of the sole finite length contribution. Systems of this kind are called *finite impulse response* systems (FIR). In all the other cases, the system is designated as *infinite impulse response* (IIR).

A.1.4 Asymptotic Behavior of the Modes

Every system mode is a sequence of the type p_i^k or $r_i^k \sin(k\theta_i + \phi)$. Behavior of each mode as $k \rightarrow +\infty$ therefore essentially depends on $|p_i|$ or r_i being larger, equal, or less than one. Under the assumption that all the roots of the characteristic equation are simple, the following conclusions hold:

- Modes of the type p_i^k, associated with the real roots of the characteristic equation, are convergent to zero for $k \rightarrow +\infty$ when $|p_i| < 1$, whereas they do not converge for $|p_i| > 1$. Condition $|p_i| = 1$ generates a constant mode when $p_i = 1$ or an oscillatory mode at the Nyquist rate when $p_i = -1$. In general, an oscillatory character—convergent or not—arises when $p_i < 0$. Examples of modes associated with real roots of the characteristic equation are shown in Fig. A.1.

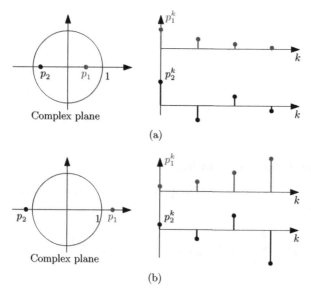

Figure A.1 Example of (a) convergent and (b) nonconvergent modes associated with real roots of the characteristic equation.

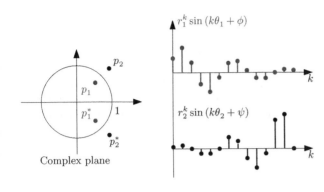

Figure A.2   Example of modes associated with complex roots of the characteristic equation.

- Modes of the type $r_i^k \sin(k\theta_i + \phi)$, associated with pairs of complex conjugate roots of the characteristic equation, converge to zero for $k \to +\infty$ when $|r_i| < 1$, whereas they do not converge when $|r_i| > 1$. Condition $r_i = 1$ originates a persistent oscillatory mode whose normalized angular frequency coincides with the pole's phase. Examples of modes associated with complex roots of the characteristic equations are shown in Fig. A.2.

For convergent modes, the rate at which they decay to zero depends on how far the corresponding root is from the unit circle in absolute value. For instance,

$$|p_i^k| = e^{k \ln |p_i|} = e^{-\frac{\ln \frac{1}{|p_i|}}{T} kT}, \tag{A.27}$$

suggesting an equivalent time constant $\tau = \frac{T}{\ln \frac{1}{|p_i|}}$, the larger the closer $|p_i|$ is to 1.

A.1.5   Further Examples

Example A.1.5:   Consider how the system of Example A.1.1 responds to a discrete-time step of amplitude U,

$$u[k] = U, \quad k \geq 0. \tag{A.28}$$

Using (A.10), one has

$$y_f[k] = \sum_{i=0}^{k} h[i]U = bU \sum_{i=0}^{k} a^i = bU \frac{1 - a^{k+1}}{1 - a}, \quad a \neq 1. \tag{A.29}$$

Hence

$$y_f[k] = \frac{b}{1 - a}U - \frac{ba^{k+1}}{1 - a}U, \quad a \neq 1. \tag{A.30}$$

When $|a| < 1$, the system mode is convergent. In this case, the first term is commonly denoted as *steady-state response*, whereas the second term is a *transient*.

Example A.1.6:   *Discrete-Time Integrator.* System

$$y[k] = y[k-1] + K_i u[k] \tag{A.31}$$

is a special case of the first-order system considered so far, with $a = 1$ and $b = K_i$. At every sampling step, the output varies by a quantity proportional to the instantaneous input. The impulse response of this system is therefore a discrete-time step of amplitude K_i,

$$h[k] = K_i, \quad k \ge 0. \tag{A.32}$$

Example A.1.7:   *Discrete-Time Differentiator.* System

$$y[k] = K_d(u[k] - u[k-1]) \tag{A.33}$$

provides an output proportional to the input variation in the last sampling step. The impulse response of such system is

$$h[k] = K_d(\delta[k] - \delta[k-1]). \tag{A.34}$$

The discrete-time differentiator is therefore an FIR system.

A.2   Z-TRANSFORM

A.2.1   Definition

The Z-transform of a discrete-time signal $h[k]$ is defined as

$$\boxed{Z[h] \triangleq \sum_{k=0}^{+\infty} h[k]z^{-k}}, \quad z \in \mathbb{C}. \tag{A.35}$$

It can be shown that the above-mentioned summation converges to a complex function $H(z)$, which is holomorphic over a set D of the type

$$D = \left\{ z \in \mathbb{C}: \quad |z| > R, \quad R \in \mathbb{R}^+ \right\}, \tag{A.36}$$

that is, on the outside of a circle of radius $R \ge 0$ and centered at the origin of the z complex plane.

Example A.2.1: Z-transform of the causal exponential sequence $h[k] = ba^k, k \ge 0$, is

$$H(z) = \sum_{k=0}^{+\infty} ba^k z^{-k} = b \sum_{k=0}^{+\infty} (az^{-1})^k. \tag{A.37}$$

This is the geometric sum of common ratio $q = az^{-1}$, which converges if and only if $|q| < 1$, that is, if and only if $|z| > a$. Sum of the series is

$$H(z) = \frac{b}{1 - az^{-1}}, \quad |z| > a. \tag{A.38}$$

In particular, the Z-transform of the discrete-time integrator impulse response examined in Example A.1.6 is

$$H(z) = \frac{K_i}{1 - z^{-1}}, \quad |z| > 1. \tag{A.39}$$

A.2.2   Properties

Main properties of the Z-transform are here summarized:

1. *Linearity.* If $h_1[k]$ and $h_2[k]$ are two sequences, the Z-transform of a superposition $h[k] = \mu h_1[k] + \lambda h_2[k]$, $\mu, \lambda \in \mathbb{R}$, is

$$H(z) = \mu H_1(z) + \lambda H_2(z), \quad z \in D_1 \cap D_2, \tag{A.40}$$

where D_1 and D_2 are the convergence regions of $H_1(z)$ and $H_2(z)$, respectively.

2. *Delay.* Given a *causal* sequence $h[k]$, that is, a sequence that is zero for all $k < 0$, the Z-transform of its delayed version $h[k - k_0]$ is

$$\begin{aligned}
Z\left[h[k - k_0]\right] &= \sum_{k=0}^{+\infty} h[k - k_0] z^{-k} \\
&= \sum_{k'=-k_0}^{+\infty} h[k'] z^{-(k'+k_0)} \\
&= z^{-k_0} \sum_{k'=0}^{+\infty} h[k'] z^{-k'} \\
&= z^{-k_0} H(z). \tag{A.41}
\end{aligned}$$

Operator z^{-1} therefore represents the basic operation of a *one-sample delay*. Delay of a noncausal sequence, on the other hand, yields

$$\begin{aligned}
Z\left[h[k - k_0]\right] &= \sum_{k=0}^{+\infty} h[k - k_0] z^{-k} \\
&= \sum_{k'=-k_0}^{+\infty} h[k'] z^{-(k'+k_0)}
\end{aligned}$$

$$= z^{-k_0} \sum_{k'=-k_0}^{-1} h[k']z^{-k'} + z^{-k_0} \sum_{k'=0}^{+\infty} h[k']z^{-k'}$$

$$= z^{-k_0} \sum_{k=-k_0}^{-1} h[k]z^{-k} + z^{-k_0} H(z). \tag{A.42}$$

3. *Discrete Convolution.* Let $h[k]$ and $u[k]$ be two signals having Z-transforms $H(z)$ and $u(z)$, converging in regions D_h and D_u, respectively. Let $y[k]$ be the discrete convolution of $h[k]$ and $u[k]$:

$$y[k] = \sum_{i=0}^{k} h[k-i]u[i]. \tag{A.43}$$

Then, the Z-transform of $y[k]$ converges into $D_y \subset D_h \cap D_u$ to

$$y(z) = H(z)u(z). \tag{A.44}$$

4. *Initial Value.* Initial value $h[0]$ of any causal sequence $h[k]$ is equal to its Z-transform evaluated at $z \to +\infty$:

$$\lim_{z \to +\infty} H(z) = h[0]. \tag{A.45}$$

5. *Final Value.* Let $h[k]$ be a sequence and $H(z)$ its Z-transform. If the poles of

$$(1 - z^{-1})H(z) \tag{A.46}$$

are located *inside* the unit circle, then

$$\lim_{z \to 1} (1 - z^{-1})H(z) = \lim_{k \to +\infty} h[k]. \tag{A.47}$$

6. *Inverse Z-Transform.* Let $H(z)$ be the Z-transform of sequence $h[k]$ and D be the region of convergence. Then

$$h[k] = \frac{1}{2\pi j} \oint_C H(z)z^{k-1} dz, \tag{A.48}$$

where C is a *closed* integration contour belonging to D and encircling the origin counterclockwise.

A.3   THE TRANSFER FUNCTION

From (A.10), it follows that the Z-transform $y_f(z)$ of the forced response of the system to an input $u(z)$ is

$$y_f(z) = H(z)u(z),$$ (A.49)

where the *transfer function* $H(z)$ is the Z-transform of the impulse response $h[k]$. $H(z)$ can be expressed in terms of the coefficients of the system difference equation by applying the Z-transform to the first and the second member of (A.1),

$$\boxed{H(z) = \frac{b_0 + b_1 z^{-1} + \ldots + b_{M-1} z^{-M+1} + b_M z^{-M}}{1 - a_1 z^{-1} - a_2 z^{-2} - \ldots - a_{N-1} z^{-N+1} - a_N z^{-N}}}.$$ (A.50)

Assuming the numerator and the denominator of $H(z)$ do not have common roots, it follows that the nonzero poles of the transfer function coincide with the roots of the system characteristic equation.

Expanding $H(z)$ in partial fractions and assuming all the poles are simple, and $M < N$,

$$H(z) = \sum_{i=1}^{N_r} \frac{A_i}{1 - p_i z^{-1}} + \sum_{i=1}^{N_c} \frac{B_i - r_i(B_i \cos\theta_i - \tilde{B}_i \sin\theta_i)z^{-1}}{1 - 2r_i \cos\theta_i z^{-1} + r_i^2 z^{-2}},$$ (A.51)

where p_i are the N_r real poles, whereas $r_i e^{\pm j\theta_i}$ are the $2N_c$ pairs of complex conjugate poles. The above-mentioned expansion is the z-domain counterpart of (A.25).

If $M \geq N$, a term appears corresponding to an initial sequence of finite duration,

$$H(z) = \sum_{i=0}^{M-N} c_i z^{-i} + \sum_{i=1}^{N_r} \frac{A_i}{1 - p_i z^{-1}} + \sum_{i=1}^{N_c} \frac{B_i - r_i(B_i \cos\theta_i - \tilde{B}_i \sin\theta_i)z^{-1}}{1 - 2r_i \cos\theta_i z^{-1} + r_i^2 z^{-2}},$$ (A.52)

which can be compared with (A.26).

Example A.3.1:   The transfer function of the discrete-time differentiator examined in Example A.1.7 is

$$H(z) = K_d(1 - z^{-1}).$$ (A.53)

A.3.1   Stability

Given the relationship between poles of the transfer function and system modes, one can draw the following conclusions:

- System is *asymptotically stable* if and only if all the poles are strictly *inside* the unit disk,

$$\begin{aligned} |p_i| &< 1, \quad \forall i = 1 \ldots N_r, \\ |r_i| &< 1, \quad \forall i = 1 \ldots N_c. \end{aligned}$$ (A.54)

- System is *unstable* when at least one of the poles lies *outside* the unit disk.
- System is *marginally stable* if all the poles are inside the unit disk, with the possible exception of a number of *simple* poles of magnitude one.

A.3.2 Frequency Response

Let $H(z)$ be the transfer function of an asymptotically stable system, and

$$u[k] = e^{jk\theta} \tag{A.55}$$

its input. The forced response of the system consists of a transient term converging to zero for $k \rightarrow +\infty$ and a steady-state term $y_{ss}[k]$ that can be expressed as

$$y_{ss}[k] = |H(e^{j\theta})|e^{j(k\theta + \angle H(e^{j\theta}))}. \tag{A.56}$$

The *frequency response* of a discrete-time system is therefore equal to the system transfer function $H(z)$ evaluated on the unit circle $z = e^{j\theta}$.

For marginally stable systems and assuming $e^{j\theta}$ does not correspond to one of the system poles, the forced response still contains a term of the type (A.56), whereas the transient term remains limited but nonconvergent.

A.4 STATE-SPACE REPRESENTATION

As with continuous-time systems, a state-space description for discrete-time systems can be developed both in the time domain and in the frequency domain. Equations of a linear, time-invariant and causal discrete-time system in state-space form are

$$\boxed{\begin{aligned} x[k+1] &= Ax[k] + Bu[k] \\ y[k] &= Cx[k] + Fu[k] \end{aligned}}, \tag{A.57}$$

where x represents the state vector, whereas u and y are the system input and output, respectively. Matrices A, B, and C are of type $\mathbb{R}^{n \times n}$, $\mathbb{R}^{n \times 1}$, and $\mathbb{R}^{1 \times n}$, respectively, where n is the number of state variables. On the other hand, for single-input, single-output systems such as those considered in this appendix, F is a scalar.

The overall system response is uniquely determined for $k \geq 0$ once the system initial state $x[0]$ and its input $u[k], k \geq 0$ are assigned,

k	$u[k]$	$x[k+1]$	$y[k]$
0	$u[0]$	$Ax[0] + Bu[0]$	$Cx[0] + Fu[0]$
1	$u[1]$	$A^2x[0] + ABu[0] + Bu[1]$	$CAx[0] + CBu[0] + Fu[1]$
2	$u[2]$	$A^3x[0] + A^2Bu[0] + ABu[1] + Bu[2]$	$CA^2x[0] + CABu[0] + CBu[1] + Fu[2]$
...	...	...	....

$$\tag{A.58}$$

In general, for $k \geq 0$,

$$\boldsymbol{x}[k+1] = \boldsymbol{A}^{(k+1)}\boldsymbol{x}[0] + \sum_{i=0}^{k} \boldsymbol{A}^{k-i}\boldsymbol{B}u[i],$$

$$y[k] = \boldsymbol{C}\boldsymbol{A}^{k}\boldsymbol{x}[0] + \sum_{i=0}^{k-1} \boldsymbol{C}\boldsymbol{A}^{k-i-1}\boldsymbol{B}u[i] + Fu[k]. \tag{A.59}$$

The latter expression makes it possible to easily distinguish between the system free response $y_o[k]$ and the forced response $y_f[k]$,

$$y_o[k] = \boldsymbol{C}\boldsymbol{A}^{k}\boldsymbol{x}[0],$$

$$y_f[k] = \sum_{i=0}^{k-1} \boldsymbol{C}\boldsymbol{A}^{k-i-1}\boldsymbol{B}u[i] + Fu[k]. \tag{A.60}$$

Observe that the forced response can be written in the form

$$y_f[k] = \sum_{i=0}^{k} h[k-i]u[i], \tag{A.61}$$

with

$$h[k] = \begin{cases} F, & k = 0, \\ \boldsymbol{C}\boldsymbol{A}^{k-1}\boldsymbol{B}, & k \geq 1. \end{cases} \tag{A.62}$$

Signal $h[k]$ is therefore the system impulse response, and the above-mentioned relationships link $h[k]$ to the system state-space matrices.

In the z-domain, (A.57) becomes

$$z\boldsymbol{x}(z) - z\boldsymbol{x}[0] = \boldsymbol{A}\boldsymbol{x}(z) + \boldsymbol{B}u(z),$$

$$y(z) = \boldsymbol{C}\boldsymbol{x}(z) + Fu(z), \tag{A.63}$$

and therefore

$$\boxed{y(z) = z\boldsymbol{C}\left(z\boldsymbol{I} - \boldsymbol{A}\right)^{-1}\boldsymbol{x}[0] + \left(\boldsymbol{C}\left(z\boldsymbol{I} - \boldsymbol{A}\right)^{-1}\boldsymbol{B} + F\right)u(z)}. \tag{A.64}$$

Letting $\boldsymbol{x}[0] = 0$ leads to an expression for the system transfer function as

$$\boxed{H(z) \triangleq \frac{y_f(z)}{u(z)} = \boldsymbol{C}\left(z\boldsymbol{I} - \boldsymbol{A}\right)^{-1}\boldsymbol{B} + F}. \tag{A.65}$$

FIXED-POINT ARITHMETIC AND HDL CODING

This appendix presents an overview of the representation of numbers in a finite precision arithmetic environment. The discussion then focuses on the binary two's complement representations and coding of fixed-point arithmetic in hardware description languages (HDLs)—VHDL and Verilog-HDL.

B.1 ROUNDING OPERATION AND ROUND-OFF ERROR

In a finite precision computing system, a limited number of digits are available to represent any given signal. *Representable numbers*, that is, those numbers having an exact representation in the considered arithmetic system, necessarily form a discrete subset $\tilde{\mathcal{I}}$ of a continuous subset $\mathcal{I}$ of the real axis.

It is first necessary to clarify how a given $c \in \mathcal{I}$ can be approximated by a suitable element $\tilde{c} \in \tilde{\mathcal{I}}$. To this end, recall that for a given integer b strictly greater than one, every nonzero real quantity c can be *uniquely* written as

$$c = \pm \mathcal{S}_b(c) \times b^{\mathcal{E}_b(c)}, \tag{B.1}$$

where

- b is called the *base* or *radix* of the representation.
- $\mathcal{E}_b(c)$ is an integer called the *exponent* of c, which defines its *order of magnitude*.
- $\mathcal{S}_b(c)$ is a real number such that $1 \leq \mathcal{S}_b(c) < b$. It is called the *significand* of c.

For instance,

$$
\begin{aligned}
\frac{3}{8} &= 3.75 \times 10^{-1} && \text{(Radix-10 representation ($b = 10$)),} \\
&= 3 \times 8^{-1} && \text{(Radix-8 representation ($b = 8$)),} \\
&= 1.5 \times 2^{-2} && \text{(Radix-2 representation ($b = 2$)).}
\end{aligned}
\tag{B.2}
$$

Digital Control of High-Frequency Switched-Mode Power Converters, First Edition.
Luca Corradini, Dragan Maksimović, Paolo Mattavelli, and Regan Zane.
© 2015 The Institute of Electrical and Electronics Engineers, Inc. Published 2015 by John Wiley & Sons, Inc.

Representation (B.1), known as *exponential notation*, suggests that, in order to define $\tilde{c}$, one can truncate or round-off $\mathcal{S}_b(c)$ to a given number of digits. Consider, for instance, number $\pi = 3.1415926535\ldots$. In a radix-10 notation, $\mathcal{E}_{10}(\pi) = 0$ and $\mathcal{S}_{10}(\pi) = \pi$. Consider then successive approximations of π in which the significand is *rounded* to its first n decimal digits,

$$
\begin{aligned}
\tilde{\pi} = & & 3 \times 10^0 & & \text{(1-digit approx.)}, \\
\tilde{\pi} = & & 3.1 \times 10^0 & & \text{(2-digits approx.)}, \\
\tilde{\pi} = & & 3.14 \times 10^0 & & \text{(3-digits approx.)}, \\
\tilde{\pi} = & & 3.141 \times 10^0 & & \text{(4-digits approx.)}, & & \text{(B.3)} \\
\tilde{\pi} = & & 3.1416 \times 10^0 & & \text{(5-digits approx.)}, \\
\tilde{\pi} = & & 3.14159 \times 10^0 & & \text{(6-digits approx.)},
\end{aligned}
$$

$\ldots$

The operation of rounding c to its first n significant digits is denoted as

$$
\boxed{\tilde{c} = \mathcal{Q}_n\,[c]}, \tag{B.4}
$$

the radix b being usually clear from the context. The notation can be simplified by eliminating the radix point "." and adjusting the exponent accordingly,

$$
\begin{aligned}
\mathcal{Q}_1\,[\pi] = & & 3 \times 10^0 & & \text{(1-digit approx.)}, \\
\mathcal{Q}_2\,[\pi] = & & 31 \times 10^{-1} & & \text{(2-digits approx.)}, \\
\mathcal{Q}_3\,[\pi] = & & 314 \times 10^{-2} & & \text{(3-digits approx.)}, \\
\mathcal{Q}_4\,[\pi] = & & 3141 \times 10^{-3} & & \text{(4-digits approx.)}, & & \text{(B.5)} \\
\mathcal{Q}_5\,[\pi] = & & 31416 \times 10^{-4} & & \text{(5-digits approx.)}, \\
\mathcal{Q}_6\,[\pi] = & & 314159 \times 10^{-5} & & \text{(6-digits approx.)}
\end{aligned}
$$

$\ldots$

From these preliminary considerations, in an n-digit, radix-b finite precision arithmetic system the representable numbers are of the form

$$
\boxed{\tilde{c} = \pm w \times b^q}, \tag{B.6}
$$

where

- The *radix* or *base* b is an integer equal or greater than 2.
- The *unsigned significand* w is a nonnegative integer, which is expressed, in positional notation, by an n-digit base-b word.

$$w = \left(d_{n-1}d_{n-2}\cdots d_1 d_0\right)_b \triangleq \sum_{i=0}^{n} d_i \times b^i, \quad d_i \in \{0, 1, \ldots b-1\}. \quad \text{(B.7)}$$

- The *exponent* q is an integer that, depending on the arithmetic system, may or may not have an explicit encoding.

Absolute and relative *round-off errors* between c and its approximation $Q_n[c]$ are denoted with

$$d_n c \triangleq Q_n[c] - c \qquad \text{(Absolute round-off error),}$$

$$\delta_n c \triangleq \frac{d_n c}{c} = \frac{Q_n[c] - c}{c} \qquad \text{(Relative round-off error).}$$

$$\text{(B.8)}$$

B.2   FLOATING-POINT VERSUS FIXED-POINT ARITHMETIC SYSTEMS

Encoding a given number in a finite precision system involves the representation of (i) the rounded significand and (ii) the exponent q. Signed numbers can be treated by devoting one additional bit to represent the sign.

The arithmetic formats where the exponent q is explicitly encoded are known as *floating-point* arithmetic systems. Consider, for instance, a radix-10 system using two digits for the significand and one digit for a signed exponent. Positive representable quantities would range from $01_{10} \times 10^{-9}$ to $99_{10} \times 10^9$, covering 20 decades with a relative round-off error never larger than $\approx 4.7\%$. The single main advantage of floating-point arithmetic, therefore, is the capability to span several orders of magnitude while maintaining a limited relative round-off error throughout the represented range. Standard IEEE Std 754™-2008 [176] defines a number of floating-point formats. For instance, in the IEEE *binary32* format, 32 total bits are available, 1 bit encoding the sign, 8 bits encoding the exponent, and the remaining 23 bits being reserved for the significand. The represented range spans approximately 83 decades.

Implementation of floating-point systems involves a significant computational overhead to carry out even the fundamental arithmetic operations, because of the need to decode and encode operands prior and after every manipulation. Furthermore, *normalization* of the represented quantities is required to make representations unique: In a three-digit system, for instance, $\tilde{c} = 8.2$ can be represented either as $082_{10} \times 10^{-1}$ or as $820_{10} \times 10^{-2}$. The representation can be made unique by requiring that $100 \le w < 1000$. More generally, in an n-digit radix-b system, one requires that $b^{n-1} \le w < b^n$.

Because of its complexity, floating-point arithmetic is nowadays implemented in most microprocessors and high-end digital signal processors (DSPs) in dedicated *floating-point units* (FPUs). On the other hand, floating-point arithmetic is typically not supported by low-cost DSPs and microcontrollers, where floating point can be *software-emulated* when absolutely needed. In general, owing to cost and complexity constraints, floating-point arithmetic is avoided in many embedded system applications, including digital controllers considered in this book.

The formats where the exponent is not explicitly encoded are called *fixed-point* arithmetic systems. Any given quantity is represented solely by a *signed significand*, whereas the exponent remains *fixed* once and for all and therefore does not require encoding. Manipulation of fixed-point quantities can be carried out much more rapidly and efficiently. In fact, it is easy to realize that *arithmetic operations in a fixed-point environment are essentially arithmetic operations between integers.* Furthermore, representations are inherently unique, with the possible exception of the zero, which may or may not have a unique encoding depending on the format used. The drawback of such simplicity is a much larger relative error with respect to a floating-point encoding using the same number of bits — or, equivalently, the need for longer word lengths to achieve the same precision. As a comparison with the previous example, suppose two integer digits are available to represent quantities over a scale of 10^2. Positive representable numbers therefore range from $01_{10} \times 10^2$ to $99_{10} \times 10^2$, that is, two decades. The worst-case relative round-off error amounts to 50%.

B.3   BINARY TWO'S COMPLEMENT (B2C) FIXED-POINT REPRESENTATION

In this book, a radix-2 fixed-point system is considered in which a *signed* significand is encoded in two's complement notation. The binary two's complement (B2C) representation is a base-2 positional system capable of encoding both positive and negative numbers, with a unique representation of the zero. It has a number of appealing features that make it easily the most commonly adopted integer arithmetic system in today's microcontrollers, DSPs, and microprocessors.

Representable numbers are of the form

$$\boxed{x = w \times 2^q}, \tag{B.9}$$

where the signed significand w is an n-bit binary word w encoded in B2C,

$$w = \left(b_{n-1} b_{n-2} \dots b_1 b_0 \right)_{\bar{2}}$$

$$\triangleq -b_{n-1} \times 2^{n-1} + \sum_{i=0}^{n-2} b_i \times 2^i, \quad b_i \in \{0, 1\}. \tag{B.10}$$

The exponent q is fixed once and for all and therefore does not have an explicit hardware encoding.

Bits b_{n-1} and b_0 of the significand are called *most significant bit* (MSB) and *least significant bit* (LSB) of the representation, respectively. The most significant bit b_{n-1} is also called the *sign bit*, as it is equal to 1 if and only if $w < 0$ and equal to 0 otherwise.

The range spanned by an n-bit B2C word is

$$\underbrace{-2^{n-1}}_{w_{min}} \leq w \leq \underbrace{2^{n-1} - 1}_{w_{max}}, \tag{B.11}$$

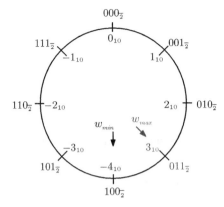

Figure B.1   Circular representation of a
3-bit B2C arithmetic system.

where w_{max} and w_{min} represent the most positive and the most negative representable numbers. For instance, the range of integers represented by a 3-bit B2C word goes from -4 to $+3$. As shown in Fig. B.1, B2C can be thought of as a circular representation: if binary one is added to the most positive number—performing the operation as if dealing with a plain base-2 representation—the most negative number is obtained.

Inset B.1 – B2C Round-off Using Matlab®

A simple Matlab® function that implements the n-digit B2C round-off operation $Q_n[x]$ of a given quantity x is given in this inset:

```
function [wk,dx]    =   Qn(x,n)

x1   =   x;

neg    =   (x<0);
x      =   abs(x);

E    =   floor(log2(x));
F    =   2^(log2(x)-E);
q    =   E-(n-2);
wd   =   round(F*2^(n-2));
if (wd==1)
     wd    =    round(F*2^(n-3));
     q     =    q+1;
end;

if (neg)
     xq    =    -wd*2^q;
     s     =    ['1',dec2bin(-wd+2^(n-1),n-1)];
     wd    =    -wd;
else
     xq    =    wd*2^q;
     s     =    ['0',dec2bin(wd,n-1)];
end;
```

```
wk.xq    =    xq;
wk.w     =    wd;
wk.q     =    q;
wk.n     =    n;
wk.s     =    s;
dx       =    xq-x1;

return;
```

The above-mentioned function accepts quantity x to be quantized and the target word length n. Its outputs wk and dx are

- wk: A structure encoding the B2C quantity. Its fields are as follows:
  - wk.xq: Rounded-off quantity $\mathcal{Q}_n[x]$.
  - wk.w: Significand w of the B2C representation of x.
  - wk.q:Scale q of the B2C representation of x.
  - wk.n: Number of bits.
  - wk.s:String representation of the n-bit B2C word w.
- dx: Absolute round-off error $d_n x$.

For instance, [wk,dx]=Qn(pi,5) produces

```
wk =

    xq: 3.2500
     w: 13
     q: -2
     n: 5
     s: '01101'

dx =

    0.1084
```

or

$$\mathcal{Q}_5[\pi] = 01101_{\bar{2}} \times 2^{-2} = 13_{10} \times 2^{-2} = 3.25_{10}. \tag{B.12}$$

B.4   SIGNAL NOTATION

Referring to (B.9), one can interpret x as a generic *signal* and word w as *representing* x over a scale 2^q. A notation is now introduced that is extensively used in Chapter 6 and that makes the relationship between x and w more explicit, without the need to rewrite (B.9) every time. Define

$$\boxed{[x]_q^n \triangleq w}, \tag{B.13}$$

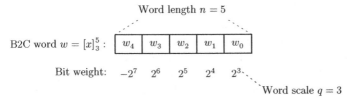

Figure B.2   Signal notation of a B2C word $w = [x]_3^5$.

so that the relationship between x and w becomes

$$x = [x]_q^n \times 2^q. \qquad (B.14)$$

In other words, $[x]_q^n$ is the *unique* n-bit B2C word that represents signal x if its least significant bit is given a weight equal to 2^q. In a sense, this notation puts emphasis on the signal x *represented* by a B2C word rather than on the word itself. As an example, Fig. B.2 shows a pictorial representation of the signal notation for a B2C word $w = [x]_3^5$ representing a signal x with 5 bits and over a scale 2^3.

B.5   MANIPULATION OF B2C QUANTITIES AND HDL EXAMPLES

This section summarizes the most common arithmetic and bitwise operations on B2C words, along with the corresponding coding in VHDL or Verilog [173, 174, 177, 178].

As far as VHDL is concerned, data types and packages standardized in the *IEEE Standard VHDL Synthesis Packages* document [172] are employed. The standard provides a description of data types, arithmetic, and logic operators, which are likely supported by any synthesis tool. B2C quantities are represented by means of the `signed` data type as defined in the `NUMERIC_STD` package, which in turn is built upon the IEEE-defined package `STD_LOGIC_1164` [171]. Hence, the following library configuration preamble is assumed for all VHDL examples in this book:

```
library IEEE;
use IEEE.STD_LOGIC_1164.ALL;
use IEEE.NUMERIC_STD.ALL;
```

An n-bit word $[x]_q^n$ representing signal x can then be defined as

```
signal x : signed(n-1 downto 0);          -- [n,q]
```

Observe that the scale q of $[x]_q^n$ is not encoded anywhere and remains *implicit*, as a fixed-point arithmetic representation is used. For such reason and for improved code readability, the comment `-- [n,q]` is included, which reports both the size and the scale of x.

The reference documentation for the Verilog language definitions and synthesizable constructs is in [175, 178]. For manipulating B2C words, the Verilog data type `signed` is employed, and words $[x]_q^n$ introduced earlier can be defined as

```
    wire signed [n-1:0] x;              // [n,q]
```

All binary and arithmetic operations described in this section implement, in hardware, purely *combinational* functions. Therefore, they are coded as either concurrent statements (if VHDL is used) or continuous assignments (in Verilog examples). Recall that the basic syntax for a VHDL concurrent statement makes use of the `<=` operator,

```
    y   <=  x;                          -- Concurrent statement
```

where x and y are two VHDL signals. A Verilog continuous assignment, on the other hand, has the basic syntax

```
    assign y = x;              // Continuous assignment
```

where x and y are declared as `wire signed` nets.

B.5.1 Sign Extension

When extending the number of binary digits from n to $n + k$, k replicas of the sign bit are to be written in the most significant portion of the word. The reason why this works is that the contribution of the sign bit can always be written as

$$-b_{n-1} \times 2^{n-1} = -b_{n-1} \times 2^n + b_{n-1} \times 2^{n-1}, \qquad (B.15)$$

which allows one to arbitrarily replicate the sign without altering the represented value.

For instance, if

$$w = \ [x]_q^5 = 10010_{\bar{2}} = -14_{10},$$

$$r = \ [y]_q^5 = 00111_{\bar{2}} = \quad 7_{10} \qquad (B.16)$$

are two 5-bit B2C words, their extensions to $5 + 3 = 8$ bits are

$$w' = \ [x]_q^8 = 11110010_{\bar{2}} = -14_{10},$$

$$r' = \ [y]_q^8 = 00000111_{\bar{2}} = \quad 7_{10}, \qquad (B.17)$$

as one can easily verify.

In signal notation, a 1-bit sign extension is simply denoted as

$$[x]_q^{n+1} \leftarrow [x]_q^n, \qquad (B.18)$$

and, more generally,

$$[x]_q^{n+k} \leftarrow [x]_q^n, \quad (k \geq 0) \qquad (B.19)$$

for a k-bit sign extension. Observe that sign extension does not modify the signal represented by the word.

As an example, Fig. B.3 shows a pictorial representation of $[x]_3^6 \leftarrow [x]_3^5$.

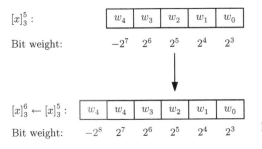

$[x]_3^5:$

Bit weight: $\quad -2^7 \quad 2^6 \quad 2^5 \quad 2^4 \quad 2^3$

$[x]_3^6 \leftarrow [x]_3^5:$

Bit weight: $\quad -2^8 \quad 2^7 \quad 2^6 \quad 2^5 \quad 2^4 \quad 2^3$

Figure B.3 Sign extension $[x]_3^6 \leftarrow [x]_3^5.$

B.5.2 Alignment

Two B2C words $[x]_q^n$ and $[y]_l^p$ are *aligned* if $q = l$, that is, if they express signals x and y over the same scale. Alignment is often necessary before arithmetic operations such as addition, subtraction, or comparisons. For instance, if

$$
\begin{aligned}
x &= \quad 01001_{\bar{2}} \times 2^2 \quad = \quad 36_{10}, \\
y &= \quad 10_{\bar{2}} \times 2^0 \quad = \quad -2_{10}
\end{aligned}
\tag{B.20}
$$

are two 5-bit and 2-bit signals represented in different scales, their aligned representation is

$$
\begin{aligned}
x &= \quad 0100100_{\bar{2}} \times 2^0 \quad = \quad 36_{10}, \\
y &= \quad 10_{\bar{2}} \times 2^0 \quad = \quad -2_{10},
\end{aligned}
\tag{B.21}
$$

where the represented values are obviously unaltered, but the significand of x is now represented on the same scale as y's and, consequently, on a larger number of bits.Therefore, alignment consists of an LSB extension of the word represented on the largest scale.

The 2-bit LSB extension of $[x]_2^5 = 01001_{\bar{2}}$ to $[x]_0^7 = 0100100_{\bar{2}}$ is indicated, in signal notation, as

$$
[x]_0^7 \leftarrow [x]_2^5.
\tag{B.22}
$$

In general, given $[x]_q^n$, one can increase the word length and correspondingly decrease the scale without altering the represented signal,

$$
[x]_{q-k}^{n+k} \leftarrow [x]_q^n, \quad (k \geq 0),
\tag{B.23}
$$

an operation that corresponds to the LSB extension mentioned earlier.

On the basis of such observation, if $[x]_q^n$ and $[y]_l^p$ are two words of different lengths and different weights, with $q > l$, alignment of $[x]_q^n$ to $[y]_l^p$ is achieved by

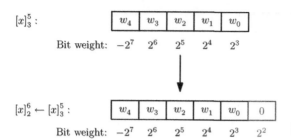

Figure B.4 One-bit LSB extension $[x]_2^6 \leftarrow [x]_3^5$.

adding and subtracting $(q - l)$ from n and q, respectively,

$$[x]_l^{n+q-l} \leftarrow [x]_q^n, \quad (q > l). \qquad (B.24)$$

Figure B.4 shows a pictorial representation of a $[x]_2^6 \leftarrow [x]_3^5$ LSB extension.

Inset B.2 – VHDL Sign Extension and Alignment

Sign extension of a word $[x]_q^n$ makes use of the VHDL concatenation operator &. Let

```
signal x     : signed(n-1 downto 0);
signal x_ext : signed(n downto 0);
```

A 1-bit sign extension of x is coded as

```
x_ext <= x(n-1)&x;
```

Similarly, for alignment, a 1-bit LSB extension of x is coded as

```
x_ext <= x&'0';
```

Inset B.3 – Verilog Sign Extension and Alignment

In a similar manner, sign extension of an n-bit wire net x to an $(n + 1)$-bit wire net x_ext, both declared as signed, is accomplished in Verilog using the concatenation operator { }:

```
wire signed [n-1:0] x;
wire signed [n:0] x_ext;
assign x_ext = {x[n-1],x};
```

For alignment, the 1-bit LSB extension of x is coded as

```
wire signed [n-1:0] x;
wire signed [n:0] x_ext;
assign x_ext = {x,1'b0};
```

Care must be taken, in general, when concatenating `signed` quantities, as *concatenate results are unsigned, regardless of the operands* [178]. The above-mentioned statements work correctly because no sign extension of $\{x[n-1],x\}$ or $\{x,1\text{'}b0\}$ is required during the assignment, but just an implicit—and irrelevant—typecasting occurs. If desired, however, the typecasting operator `$signed` can be explicitly invoked. For instance, the above-mentioned 1-bit LSB extension would become

```
assign  x_ext = $signed({x,1'b0});
```

B.5.3   Sign Reversal

In an n-bit B2C system, every number has its additive inverse except for the most negative number. Therefore, an $(n+1)$-bit word is required when changing the sign of an n-bit B2C quantity.

Operatively, the sign of an n-bit word w can be changed by first extending its representation to $n+1$ bits, then bit-wise negating all the bits, and finally adding one. For instance, if $w = 0110_{\bar{2}} = 6_{10}$, then -6_{10} is calculated as

$$w = 0110_{\bar{2}} \qquad \text{(original word)}$$

$$\rightarrow 00110_{\bar{2}} \qquad \text{(sign extension)}$$

$$\rightarrow 11001_{\bar{2}} \qquad \text{(bit-wise negation)}$$

$$\rightarrow 11010_{\bar{2}} = -6_{10} \quad \text{(add } 00001_{\bar{2}}\text{).} \tag{B.25}$$

In signal notation, sign reversal is indicated as

$$[-x]_q^{n+1} \longleftarrow -[x]_q^n . \tag{B.26}$$

$$\boxed{\textbf{Inset B.4 – VHDL Sign Reversal}}$$

Define

```
signal x     : signed(n-1 downto 0);    --  [n,q]
signal x_ext : signed(n downto 0);      --  [n+1,q]
signal z     : signed(n downto 0);      --  [n+1,q]
```

VHDL sign reversal of x is accomplished by first sign-extending x, then employing the "-" unary operator defined in package NUMERIC_STD:

```
x_ext <= x(n-1)&x;
z     <= -x_ext;
```

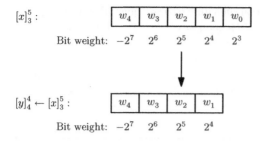

Figure B.5 One-bit LSB truncation $[y]_4^4 \leftarrow [x]_3^5$.

Inset B.5 – Verilog Sign Reversal

Verilog coding of sign reversal of an n-bit signal x is accomplished by first sign-extending x, then employing the "-" unary operator:

```
wire signed [n-1:0] x;              //  [n,q]
wire signed [n:0]   x_ext;          //  [n+1,q]
wire signed [n:0] z;                //  [n+1,q]
assign  x_ext = {x[n-1],x};
assign  z     = -x_ext;
```

where z is a $(n+1)$-bit signal.

B.5.4 LSB and MSB Truncation

Truncation—that is, removal—one or more LSBs or MSBs from an n-bit B2C word destroys, in general, the represented number. This operation is nonetheless discussed as it is frequently employed during bit manipulation. In signal notation, a 1-bit LSB truncation is denoted as

$$[y]_{q+1}^{n-1} \leftarrow [x]_q^n . \tag{B.27}$$

In general, $y = x$ if and only if the least significant bit of $[x]_q^n$ is zero, that is, if and only if x is a multiple of 2^q. Otherwise, $y = x - 2^q$.

More generally, truncation of the first k least significant bits of a word $[x]_q^n$ is denoted as

$$[y]_{q+k}^{n-k} \leftarrow [x]_q^n , \quad (0 \le k \le n-1), \tag{B.28}$$

and $y = x$ if and only if x is a multiple of 2^{q+k}. Figure B.5 illustrates a pictorial representation of a 1-bit LSB truncation $[y]_4^4 \leftarrow [x]_3^5$.

A 1-bit MSB truncation, on the other hand, is denoted as

$$[y]_q^{n-1} \leftarrow [x]_q^n , \tag{B.29}$$

and, more generally for a k-bit MSB truncation, one has

$$[y]_q^{n-k} \leftarrow [x]_q^n , \quad (0 \le k \le n-1). \tag{B.30}$$

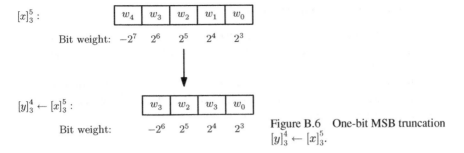

Figure B.6 One-bit MSB truncation $[y]_3^4 \leftarrow [x]_3^5$.

Figure B.6 illustrates a pictorial representation of a 1-bit MSB truncation $[y]_3^4 \leftarrow [x]_3^5$.

Inset B.6 – VHDL Truncation of a B2C Quantity

Truncation of a signal x is simply coded, in VHDL, by assigning the proper portion of x to a signal of smaller length. Focusing, for definiteness, on a 1-bit LSB truncation, define x and y as

```
signal  x : signed(n-1 downto 0);    --   [n,q]
signal  y : signed(n-2 downto 0);    --   [n-1,q+1]
```

Truncation of x is then coded as

```
y <= x(n-1 downto 1);
```

Inset B.7 – Verilog Truncation of a B2C Quantity

The Verilog construct for a 1-bit LSB truncation is by all means analogous to the VHDL one:

```
wire signed [n-1:0] x;    //   [n,q]
wire signed [n-2:0] y;    //   [n-1,q+1]
assign  y = x[n-1:1];
```

where y is a $(n-1)$-bit Verilog signal. Note that, as with the concatenation operator, *part-select results are unsigned, regardless of the operands even if the part-select specifies the entire vector* [175]. In other words, x[n-1:1] on the assignment right-hand side is of unsigned type. Nonetheless, the above-mentioned example works correctly, as no sign extension of x[n-1:1] is required during the assignment and just a conversion back to signed occurs. If desired, however, the $signed typecasting operator can be explicitly used:

```
assign  y = $signed(x[n-1:1]);
```

B.5.5 Addition and Subtraction

Representing the sum of two n-bit B2C words requires an $(n+1)$-bit word. For instance, in a 3-bit B2C system in which numbers range from -4_{10} to $+3_{10}$, possible sums of its elements range between -8_{10} and $+6_{10}$.

The simplest way to handle the need for an extended range is a preliminary sign extension of the addends. Addition between two B2C words is then accomplished with the same rules of plain base-2 addition. For instance, let

$$w = [x]_q^3 = 011_{\bar{2}} = 3_{10},$$
$$r = [y]_q^3 = 111_{\bar{2}} = -1_{10} \tag{B.31}$$

be the two 3-bit addends. Their sum is then accomplished over 4 bits as

$$
\begin{array}{llll}
0011_{\bar{2}} & + & 3_{10} & + \\
1111_{\bar{2}} & = \quad \text{i.e.,} & -1_{10} & = \\
0010_{\bar{2}} & & 2_{10}.
\end{array}
\tag{B.32}
$$

In a similar manner, subtraction of two n-bit B2C quantities can be exactly represented over $n+1$ bits. Difference between words w and r defined earlier can be accomplished as the B2C sum of w with $-r$. The latter is obtained, according to the discussion in the previous section, with a preliminary sign extension of r, followed by its bit-wise negation and a unit increment,

$$-r = 0000_{\bar{2}} + 0001_{\bar{2}} = 0001_{\bar{2}}. \tag{B.33}$$

Hence, $w - r$ becomes

$$
\begin{array}{llll}
0011_{\bar{2}} & + & 3_{10} & + \\
0000_{\bar{2}} & + & 0_{10} & + \\
0001_{\bar{2}} & = \quad \text{i.e.,} & 1_{10} & = \\
0100_{\bar{2}} & & 4_{10}.
\end{array}
\tag{B.34}
$$

In signal notation, addition or subtraction between two n-bit words and storage into a $(n+1)$-bit word is denoted with

$$[x \pm y]_q^{n+1} \leftarrow [x]_q^n \pm [y]_q^n. \tag{B.35}$$

Observe that the operation only makes sense *as long as the two addends are aligned*. If not, a preliminary alignment is required.

When the two addends are aligned but have different lengths n and p, their sum or difference can always be correctly represented by a $(\max(n, p) + 1)$-bit word,

$$[x \pm y]_q^{\max(n,p)+1} \leftarrow [x]_q^n \pm [y]_q^p. \tag{B.36}$$

Inset B.8 – VHDL Addition of B2C Quantities

Addition of two *aligned* words $[x]_q^n$ and $[y]_q^p$ can be VHDL-coded as follows. Consider the signal declarations

```
signal x : signed(n-1 downto 0);        -- [n,q]
signal y : signed(p-1 downto 0);        -- [p,q]
signal z : signed(max(n,p) downto 0);   -- [max(n,p)+1,q]
```

In the above-mentioned code, it is assumed that a function max is defined that returns the largest of its two arguments. Addition between x and y can be accomplished by first sign extending both signals, then using the + operator defined on signed data types to store the result into z:

```
z <= (x(n-1)&x)+(y(p-1)&y);
```

Note that operator +, defined in the NUMERIC_STD package, evaluates the result to a vector whose length is the largest between the lengths of the operands [172]. Therefore, both sides of the foregoing concurrent statement have the same length.

Subtraction of two B2C quantities is accomplished with the "-" operator, which automatically implements a signed difference as discussed earlier. All considerations and constructs examined earlier apply with no other modifications.

Inset B.9 – Verilog Addition of B2C Quantities

Addition of two aligned words $[x]_q^n$ and $[y]_q^p$ can be Verilog-coded as follows. Consider the signal declarations

```
wire    signed  [n-1:0] x;      //  [n,q]
wire    signed  [p-1:0] y;      //  [p,q]
wire    signed  [max(n,p):0] z; //  [max(n+p)+1,q]
```

As in VHDL, addition between x and y can be accomplished by first sign-extending both signals, then using the + operator defined on signed data types to store the result into z:

```
wire signed [n:0] x_ext = {x[n-1],x};
wire signed [p:0] y_ext = {y[p-1],y};
assign z              = x_ext + y_ext;
```

Subtraction is accomplished in a similar manner with the use of Verilog operator "-".

B.5.6 Multiplication

The product of an n-bit B2C word with a p-bit B2C word can be exactly represented by a $(n + p)$-bit B2C word. In particular, multiplication of two n-bit words requires a $(2n)$-bit word to store the product. For instance, in a 3-bit B2C system, possible products between its elements range between -12_{10} and $+16_{10}$.

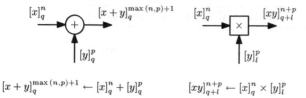

$$[x + y]_q^{\max(n,p)+1} \leftarrow [x]_q^n + [y]_q^p \qquad\qquad [xy]_{q+l}^{n+p} \leftarrow [x]_q^n \times [y]_l^p$$

Figure B.7   Block diagram symbols for addition and multiplication.

In signal notation, multiplication between two words and storage into an output word is denoted as

$$[xy]_{q+l}^{n+p} \leftarrow [x]_q^n \times [y]_l^p . \tag{B.37}$$

Block diagram symbols for addition and multiplication are depicted in Fig. B.7. It should be noted that *multiplication changes the scale of the represented quantity*.

Inset B.10 – VHDL Multiplication of B2C Quantities

VHDL multiplication of two B2C quantities is simply accomplished via the * operator defined for signed data types. Let

```
signal x : signed(n-1 downto 0);       -- [n,q]
signal y : signed(p-1 downto 0);       -- [p,l]
signal z : signed(n+p-1 downto 0);     -- [n+p,q+l]
```

be signals representing B2C words of lengths n, p, and $n + p$, respectively. B2C multiplication between x and y is simply coded as

```
z <= x*y;
```

Inset B.11 – Verilog Multiplication of B2C Quantities

Verilog multiplication of two B2C quantities is simply accomplished via the * operator defined for signed data types. Let

```
wire    signed  [n-1:0] x;       //  [n,q]
wire    signed  [p-1:0] y;       //  [p,l]
wire    signed  [n+p-1:0] z;     //  [n+p,q+l]
```

be signals representing B2C words of lengths n, p, and $n + p$, respectively. B2C multiplication between x and y is coded as

```
assign z = x*y;
```

B.5.7 Overflow Detection and Saturated Arithmetic

If result (B.34) is to be stored in a 3-bit word, an *overflow* would occur, as 4_{10} is not representable in a 3-bit B2C system. Simply dropping the most significant bit from the $(3 + 1)$-bit result would yield $100_{\bar{2}} = -4_{10}$.

Given an $(n + 1)$-bit B2C word $w = [x]_q^{n+1}$, it is therefore of interest to determine whether x could be represented in n bits, a check often referred to as *overflow detection*. Overflow detection can be accomplished in a variety of ways. If

$$w = [x]_q^{n+1} = (w_n \ldots w_0)_{\bar{2}}, \tag{B.38}$$

is a generic $(n + 1)$-bit B2C word, the range overflow occurs if and only if the two most significant bits of w differ,

$$OV = w_n \text{ XOR } w_{n-1}. \tag{B.39}$$

Whenever $OV = 0$, truncation of the MSB does not alter the represented quantity. It can be verified that the above-mentioned criterion would correctly detect an overflow condition in the case of sum (B.34).

The above-mentioned result can be generalized to an $(n + l)$-bit word

$$w = [x]_q^{n+l} = (w_{n+l-1} \ldots w_0)_{\bar{2}}. \tag{B.40}$$

Word w can be exactly stored in an n-bit word if and only the $l + 1$ most significant bits of w are equal,

$$OV = \text{NOT} \left(w_{n+l-1} = w_{n+l-2} = \ldots = w_{n-1} \right). \tag{B.41}$$

The action to be undertaken when an overflow occurs depends on the system in which the B2C arithmetic is implemented. A frequent provision *saturates* an overflowed result to the most positive or the most negative representable number. In such *saturated arithmetic*, the result of (B.34) would be $011_{\bar{2}} = 3_{10}$, the most positive representable number in a 3-bit B2C system.

In general, a saturated assignment involving truncation of l MSBs is indicated as

$$[y]_q^n \Leftarrow [x]_q^{n+l}, \tag{B.42}$$

The above-mentioned assignment is to be interpreted as follows: whenever the result of the operation on the right-hand side of the assignment overflows, the left-hand side is set to the most positive or the most negative values representable on n bits, depending on the overflow direction. If no overflow occurs, a simple MSB truncation of the result is accomplished. For instance, a saturated sign reversal is denoted as

$$[z]_q^n \Leftarrow -[x]_q^n, \tag{B.43}$$

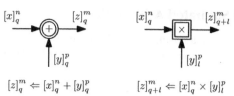

$$[z]_q^m \Leftarrow [x]_q^n + [y]_q^p \qquad [z]_{q+l}^m \Leftarrow [x]_q^n \times [y]_l^p$$

$$m \le \max(n,p) \qquad\qquad m < n + p$$

Figure B.8   Block diagram symbols for saturated addition and saturated multiplication.

whereas signal notation for saturated addition/subtraction and multiplication becomes

$$[z]_q^m \Leftarrow [x]_q^n \pm [y]_q^p \quad (m \le \max(n,p)) \tag{B.44}$$

and

$$[z]_{q+l}^m \Leftarrow [x]_q^n \times [y]_l^p \quad (m < n + p). \tag{B.45}$$

Block diagram symbols used for saturated addition and multiplication are shown in Fig. B.8.

Inset B.12 – VHDL Saturated Addition and Multiplication

A combinational saturated adder can be VHDL-coded as follows. Consider the entity declaration first:

```
entity saturated_adder is
    generic (
        n, p, m : integer              -- m <= max(n,p)+1
        );

    port (
        x      : in signed(n-1 downto 0);   --  [n,q]
        y      : in signed(p-1 downto 0);   --  [p,q]
        z      : out signed(m-1 downto 0);  --  [m,q]
        OV     : out std_logic;
        op     : in std_logic
        );
end saturated_adder;
```

Entity `saturated_adder` operates on two inputs `x` and `y`, n-bit and p-bit long, respectively, and outputs their saturated sum as a m-bit word `z`, with $m \le \max(n,p) + 1$. Input flag `op` specifies whether the sum is actually an addition (if `op='1'`), or a subtraction (if `op='0'`). Output flag `OV` is asserted in the presence of a range overflow with respect to the target word length m.

A possible implementation of the saturated adder is as follows:

```
1   architecture saturated_adder_arch of saturated_adder is
2
3       function MAX(LEFT, RIGHT: INTEGER) return INTEGER is
4       begin
```

```
5          if LEFT > RIGHT then return LEFT;
6          else return RIGHT;
7              end if;
8          end;
9
10    signal zx          :    signed(max(n,p) downto 0);
11    signal OVi         :    std_logic;
12    constant wmax      :    signed(m-2 downto 0)    :=    (others=>'1');
13    constant wmin      :    signed(m-2 downto 0)    :=    (others=>'0');
14
15  begin
16
17    zx      <=    (x(n-1)&x) + (y(p-1)&y) when op='1' else
18                  (x(n-1)&x) - (y(p-1)&y);
19
20    overflow_detect :    process(zx)
21          variable temp  :   std_logic;
22          begin
23              temp      :=   '0';
24              for I in m to max(n,p) loop
25                  if ((zx(I) xor zx(m-1))='1') then
26                      temp      :=   '1';
27                  end if;
28              end loop;
29              OVi <=  temp;
30          end process;
31
32    z     <=    ('0'&wmax)   when     OVi='1' AND zx(max(n,p))='0'   else
33                ('1'&wmin)   when     OVi='1' AND zx(max(n,p))='1'   else
34                zx(m-1 downto 0);
35
36    OV    <=OVi;
37
38  end saturated_adder_arch;
```

In the preceding example, an overflow check is accomplished by process `overflow_detect` defined in line 20.
Entity declaration for a combinational saturated multiplier is

```
entity saturated_multiplier is
    generic (
        n, p, m :    integer      -- m<=n+p
        );

    port (
        x       :    in signed(n-1 downto 0);       -- [n,q]
        y       :    in signed(p-1 downto 0);       -- [p,l]
        z       :    out signed(m-1 downto 0);      -- [m,q+l]
        OV      :    out std_logic
        );
end saturated_multiplier;
```

In this case, the word length of the saturated product is $m \leq n + p$. An implementation example of the foregoing entity is

```
1   architecture saturated_multiplier_arch of saturated_multiplier is
2
3       signal zx        :   signed(n+p-1 downto 0);
4       signal OVi       :   std_logic;
5       constant wmax    :   signed(m-2 downto 0)   :=  (others=>'1');
6       constant wmin    :   signed(m-2 downto 0)   :=  (others=>'0');
7
8   begin
9
10      zx      <=  x*y;
11
12      overflow_detect :   process(zx)
13          variable temp   :   std_logic;
14          begin
15              temp    :=  '0';
16              for I in m to n+p-1 loop
17                  if ((zx(I) xor zx(m-1))='1') then
18                      temp    :=  '1';
19                  end if;
20              end loop;
21              OVi <=  temp;
22          end process;
23
24      z    <=  ('0'&wmax)   when   OVi='1' AND zx(n+p-1)='0'   else
25               ('1'&wmin)   when   OVi='1' AND zx(n+p-1)='1'   else
26               zx(m-1 downto 0);
27
28      OV  <=OVi;
29
30  end saturated_multiplier_arch;
```

Inset B.13 – Verilog Saturated Addition and Multiplication

Following the previous VHDL example, a combinational saturated adder can be Verilog-coded as follows:

```
1   module saturated_adder(x,y,z,OV,op);
2
3       function integer max;
4           input integer left, right;
5           if (left>right)
6               max = left;
7           else
8               max = right;
9       endfunction
10
11      parameter n;
12      parameter p;
```

```
13      parameter m;                        // Assuming m <= max(n,p)+1
14      parameter mx = max(n,p)+1;
15
16      input   signed [n-1:0]   x;
17      input signed [p-1:0]     y;
18      output reg signed    [m-1:0]     z;
19      input op;
20
21      output reg OV;
22
23      wire signed [n:0]    xx = {x[n-1],x};
24      wire signed [p:0]    yx = {y[p-1],y};
25      wire signed [mx-1:0]    zx;
26
27      assign zx = (op==1'b1) ? xx+yx : xx-yx;
28
29      reg temp;
30      integer I;
31      always @(zx)
32          begin
33              temp = 1'b0;
34              for (I=m;I<=mx-1;I=I+1)
35                  begin
36                      if ((zx[I]^zx[m-1])==1'b1)
37                          temp = 1'b1;
38                  end
39              OV = temp;
40          end
41
42      always @(OV,zx)
43          case (OV)
44              1'b0:    z = zx[m-1:0];
45              1'b1:
46                  begin
47                      if (zx[mx-1]==1'b0)
48                          z = {1'b0,{(m-1){1'b1}}};
49                      else
50                          z = {1'b1,{(m-1){1'b0}}};
51                  end
52          endcase
53
54  endmodule
```

Overflow check is implemented by the `for` loop contained in the `always` statement in line 42.

Similarly, Verilog code for a combinational saturated multiplier is

```
1   module saturated_multiplier(x,y,z,OV);
2
3       parameter n;
4       parameter p;
```

```
5        parameter m;                    // Assuming m <= n+p
6
7        input   signed [n-1:0]  x;      //  [n,q]
8        input   signed [p-1:0]  y;      //  [p,1]
9        output  reg [m-1:0] z;          //  [m,q+1]
10
11       output reg OV;
12
13       wire signed [n+p-1:0]   zx;
14       assign zx = x*y;
15
16       reg temp;
17       integer I;
18       always @(zx)
19           begin
20               temp = 1'b0;
21               for (I=m;I<=n+p-1;I=I+1)
22                   begin
23                       if ((zx[I]^zx[m-1])==1'b1)
24                           temp = 1'b1;
25                   end
26               OV = temp;
27           end
28
29       always @(OV,zx)
30           case (OV)
31               1'b0:   z = zx[m-1:0];
32               1'b1:
33                   begin
34                       if (zx[n+p-1]==1'b0)
35                           z = {1'b0,{(m-1){1'b1}}};
36                       else
37                           z = {1'b1,{(m-1){1'b0}}};
38                   end
39           endcase
40
41   endmodule
```

SMALL-SIGNAL PHASE LAG OF UNIFORMLY SAMPLED PULSE WIDTH MODULATORS

The small-signal delay associated with uniformly sampled pulse width modulators (USPWMs), which are typically employed in digital controllers, is introduced in Section 2.5.2. This modulation delay is an important contribution to the total loop delay in digitally controlled converters. This appendix presents a proof for the results given in (2.27) and summarized in Table 2.1.

C.1 TRAILING-EDGE MODULATORS

Referring to a trailing-edge uniformly sampled modulator (TE-USPWM), the proof closely follows the derivation originally presented in [126]. A more general calculation of the PWM spectrum for both naturally sampled and uniformly sampled modulators can be found in [186].

Figure C.1 reports the main waveforms of a TE-USPWM. The input modulating signal $u[k]$ has sampling period T_s equal to the switching period and is assumed to be updated at the beginning of every switching cycle.

In steady-state operation with a constant input modulating signal U, the modulator output $c_s(t)$ is a square wave having duty cycle

$$D = \frac{U}{N_r},$$ (C.1)

where N_r denotes the carrier amplitude.

Consider now a sinusoidal perturbation $\hat{u}[k]$ superimposed to U,

$$u[k] = U + \hat{u}[k] = U + \hat{u}_m \sin\left(\omega k T_s + \varphi\right),$$ (C.2)

which produces a corresponding perturbation $\hat{c}(t)$ in the output PWM command,

$$c(t) = c_s(t) + \hat{c}(t).$$ (C.3)

Digital Control of High-Frequency Switched-Mode Power Converters, First Edition.
Luca Corradini, Dragan Maksimović, Paolo Mattavelli, and Regan Zane.

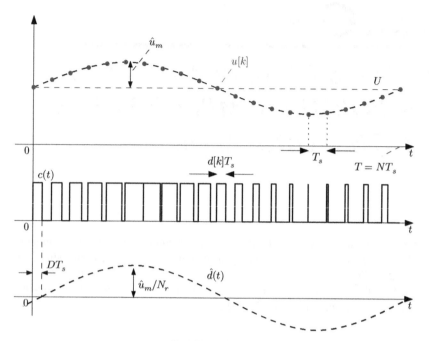

Figure C.1 Waveforms of a TE-USPWM for $N = 20$, $L = 1$.

Suppose also that the perturbation frequency ω and the switching rate ω_s are commensurable,

$$\frac{\omega_s}{\omega} = \frac{f_s}{f} = \frac{T}{T_s} = \frac{N}{L}, \qquad L, N \in \mathbb{Z}^+. \tag{C.4}$$

This assumption is equivalent to assuming that $\hat{u}[k]$ is periodic with period $T_p = NT_s = LT$. Although not strictly necessary, such assumption allows the derivation to proceed using Fourier series and summations rather than transforms and integrals. Furthermore, given the density of the rational set $\mathbb{Q}$ within the real set $\mathbb{R}$, any ratio f_s/f can be approximated arbitrarily well by a fraction. In Fig. C.1, the case $N = 20$, $L = 1$ is exemplified.

The small-signal frequency response $G_{PWM,TE}(j\omega)$ to be determined is defined as the ratio between the Fourier components $c(\omega)$ and $u(\omega)$ of c and u at the perturbation frequency ω, in the small-signal limit [126, 187],

$$\boxed{G_{PWM,TE}(j\omega) \triangleq \lim_{\hat{u}_m \to 0} \frac{c(\omega)}{u(\omega)}.} \tag{C.5}$$

The time-domain counterpart of $c(\omega)$ is illustrated in Fig. C.1 and is denoted with $\hat{d}(t)$.

The quantity $u(\omega)$ can be derived directly from the definition (C.2),

$$u(\omega) = \frac{\hat{u}_m}{2j}e^{j\varphi}. \tag{C.6}$$

Evaluation of $c(\omega)$, on the other hand, is carried out via a Fourier analysis of $c(t)$. During the kth switching interval, the modulated signal $c(t)$ is defined as

$$c(t) = \begin{cases} 1, & kT_s < t < kT_s + d[k]T_s, \\ 0, & kT_s + d[k]T_s < t < (k+1)T_s, \end{cases} \tag{C.7}$$

where $d[k] = u[k]/N_r$ is the duty cycle. Observe that $c(t)$ has the same periodicity $T_p = NT_s = LT$ as $u[k]$ and can therefore be expanded as

$$c(t) = \sum_{n=-\infty}^{+\infty} c(n\omega_p)e^{jn\omega_p t}, \tag{C.8}$$

where $\omega_p = 2\pi/T_p$ and where $c(n\omega_p)$ is the Fourier coefficient of c,

$$c(n\omega_p) = \frac{1}{T_p}\int_0^{T_p} c(\tau)e^{-jn\omega_p\tau}d\tau. \tag{C.9}$$

As $f = L/T_p = Lf_p$, the harmonic component of $c(t)$ to be evaluated is that of order L,[1]

$$c(\omega) = c(L\omega_p) = \frac{1}{T_p}\int_0^{T_p} c(\tau)e^{-jL\omega_p\tau}d\tau = \frac{1}{NT_s}\int_0^{NT_s} c(\tau)e^{-j\omega\tau}d\tau. \tag{C.10}$$

The above-mentioned integral can be expressed in terms of the contributions from each switching cycle,

$$c(\omega) = \frac{1}{NT_s}\sum_{k=0}^{k=N-1}\int_{kT_s}^{(k+1)T_s} c(\tau)e^{-j\omega\tau}d\tau. \tag{C.11}$$

By considering (C.7),

$$c(\omega) = \frac{1}{NT_s}\sum_{k=0}^{k=N-1}\int_{kT_s}^{(k+d[k])T_s} e^{-j\omega\tau}d\tau. \tag{C.12}$$

---

[1] It can be shown that, in the limit as $\hat{u}_m \to 0$ considered here, the Lth harmonic is the only nonvanishing Fourier component in the range $[0, \omega_s/2]$.

Evaluating the integrals explicitly yields

$$c(\omega) = \frac{1}{NT_s} \sum_{k=0}^{k=N-1} \left[\frac{e^{-j\omega\tau}}{-j\omega} \right]_{\tau=kT_s}^{\tau=(k+d[k])T_s}$$

$$= \frac{1}{j\omega NT_s} \sum_{k=0}^{k=N-1} e^{-j\omega kT_s} \left(1 - e^{-j\omega d[k]T_s} \right), \qquad (C.13)$$

which is an exact large-signal result.

Introduce, at this point, the small-signal assumption by substituting the exponential term containing $d[k]$ with its first-order Taylor approximation around $d = D$,

$$e^{-j\omega dT_s} \approx e^{-j\omega DT_s} + \left. \frac{\partial e^{-j\omega dT_s}}{\partial d} \right|_{d=D} (d - D)$$

$$= e^{-j\omega DT_s} - j\omega T_s e^{-j\omega DT_s} \hat{d}, \qquad (C.14)$$

with $\hat{d} \triangleq d - D$. Substituting such approximation into (C.13) yields

$$c(\omega) = \frac{1}{j\omega NT_s} \sum_{k=0}^{k=N-1} e^{-j\omega kT_s} \left(1 - e^{-j\omega DT_s} \left(1 - j\omega T_s \hat{d}[k] \right) \right). \qquad (C.15)$$

Write now the above-mentioned result as the sum of two terms,

$$c(\omega) = \frac{1}{j\omega NT_s} \left(1 - e^{-j\omega DT_s} \right) \sum_{k=0}^{k=N-1} e^{-j\omega kT_s}$$

$$+ \frac{1}{j\omega NT_s} j\omega T_s e^{-j\omega DT_s} \sum_{k=0}^{k=N-1} e^{-j\omega kT_s} \hat{d}[k], \qquad (C.16)$$

and examine them separately. The first term vanishes, as

$$\sum_{k=0}^{k=N-1} e^{-j\omega kT_s} = 0 \qquad (C.17)$$

for any ω that is not a multiple of ω_s. The expression for $c(\omega)$ therefore simplifies to

$$c(\omega) = \frac{e^{-j\omega DT_s}}{N} \sum_{k=0}^{k=N-1} e^{-j\omega kT_s} \hat{d}[k]. \qquad (C.18)$$

From (C.2) and $d[k] = u[k]/N_r$, it follows that

$$\hat{d}[k] = \frac{\hat{u}_m}{N_r} \sin(k\omega T_s + \varphi) = \frac{\hat{u}_m}{N_r} \left(\frac{e^{j(k\omega T_s + \varphi)} - e^{-j(k\omega T_s + \varphi)}}{2j} \right), \qquad (C.19)$$

which leads to

$$u(\omega) = \frac{e^{-j\omega DT_s}}{N} \frac{\hat{u}_m}{2jN_r} e^{j\varphi} \sum_{k=0}^{k=N-1} \left(1 - e^{-2j(\omega kT_s + \varphi)}\right). \tag{C.20}$$

As the summation

$$\sum_{k=0}^{k=N-1} e^{-2j\omega kT_s} \tag{C.21}$$

vanishes for any ω that is not a multiple of $\omega_s/2$, the expression of $c(\omega)$ reduces to

$$c(\omega) = \frac{e^{-j\omega DT_s}}{N_r} \frac{\hat{u}_m}{2j} e^{j\varphi}. \tag{C.22}$$

Finally, the desired result is obtained from (C.5), (C.6), and (C.22),

$$\boxed{G_{PWM,TE}(j\omega) = \frac{1}{N_r} e^{-j\omega DT_s}}, \tag{C.23}$$

which is valid for $0 < \omega < \omega_s/2$. As anticipated, such frequency response describes a transport delay $t_{DPWM} = DT_s$. The result is the TE-USPWM entry in Table 2.1.

As a final remark, and according to (1.87) of Section 1.6, the duty cycle $d(t)$, which determines the averaged dynamics of the converter, is just the baseband component of $c(\omega)$,

$$d(t) = \underbrace{\frac{U}{N_r}}_{D} + \underbrace{\frac{\hat{u}_m}{N_r} \sin(\omega\,(t - DT_s) + \varphi)}_{\hat{d}(t)}. \tag{C.24}$$

C.2   LEADING-EDGE MODULATORS

The above-mentioned calculation, developed for a trailing-edge modulator example, can be readily adapted to the leading-edge case by replacing (C.7) with

$$c(t) \;=\; \begin{cases} 0, & kT_s < t < kT_s + (1 - d[k])\,T_s, \\ 1, & kT_s + (1 - d[k])\,T_s < t < (k+1)T_s, \end{cases} \tag{C.25}$$

which expresses the relationship between $c(t)$ and $d[k]$ in a leading-edge modulator. With this modification, the modulator small-signal frequency response is found to be

$$\boxed{G_{PWM,LE}(j\omega) = \frac{1}{N_r} e^{-j\omega(1-D)T_s}}, \tag{C.26}$$

which is associated with a small-signal transport delay $t_{DPWM} = (1 - D)\,T_s$. The result is reported in the second entry of Table 2.1.

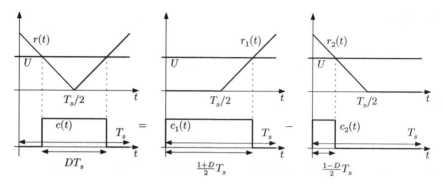

Figure C.2   Decomposition of a symmetrical modulation into trailing-edge modulations.

C.3   SYMMETRICAL MODULATORS

For a symmetrical modulator, dependence of $c(t)$ on $d[k]$ is expressed as

$$
c(t) = \begin{cases}
0, & kT_s < t < kT_s + (1 - d[k])\,\frac{T_s}{2}, \\
1, & kT_s + (1 - d[k])\,\frac{T_s}{2} < t < kT_s + (1 + d[k])\,\frac{T_s}{2}, \\
0, & kT_s + (1 + d[k])\,\frac{T_s}{2} < t < (k + 1)T_s.
\end{cases} \tag{C.27}
$$

Once (C.7) is replaced with (C.27), the small-signal frequency response of the symmetrical modulator can be derived.

A quicker approach to the derivation of the symmetrical modulator frequency response, which is applicable once the small-signal result (C.23) is known, is to think of the symmetrical modulation as a combination of two trailing-edge submodulations, as depicted in Fig. C.2. The symmetrical carrier $r(t)$ is decomposed into two trailing-edge carriers $r_1(t)$ and $r_2(t)$, against which the modulating signal $u[k]$ is compared to produce modulated signals $c_1(t)$ and $c_2(t)$. Symmetrically modulated signal $c(t)$ is then evaluated as

$$
c(t) = c_1(t) - c_2(t). \tag{C.28}
$$

Because of the linearity of the foregoing equation, one necessarily has that

$$
G_{PWM,Sym}(j\omega) = G_{PWM,1}(j\omega) - G_{PWM,2}(j\omega), \tag{C.29}
$$

where $G_{PWM,1}(j\omega)$ and $G_{PWM,2}(j\omega)$ frequency responses associated with $c_1(t)$ and $c_2(t)$, respectively. These can be both derived from (C.23) after noting that

1. Slopes of $r_1(t)$ and $r_2(t)$ are, in absolute value, equal to $2N_r/T_s$, whereas in the standard trailing-edge modulation, the carrier slope is N_r/T_s. Hence, the static small-signal gain associated with the two submodulations is halved with respect to the standard trailing-edge case.

2. The static differential gain associated with $G_{PWM,2}(j\omega)$ is *negative*, as duty cycle of $c_2(t)$ decreases as the modulating signal increases.

With the above-mentioned remarks in mind, (C.29) becomes

$$
G_{PWM,Sym}(j\omega) = \frac{1}{2N_r}e^{-j\omega\frac{1+D}{2}T_s} - \left(-\frac{1}{2N_r}\right)e^{-j\omega\frac{1-D}{2}T_s}
$$

$$
= \frac{1}{2N_r}\left(e^{j\omega D\frac{T_s}{2}} + e^{j\omega D\frac{T_s}{2}}\right)e^{-j\omega\frac{T_s}{2}}, \tag{C.30}
$$

and therefore

$$
\boxed{G_{PWM,Sym}(j\omega) = \frac{1}{N_r}\cos\left(\frac{\omega D T_s}{2}\right)e^{-j\omega\frac{T_s}{2}}}, \tag{C.31}
$$

which is the result reported in the last entry of Table 2.1. The small-signal transport delay associated with (C.31) is $t_{DPWM} = T_s/2$, independent of the operating point.

REFERENCES

[1] R. W. Erickson and D. Maksimović, *Fundamentals of Power Electronics*, 2nd ed. Springer, 2001.

[2] J. Kassakian, M. Schlecht and G. Verghese, *Principles of Power Electronics*, 1st ed. Addison-Wesley, 1991.

[3] P. Krein, *Elements of Power Electronics*, 1st ed. Oxford University Press, 1997.

[4] N. Mohan, T. Undeland and W. Robbins, *Power Electronics: Converters, Applications, and Design*, 3rd ed. John Wiley & Sons, Inc., 2002.

[5] M. H. Rashid, *Power Electronics: Circuits, Devices and Applications*, 3rd ed. Prentice-Hall, 2004.

[6] G. F. Franklin, J. D. Powell and A. Emami-Naeni, *Feedback Control of Dynamic Systems*, 6th ed. Prentice-Hall, 2010.

[7] G. F. Franklin, J. D. Powell and M. L. Workman, *Digital Control of Dynamic Systems*, 3rd ed. Prentice-Hall, 1998.

[8] A. V. Oppenheim, A. V. Schafer and J. R. Buck, *Discrete-Time Signal Processing*, 2nd ed. Prentice-Hall, 1999.

[9] PMBus. PMBusTM: Power Management Defined [Online]. Available: http://pmbus.org/ (accessed 17 October 2014).

[10] R. V. White and D. Durant, "Understanding and using PMBus data formats," in *Proceedings of the 21st IEEE Applied Power Electronics Conference and Exposition (APEC)*, Mar. 2006.

[11] D. Freeman and B. McDonald, "Parameterization for power solutions," in *Proceedings of the 12th IEEE Workshop on Control and Modeling for Power Electronics (COMPEL)*, Jun. 2010, pp. 1–6.

[12] D. Maksimović, R. Zane and R. W. Erickson, "Impact of digital control in power electronics," in *Proceedings of the 16th IEEE International Symposium on Power Semiconductor Devices*, May 2004, pp. 13–22.

[13] D. Maksimović, R. Zane and L. Corradini, "Advances in digital control for high-frequency switched-mode power converters," *Power Electron. Mon.*, vol. 44, no. 12, pp. 2–19, Dec. 2010, serial no. 217, sponsored by Xi'an Power Electronics Research Institute, China.

[14] S. Buso and P. Mattavelli, *Digital Control in Power Electronics*, 1st ed. Morgan & Claypool, 2006.

[15] J. Morroni, A. Dolgov, M. Shirazi, R. Zane and D. Maksimović, "Online health monitoring in digitally controlled power converters," in *Proceedings of IEEE Power Electronics Specialists Conference (PESC)*, Jun. 2007, pp. 112–118.

Digital Control of High-Frequency Switched-Mode Power Converters, First Edition.
Luca Corradini, Dragan Maksimović, Paolo Mattavelli, and Regan Zane.
© 2015 The Institute of Electrical and Electronics Engineers, Inc. Published 2015 by John Wiley & Sons, Inc.

321

[16] B. Mather, D. Maksimović and I. Cohen, "Input power measurement techniques for single-phase digitally controlled PFC rectifiers," in *Proceedings of the 24th IEEE Applied Power Electronics Conference and Exposition (APEC)*, 2009, pp. 767–773.

[17] S. Buso, P. Mattavelli, L. Rossetto and G. Spiazzi, "Simple digital control improving dynamic performance of power factor preregulators," *IEEE Trans. Power Electron.*, vol. 13, no. 5, pp. 814–823, Sept. 1998.

[18] U. Meyer-Baese, *Digital Signal Processing with Field Programmable Gate Arrays*, 3rd ed. Springer, 2007.

[19] S. Brown and Z. Vranesic, *Fundamentals of Digital Logic with Verilog Design*, 1st ed. McGraw-Hill, 2003.

[20] A. M. Wu, X. Jinwen, D. Marković and S. R. Sanders, "Digital PWM control: application in voltage regulation modules," in *Proceedings of the 30th IEEE Power Electronics Specialists Conference (PESC)*, vol. 1, Jul. 1999, pp. 77–83.

[21] B. J. Patella, "Implementation of a high frequency, low-power digital pulse width modulation controller chip," Master's thesis, University of Colorado at Boulder, Dec. 2000.

[22] B. J. Patella, A. Prodić, A. Zirger and D. Maksimović, "High-frequency digital PWM controller IC for DC-DC converters," *IEEE Trans. Power Electron.*, vol. 18, no. 1, pp. 438–446, Mar. 2003.

[23] A. Prodić and D. Maksimović, "Design of a digital PID regulator based on look-up tables for control of high-frequency DC-DC converters," in *Proceedings of the 8th IEEE Workshop on Computers in Power Electronics (COMPEL)*, Jun. 2002, pp. 18–22.

[24] A. Syed, E. Ahmed and D. Maksimović, "Digital PWM controller with feed-forward compensation," in *Proceedings of the 19th IEEE Applied Power Electronics Conference and Exposition (APEC)*, vol. 1, 2004, pp. 60–66.

[25] A. Prodić, D. Maksimović and R. W. Erickson, "Digital controller chip set for isolated DC power supplies," in *Proceedings of the 18th IEEE Applied Power Electronics Conference and Exposition (APEC)*, vol. 2, Feb. 2003, pp. 866–872.

[26] A. V. Peterchev, J. Xiao and S. R. Sanders, "Architecture and IC implementation of a digital VRM controller," *IEEE Trans. Power Electron.*, vol. 18, no. 1, pp. 356–364, Jan. 2003.

[27] J. Xiao, A. V. Peterchev, J. Zhang and S. Sanders, "A $4\mu A$ quiescent current dual-mode digitally controlled buck converter IC for cellular phone applications," *IEEE J. Solid-State Circuits*, vol. 39, no. 12, pp. 2342–2348, Dec. 2004.

[28] A. V. Peterchev, "Digital pulse-width modulation control in power electronic circuits: theory and applications," Ph.D. dissertation, University of California at Berkeley, 2005.

[29] K. Wang, N. Rahman, Z. Lukić and A. Prodić, "All-digital DPWM/DPFM controller for low-power DC-DC converters," in *Proceedings of the 21st IEEE Applied Power Electronics Conference and Exposition (APEC)*, Mar. 2006.

[30] J. Zhang and S. R. Sanders, "A digital multi-mode multi-phase IC controller for voltage regulator application," in *Proceedings of the 22nd IEEE Applied Power Electronics Conference and Exposition (APEC)*, Mar. 2007, pp. 719–726.

[31] Z. Lukić, N. Rahman and A. Prodić, "Multibit Σ-Δ PWM digital controller IC for DC-DC converters operating at switching frequencies beyond 10 MHz," *IEEE Trans. Power Electron.*, vol. 22, no. 5, pp. 1693–1707, Sept. 2007.

[32] A. Parayandeh and A. Prodić, "Programmable analog-to-digital converter for low-power DC-DC smps," *IEEE Trans. Power Electron.*, vol. 23, no. 1, pp. 500–505, Jan. 2008.

[33] T. Takayama and D. Maksimović, "Digitally controlled 10 MHz monolithic buck converter," in *Proceedings 10th IEEE Workshop on Computers in Power Electronics (COMPEL)*, Jul. 2006, pp. 154–158.

[34] L. Corradini, A. Costabeber, P. Mattavelli and S. Saggini, "Parameter-independent time-optimal digital control for point-of-load converters," *IEEE Trans. Power Electron.*, vol. 24, no. 10, pp. 2235–2248, Oct. 2009.

[35] L. Corradini, E. Orietti, P. Mattavelli and S. Saggini, "Digital hysteretic voltage-mode control for DC-DC converters based on asynchronous sampling," *IEEE Trans. Power Electron.*, vol. 24, no. 1, pp. 201–211, Jan. 2009.

[36] D. Maksimović and R. Zane, "Small-signal discrete-time modeling of digitally controlled PWM converters," *IEEE Trans. Power Electron.*, vol. 22, no. 6, pp. 2552–2556, Nov. 2007.

[37] A. V. Peterchev and S. R. Sanders, "Quantization resolution and limit cycling in digitally controlled PWM converters," *IEEE Trans. Power Electron.*, vol. 18, no. 1, pp. 301–308, Jan. 2003.

[38] H. Peng, A. Prodić, E. Alarcon and D. Maksimović, "Modeling of quantization effects in digitally controlled DC-DC converters," *IEEE Trans. Power Electron.*, vol. 22, no. 1, pp. 208–215, Jan. 2007.

[39] W. W. Burns and T. G. Wilson, "A state-trajectory control law for DC-to-DC converters," *IEEE Trans. Aerosp.*, vol. AES-14, no. 1, pp. 2–20, Jan. 1978.

[40] D. Biel, L. Martinez, J. Tenor, B. Jammes and J.-C. Marpinard, "Optimum dynamic performance of a buck converter," in *Proceedings of IEEE International Symposium on Circuits and Systems (ISCAS)*, vol. 1, 1996, pp. 589–592.

[41] G. Feng, E. Meyer and Y. F. Liu, "High performance digital control algorithms for DC-DC converters based on the principle of capacitor charge balance," in *Proceedings of the 37th IEEE Power Electronics Specialists Conference (PESC)*, 2006, pp. 1–7.

[42] G. Feng, E. Meyer and Y. F. Liu, "A new digital control algorithm to achieve optimal dynamic performance in DC-to-DC Converters," *IEEE Trans. Power Electron.*, vol. 22, no. 4, pp. 1489–1498, 2007.

[43] E. Meyer and Y. F. Liu, "A quick capacitor charge balance control method to achieve optimal dynamic response for buck converters," in *Proceedings of the 38th IEEE Power Electronics Specialists Conference, (PESC)*, 2007, pp. 1549–1555.

[44] E. Meyer and Y. F. Liu, "A practical minimum time control method for buck converters based on capacitor charge balance," in *Proceedings of the 23rd IEEE Applied Power Electronics Conference and Exposition (APEC)*, 2008, pp. 10–16.

[45] E. Meyer, Z. Zhang and Y. F. Liu, "An optimal control method for buck converters using a practical capacitor charge balance technique," *IEEE Trans. Power Electron.*, vol. 23, no. 4, pp. 1802–1812, 2008.

[46] E. Meyer, D. Wang, L. Jia and Y. F. Liu, "Digital charge balance controller with an auxiliary circuit for superior unloading transient performance of buck converters," in *Proceedings of the 25th IEEE Applied Power Electronics Conference and Exposition (APEC)*, 2010, pp. 124–131.

[47] W. Fang, Y. J. Qiu, X. Liu and Y. F. Liu, "A new digital capacitor charge balance control algorithm for boost DC/DC Converter," in *Proceedings of the 2nd IEEE Energy Conversion Conference and Exposition (ECCE)*, 2010, pp. 2035–2040.

[48] X. Liu, L. Ge, W. Fang and Y. F. Liu, "An algorithm for buck-boost converter based on the principle of capacitor charge balance," in *Proceedings of the 6th IEEE Conference on Industrial Electronics and Applications (ICIEA)*, 2011, pp. 1365–1369.

[49] L. Jia, D. Wang, Y. F. Liu and P. C. Sen, "A novel analog implementation of capacitor charge balance controller with a practical extreme voltage detector," in *Proceedings of the 26th IEEE Applied Power Electronics Conference and Exposition (APEC)*, 2011, pp. 245–252.

[50] E. Meyer, Z. Zhang and Y. F. Liu, "Digital charge balance controller to improve the loading/unloading transient response of buck converters," *IEEE Trans. Power Electron.*, vol. 27, no. 3, pp. 1314–1326, 2012.

[51] L. Jia and Y. F. Liu, "Voltage-based charge balance controller suitable for both digital and analog implementations," *IEEE Trans. Power Electron.*, vol. 28, no. 2, pp. 930–944, 2013.

[52] A. Costabeber, L. Corradini, S. Saggini and P. Mattavelli, "Time-optimal, parameters-insensitive digital controller for DC-DC buck converters," in *Proceedings of the 39th IEEE Power Electronics Specialists Conference (PESC)*, 2008, pp. 1243–1249.

[53] L. Corradini, A. Costabeber, P. Mattavelli and S. Saggini, "Time optimal, parameters-insensitive digital controller for VRM applications with adaptive voltage positioning," in *Proceedings of the 11th IEEE Workshop on Control and Modeling for Power Electronics (COMPEL)*, 2008, pp. 1–8.

[54] V. Yousefzadeh, A. Babazadeh, R. Ramachandran, E. Alarcon, L. Pao and D. Maksimović, "Proximate time-optimal digital control for synchronous buck DC-DC converters," *IEEE Trans. Power Electron.*, vol. 23, no. 4, pp. 2018–2026, Jul. 2008.

[55] G. Pitel and P. Krein, "Minimum-time transient recovery for DC-DC converters using raster control surfaces," *IEEE Trans. Power Electron.*, vol. 24, no. 12, pp. 2692–2703, Dec. 2009.

[56] A. Babazadeh and D. Maksimović, "Hybrid digital adaptive control for fast transient response in synchronous buck DC-DC converters," *IEEE Trans. Power Electron.*, vol. 24, no. 4, pp. 2625–2638, Nov. 2009.

[57] L. Corradini, A. Babazadeh, A. Bjeletić and D. Maksimović, "Current-limited time-optimal response in digitally-controlled DC-DC converters," *IEEE Trans. Power Electron.*, vol. 25, no. 11, pp. 2869–2880, Nov. 2010.

[58] A. Costabeber, P. Mattavelli and S. Saggini, "Digital time-optimal phase shedding in multiphase buck converters," *IEEE Trans. Power Electron.*, vol. 25, no. 9, pp. 2242–2247, Sept. 2010.

[59] A. Radić, Z. Lukić, A. Prodić and R. H. de Nie, "Minimum-deviation digital controller IC for DC-DC switch-mode power supplies," *IEEE Trans. Power Electron.*, vol. 28, no. 9, pp. 4281–4298, Sept. 2013.

[60] J. M. Galvez, M. Ordonez, P. Luchino and J. E. Quaicoe, "Improvements in boundary control of boost converters using the natural switching surface," *IEEE Trans. Power Electron.*, vol. 26, no. 11, pp. 3367–3376, Nov. 2011.

[61] L. Corradini, P. Mattavelli, E. Tedeschi and D. Trevisan, "High-bandwidth multisampled digitally controlled DC-DC converters using ripple compensation," *IEEE Trans. Ind. Electron.*, vol. 55, no. 4, pp. 1501–1508, Apr. 2008.

[62] L. Corradini and P. Mattavelli, "Modeling of multisampled pulse width modulators for digitally controlled DC-DC converters," *IEEE Trans. Power Electron.*, vol. 23, no. 4, pp. 1839–1847, Jul. 2008.

[63] Z. Lukić, A. Radić, A. Prodić and S. Effler, "Oversampled digital controller IC based on successive load-change estimation for DC-DC converters," in *Proceedings of the 25th IEEE Applied Power Electronics Conference and Exposition (APEC)*, Feb. 2010, pp. 315–320.

[64] S. Saggini, D. Trevisan, P. Mattavelli and M. Ghioni, "Synchronous-asynchronous digital voltage-mode control for DC-DC converters," *IEEE Trans. Power Electron.*, vol. 22, no. 4, pp. 1261–1268, Jul. 2007.

[65] Z. Zhao and A. Prodić, "Continuous-time digital controller for high-frequency DC-DC converters," *IEEE Trans. Power Electron.*, vol. 23, no. 2, pp. 564–573, Mar. 2008.

[66] S. Saggini, M. Ghioni and A. Geraci, "An innovative digital control architecture for low-voltage high-current DC-DC converters with tight voltage regulation," *IEEE Trans. Power Electron.*, vol. 19, no. 1, pp. 210–218, Jan. 2004.

[67] S. Saggini, P. Mattavelli, M. Ghioni and M. Redaelli, "Mixed-signal voltage-mode control for DC-DC converters with inherent analog derivative action," *IEEE Trans. Power Electron.*, vol. 23, no. 3, pp. 1485–1493, May 2008.

[68] J. Li, F. C. Lee and Y. Qiu, "New digital control architecture eliminating the need for high resolution DPWM," in *Proceedings of the 37th IEEE Power Electronics Specialists Conference (PESC)*, 2007, pp. 814–819.

[69] J. Li and F.C. Lee, "Digital current mode control architecture with improved performance for DC-DC converters," in *Proceedings of the 23rd IEEE Applied Power Electronics Conference and Exposition (APEC)*, 2008, pp. 1087–1092.

[70] K. Y. Cheng, F. Yu, F. C. Lee and P. Mattavelli, "Digital enhanced V2-type constant on-time control using inductor current ramp estimation for a buck converter with low-ESR capacitors," *IEEE Trans. Power Electron.*, vol. 28, no. 3, pp. 1241–1252, Mar. 2013.

[71] H. Hu, V. Yousefzadeh and D. Maksimović, "Nonuniform A/D quantization for improved dynamic responses of digitally controlled DC-DC converters," *IEEE Trans. Power Electron.*, vol. 23, no. 4, pp. 1998–2005, Jul. 2008.

[72] A. Soto, P. Alou and J. A. Cobos, "Nonlinear digital control breaks bandwidth limitations," in *Proceedings of the 21st IEEE Applied Power Electronics Conference and Exposition (APEC)*, Mar. 2006, pp. 724–730.

[73] A. Prodić, D. Maksimović and R. W. Erickson, "Dead-zone digital controllers for improved dynamic response of low harmonic rectifiers," *IEEE Trans. Power Electron.*, vol. 21, no. 1, pp. 173–181, Jan. 2006.

[74] X. Zhang, Y. Zhang, R. Zane and D. Maksimović, "Design and implementation of a wide-bandwidth digitally controlled 16-phase converter," in *Proceedings of the 10th IEEE Workshop on Computers in Power Electronics (COMPEL)*, 2006, pp. 106–111.

[75] Y. Zhang, X. Zhang, R. Zane and D. Maksimović, "Wide-bandwidth digital multi-phase controller," in *Proceedings of the 36th IEEE Power Electronics Specialists Conference (PESC)*, Jun. 2006, pp. 1–7.

[76] A. Stupar, Z. Lukić and A. Prodić, "Digitally-controlled steered-inductor buck converter for improving heavy-to-light load transient response," in *Proceedings of IEEE Power Electronics Specialists Conference (PESC)*, Jun. 2008, pp. 3950–3954.

[77] S. S. Ahsanuzzaman, A. Parayandeh, A. Prodić and D. Maksimović, "Load-interactive steered-inductor DC-DC converter with minimized output filter capacitance," in *Proceedings of the 25th IEEE Applied Power Electronics Conference and Exposition (APEC)*, Feb. 2010, pp. 980–985.

[78] R. D. Middlebrook, "Measurement of loop gain in feedback systems," *Int. J. Electron.*, vol. 38, no. 4, pp. 485–512, 1975.

[79] B. Miao, R. Zane and D. Maksimović, "System identification of power converters with digital control through cross-correlation methods," *IEEE Trans. Power Electron.*, vol. 20, no. 5, pp. 1093–1099, Sept. 2005.

[80] M. Shirazi, J. Morroni, A. Dolgov, R. Zane and D. Maksimović, "Integration of frequency response measurement capabilities in digital controllers for DC-DC converters," *IEEE Trans. Power Electron.*, vol. 23, no. 5, pp. 2524–2535, Sept. 2008.

[81] M. Shirazi, R. Zane, D. Maksimović, L. Corradini and P. Mattavelli, "Autotuning techniques for digitally-controlled point-of-load converters with wide range of capacitive loads," in *Proceedings of the 22nd IEEE Applied Power Electronics Conference and Exposition (APEC)*, 2007, pp. 14–20.

[82] J. G. Ziegler and N. B. Nichols, "Optimum settings for automatic controllers," *Trans. ASME*, vol. 64, pp. 759–768, 1942.

[83] K. Åström and T. Hägglund, "Automatic tuning of simple regulators with specifications on phase and amplitude margins," *Automatica*, vol. 20, no. 5, pp. 645–651, 1984.

[84] A. Leva, "PID autotuning algorithm based on relay feedback," *IEE Proc. Control Theory Appl.*, vol. 140, no. 5, pp. 328–338, Sept. 1993.

[85] K. Åström and B. Wittenmark, *Adaptive Control*, 2nd ed. Addison-Wesley, 1995.

[86] W. Stefanutti, P. Mattavelli, S. Saggini and M. Ghioni, "Autotuning of digitally controlled buck converters based on relay feedback," in *Proceedings*

of the 36th IEEE Power Electronics Specialists Conference (PESC), 2005, pp. 2140–2145.

[87] Z. Zhao, A. Prodić and P. Mattavelli, "Self-programmable PID compensator for digitally controlled SMPS," in *Proceedings of the 10th IEEE Workshop on Computers in Power Electronics (COMPEL)*, Jul. 2006, pp. 112–116.

[88] Z. Zhao, A. Prodić and P. Mattavelli, "Limit-cycle oscillations based auto-tuning system for digitally controlled DC-DC power supplies," *IEEE Trans. Power Electron.*, vol. 22, no. 6, pp. 2211–2222, Nov. 2007.

[89] W. Stefanutti, P. Mattavelli, S. Sagginia and M. Ghioni, "Autotuning of digitally controlled buck converters based on relay feedback," *IEEE Trans. Power Electron.*, vol. 22, no. 1, pp. 199–207, Jan. 2007.

[90] L. Corradini, P. Mattavelli and D. Maksimović, "Robust relay-feedback based autotuning for DC-DC converters," in *Proceedings of the 38th IEEE Power Electronics Specialists Conference (PESC)*, 2007, pp. 2196–2202.

[91] W. Stefanutti, S. Saggini, E. Tedeschi, P. Mattavelli and P. Tenti, "Simplified model reference tuning of PID regulators of digitally controlled DC-DC converters based on crossover frequency analysis," in *Proceedings of the 38th IEEE Power Electronics Specialists Conference (PESC)*, 2007, pp. 785–791.

[92] W. Stefanutti, S. Saggini, L. Corradini, E. Tedeschi, P. Mattavelli and D. Trevisan, "Closed-loop model-reference tuning of PID regulators for digitally controlled DC-DC converters based on duty-cycle perturbation," in *Proceedings of the 33th IEEE Conference of the Industrial Electronics Society (IECON)*, Nov. 2007, pp. 1553–1558.

[93] L. Corradini, P. Mattavelli, W. Stefanutti and S. Saggini, "Simplified model reference-based autotuning for digitally controlled SMPS," *IEEE Trans. Power Electron.*, vol. 23, no. 4, pp. 1956–1963, Jul. 2008.

[94] Z. Lukić, Z. Zhao, S. Ahsanuzzaman and A. Prodić, "Self-tuning digital current estimator for low-power switching converters," in *Proceedings of the 23rd IEEE Applied Power Electronics Conference and Exposition (APEC)*, Feb. 2008, pp. 529–534.

[95] J. Morroni, R. Zane and D. Maksimović, "Design and implementation of an adaptive tuning system based on desired phase margin for digitally controlled DC-DC converters," *IEEE Trans. Power Electron.*, vol. 24, no. 2, pp. 559–568, Feb. 2009.

[96] M. Shirazi, R. Zane and D. Maksimović, "An autotuning digital controller for DC-DC power converters based on on-line frequency response measurement," *IEEE Trans. Power Electron.*, vol. 24, no. 11, pp. 2578–2588, Nov. 2009.

[97] J. Morroni, L. Corradini, R. Zane and D. Maksimović, "Adaptive tuning of switched-mode power supplies operating in discontinuous and continuous conduction modes," *IEEE Trans. Power Electron.*, vol. 24, no. 11, pp. 2603–2611, Nov. 2009.

[98] Z. Lukić, S. Ahsanuzzaman, A. Prodić and Z. Zhao, "Self-tuning sensorless digital current-mode controller with accurate current sharing for multi-phase DC-DC converters," in *Proceedings of the 24th IEEE Applied Power Electronics Conference and Exposition (APEC)*, Feb. 2009, pp. 264–268.

[99] S. Moon, L. Corradini and D. Maksimović, "Accurate mode boundary detection in digitally controlled boost power factor correction rectifiers," in *Proceedings of the 2nd IEEE Energy Conversion Conference and Exposition (ECCE)*, 2010, pp. 1212–1217.

[100] S. Moon, L. Corradini and D. Maksimović,, "Auto-tuning of digitally controlled boost power factor correction rectifiers operating in continuous conduction mode," in *Proceedings of the 12th IEEE Workshop on Control and Modeling for Power Electronics (COMPEL)*, 2010, pp. 1–8.

[101] S. Moon, L. Corradini and D. Maksimović, "Autotuning of digitally controlled boost power factor correction rectifiers," *IEEE Trans. Power Electron.*, vol. 26, no. 10, pp. 3006–3018, Oct. 2011.

[102] V. Yousefzadeh and D. Maksimović, "Sensorless optimization of dead times in DC-DC converters with synchronous rectifiers," *IEEE Trans. Power Electron.*, vol. 21, no. 4, pp. 994–1002, Jul. 2006.

[103] S. H. Kang, D. Maksimović and I. Cohen, "Efficiency optimization in digitally controlled Flyback DC-DC converters over wide ranges of operating conditions," *IEEE Trans. Power Electron.*, vol. 27, no. 8, pp. 3734–3748, Aug. 2012.

[104] F. Z. Chen and D. Maksimović, "Digital control for improved efficiency and reduced harmonic distortion over wide load range in boost PFC rectifiers," in *Proceedings of the 24th IEEE Applied Power Electronics Conference and Exposition (APEC)*, 2009, pp. 760–766.

[105] F. Z. Chen and D. Maksimović,, "Digital control for efficiency improvements in interleaved boost PFC rectifiers," in *Proceedings of the 25th IEEE Applied Power Electronics Conference and Exposition (APEC)*, 2010, pp. 188–195.

[106] F. Z. Chen and D. Maksimović, "Digital control for improved efficiency and reduced harmonic distortion over wide load range in boost PFC rectifiers," *IEEE Trans. Power Electron.*, vol. 25, no. 10, pp. 2683–2692, Oct. 2010.

[107] W. Feng, F. C. Lee, P. Mattavelli and D. Huang, "A universal adaptive driving scheme for synchronous rectification in LLC resonant converters," *IEEE Trans. Power Electron.*, vol. 27, no. 8, pp. 3775–3781, Aug. 2012.

[108] O. Trescases, G. Wei, A. Prodić and W.T. Ng, "Predictive efficiency optimization for DC-DC converters with highly dynamic digital loads," *IEEE Trans. Power Electron.*, vol. 23, no. 4, pp. 1859–1869, Jul. 2008.

[109] A. Parayandeh, C. Pang and A. Prodić, "Digitally controlled low-power DC-DC converter with instantaneous on-line efficiency optimization," in *Proceedings of the 24th IEEE Applied Power Electronics Conference and Exposition (APEC)*, 2009, pp. 159–163.

[110] A. Parayandeh and A. Prodić, "Digitally controlled low-power DC-DC converter with segmented output stage and gate charge based instantaneous efficiency optimization," in *Proceedings of the 1st IEEE Energy Conversion Conference and Exposition (ECCE)*, 2009, pp. 3870–3875.

[111] S. Effler, M. Halton and K. Rinne, "Efficiency-based current distribution scheme for scalable digital power converters," *IEEE Trans. Power Electron.*, vol. 26, no. 4, pp. 1261–1269, Apr. 2011.

[112] Z. Lukić, Z. Zhenyu, A. Prodić and D. Goder, "Digital controller for multi-phase DC-DC converters with logarithmic current sharing," in *Proceedings of IEEE Power Electronics Specialists Conference (PESC)*, Jun. 2007, pp. 119–123.

[113] A. Parayandeh, B. Mahdavikkhah, S. S. Ahsanuzzaman, A. Radić and A. Prodić, "A 10 MHz mixed-signal CPM controlled DC-DC converter IC with novel gate swing circuit and instantaneous efficiency optimization," in *Proceedings of the 3rd IEEE Energy Conversion Conference and Exposition (ECCE)*, 2011, pp. 1229–1235.

[114] A. V. Peterchev and S. R. Sanders, "Digital multimode buck converter control with loss-minimizing synchronous rectifier adaptation," *IEEE Trans. Power Electron.*, vol. 21, no. 6, pp. 1588–1599, Nov. 2006.

[115] W. Al-Hoor, J. Abu-Qahouq, L. Huang, C. Ianello, W. Mikhael and I. Batarseh, "Multivariable adaptive efficiency optimization digital controller," in *Proceedings of IEEE Power Electronics Specialists Conference (PESC)*, Jun. 2008, pp. 4590–4596.

[116] S. H. Kang, H. Nguyen, D. Maksimović and I. Cohen, "Efficiency characterization and optimization in flyback DC-DC converters," in *Proceedings of the 2nd IEEE Energy Conversion Congress and Exposition (ECCE)*, 2010, pp. 527–534.

[117] S. H. Kang, D. Maksimović and I. Cohen, "On-line efficiency optimization in Flyback DC-DC converters over wide ranges of operating conditions," in *Proceedings of the 26th IEEE Applied Power Electronics Conference and Exposition (APEC)*, Feb. 2011, pp. 1417–1424.

[118] G. W. Wester, "Low-frequency characterization of switched DC-DC converters," Ph.D. dissertation, California Institute of Technology, May 1972.

[119] S. M. Ćuk, "Modelling, analysis, and design of switching converters," Ph.D. dissertation, California Institute of Technology, Nov. 1976.

[120] R. D. Middlebrook and S. Ćuk, "A general unified approach to modeling switching-converter power stages," *Int. J. Electron.*, vol. 42, pp. 521–550, Jun. 1977.

[121] S. Ćuk and R. D. Middlebrook, "A general unified approach to modeling switching DC-to-DC converters in discontinuous conduction mode," in *Proceedings of IEEE Power Electronics Specialists Conference (PESC)*, 1977, pp. 36–57.

[122] D. Maksimović and S. Ćuk, "A unified analysis of PWM converters in discontinuous modes," *IEEE Trans. Power Electron.*, vol. 6, no. 3, pp. 476–490, 1991.

[123] J. Sun, D. M. Mitchell, M. F. Greuel, P. T. Krein and R. M. Bass, "Averaged modeling of PWM converters operating in discontinuous conduction mode," *IEEE Trans. Power Electron.*, vol. 16, no. 1, pp. 482–492, Jul. 2001.

[124] Vatché Vorpérian, "Simplified analysis of PWM converters using model of PWM switch Part I: continuous conduction mode," *IEEE Trans. Aerosp. Appl.*, vol. 26, no. 3, pp. 490–496, May 1990.

[125] A. R. Brown and R. D. Middlebrook, "Sampled-data modeling of switching regulators," in *Proceedings of IEEE Power Electronics Specialists Conference (PESC)*, 1981, pp. 349–369.

[126] R. D. Middlebrook, "Predicting modulator phase lag in PWM converter feedback loops," in *Proceedings of the 8th Int. Solid-State Power Conversion Conference (POWERCON)*, Apr. 1981.

[127] K. Åström and T. Hägglund, *PID Controllers: Theory, Design, and Tuning*, 2nd ed. The Instrumentation, Systems, and Automation Society, 1995.

[128] D. J. Packard, "Discrete modeling and analysis of switching regulators," Ph.D. dissertation, California Institute of Technology, Nov. 1976.

[129] C.-C. Fang and E. Abed, "Sampled-data modeling and analysis of the power stage of PWM DC-DC converters," *Int. J. Electron.*, vol. 88, no. 3, pp. 347–369, Mar. 2001.

[130] C.-C. Fang, "Sampled-data poles and zeros of buck and boost converters," in *Proceedings of IEEE International Symposium on Circuits and Systems (ISCAS)*, vols. 2 and 3, May 2001, pp. 731–734.

[131] D. M. Van de Sype, K. De Gusseme, F. M. L. L. De Belie, A. P. Van den Bossche and J. A. Melkebeek, "Small-signal z-domain analysis of digitally controlled converters," *IEEE Trans. Power Electron.*, vol. 21, no. 2, pp. 470–478, Mar. 2006.

[132] V. Yousefzadeh, M. Shirazi and D. Maksimović, "Minimum phase response in digitally controlled boost and flyback converters," in *Proceedings of the 22nd IEEE Applied Power Electronics Conference and Exposition (APEC)*, Feb. 2007, pp. 865–870.

[133] F. Krismer and J. Kolar, "Accurate small-signal model for the digital control of an automotive bidirectional dual active bridge," *IEEE Trans. Power Electron.*, vol. 24, no. 12, pp. 2756–2768, Dec. 2009.

[134] J. Hefferon, *Linear Algebra*, 2012. [Online]. Available: http://joshua.smcvt .edu/linearalgebra/ (accessed 17 October 2014).

[135] S. Bibian and J. Hua, "High performance predictive dead-beat digital controller for DC power supplies," *IEEE Trans. Power Electron.*, vol. 17, no. 3, pp. 420–427, May 2002.

[136] J. Chen, A. Prodić, R. W. Erickson and D. Maksimović, "Predictive digital current programmed control," *IEEE Trans. Power Electron.*, vol. 18, no. 1, pp. 411–419, Jan. 2003.

[137] H. Peng and D. Maksimović, "Digital current-mode controller for DC-DC converters," in *Proceedings of the 20th IEEE Applied Power Electronics Conference and Exposition (APEC)*, vol. 2, Mar. 2005, pp. 899–905.

[138] B. Mather, B. Ramachandran and D. Maksimović, "A digital PFC controller without input voltage sensing," in *Proceedings of the 22nd IEEE Applied Power Electronics Conference and Exposition (APEC)*, 2007, pp. 198–204.

[139] B. Mather and D. Maksimović, "A simple digital power factor correction rectifier controller," *IEEE Trans. Power Electron.*, vol. 26, no. 1, pp. 9–19, Jan. 2011.

[140] V. M. Lopez, F. J. Azcondo, A. de Castro and R. Zane, "Universal digital controller for boost CCM power factor correction stages based on current rebuilding concept," *IEEE Trans. Power Electron.*, vol. 29, no. 7, pp. 3818–3829, Jul. 2014.

[141] D. Costinett, R. Zane and D. Maksimović, "Discrete time modeling of output disturbances in the dual active bridge converter," in *Proceedings of the 29th IEEE Applied Power Electronics Conference and Exposition (APEC)*, Mar. 2014.

[142] J. Öhr, "Anti-windup and control of systems with multiple input saturations – tools, solutions and case studies," Ph.D. dissertation, Uppsala University, 2003.

[143] S. Saggini, W. Stefanutti, D. Trevisan, P. Mattavelli and G. Garcea, "Prediction of limit-cycles oscillations in digitally controlled DC-DC converters using statistical approach," in *Proceedings of the 31st Annual Conference of IEEE Industrial Electronics Society (IECON)*, Nov. 2005, pp. 561–566.

[144] W. Stefanutti, P. Mattavelli, S. Saggini and G. Garcea, "Energy-based approach for predicting limit cycle oscillations in voltage-mode digitally-controlled DC-DC converters," in *Proceedings of the 21st IEEE Applied Power Electronics Conference and Exposition (APEC)*, Mar. 2006, pp. 1148–1154.

[145] B. Mather and D. Maksimović, "Quantization effects and limit cycling in digitally controlled single-phase PFC rectifiers," in *Proceedings of the 39th IEEE Power Electronics Specialists Conference (PESC)*, 2008, pp. 1297–1303.

[146] M. Bradley, E. Alarcon and O. Feely, "Analysis of limit cycles in a PI digitally controlled buck converter," in *2012 IEEE International Symposium on Circuits and Systems (ISCAS)*, May 2012, pp. 628–631.

[147] M. Bradley, E. Alarcon and O. Feely, "Design-oriented analysis of quantization-induced limit cycles in a multiple-sampled digitally controlled buck converter," *IEEE Trans. Circuits Syst. Regul. Pap.*, vol. 61, no. 4, pp. 1192–1205, Apr. 2014.

[148] A. Syed, E. Ahmed, D. Maksimović and E. Alarcon, "Digital pulse width modulator architectures," in *Proceedings of the 35th IEEE Power Electronics Specialists Conference (PESC)*, vol. 6, 2004, pp. 4689–4695.

[149] A. Dancy and A. Chandrakasan, "Ultra low power control circuits for PWM converters," in *Proceedings of the 28th IEEE Power Electronics Specialists Conference (PESC)*, vol. 1, Jun. 1997, pp. 21–27.

[150] A. Dancy, R. Amirtharajah and A. Chandrakasan, "High-efficiency multiple-output DC-DC conversion for low-voltage systems," *IEEE Trans. VLSI Syst.*, vol. 8, no. 3, pp. 252–263, Jun. 2000.

[151] E. O'Malley and K. Rinne, "A programmable digital pulse width modulator providing versatile pulse patterns and supporting switching frequencies beyond 15 MHz," in *Proceedings of the 19th IEEE Applied Power Electronics Conference and Exposition (APEC)*, vol. 1, 2004, pp. 53–59.

[152] R. F. Foley, R. C. Kavanagh, W. P. Marnane and M. G. Egan, "An area-efficient digital pulsewidth modulation architecture suitable for FPGA implementation," in *Proceedings of the 20th IEEE Applied Power Electronics Conference and Exposition (APEC)*, vol. 3, Mar. 2005, pp. 1412–1418.

[153] V. Yousefzadeh, T. Takayama and D. Maksimović, "Hybrid DPWM with digital delay-locked loop," in *Proceedings of the 10th IEEE Workshop on Computers in Power Electronics (COMPEL)*, Jul. 2006, pp. 142–148.

[154] S. C. Huerta, A. De Castro, O. Garcia and J. A. Cobos, "FPGA-based digital pulse-width modulator with time resolution under 2 ns," *IEEE Trans. Power Electron.*, vol. 23, no. 6, pp. 3135–3141, Nov. 2008.

[155] L. S. Ge, Z. X. Chen, Z. J. Chen and Y. F. Liu, "Design and implementation of a high resolution DPWM based on a low-cost FPGA," in *Proceedings of IEEE Energy Conversion Congress and Exposition (ECCE)*, Sept. 2010, pp. 2306–2311.

[156] D. Costinett, M. Rodriguez and D. Maksimović, "Simple digital pulse width modulator under 100 ps resolution using general-purpose FPGAs," *IEEE Trans. Power Electron.*, vol. 28, no. 10, pp. 4466–4472, Oct. 2013.

[157] T. Carosa, R. Zane and D. Maksimović, "Implementation of a 16 phase digital modulator in a 0.35μm process," in *Proceedings of the 10th IEEE Workshop on Computers in Power Electronics (COMPEL)*, 2006, pp. 159–165.

[158] T. Carosa, R. Zane and D. Maksimović, "Scalable digital multiphase modulator," *IEEE Trans. Power Electron.*, vol. 23, no. 4, pp. 2201–2205, Jul. 2008.

[159] M. Scharrer, M. Halton, T. Scanlan and K. Rinne, "FPGA-based multi-phase digital pulse width modulator with dual-edge modulation," in *Proceedings of the 25th IEEE Applied Power Electronics Conference and Exposition (APEC)*, Feb. 2010, pp. 1075–1080.

[160] A. De Castro and E. Todorovich, "High resolution FPGA DPWM based on variable clock phase shifting," *IEEE Trans. Power Electron.*, vol. 25, no. 5, pp. 1115–1119, May 2010.

[161] L. Corradini, A. Bjeletić, R. Zane and D. Maksimović, "Fully digital hysteretic modulator for DC-DC switching converters," *IEEE Trans. Power Electron.*, vol. 26, no. 10, pp. 2969–2979, Oct. 2011.

[162] D. Navarro, O. Lucia, L.A. Barragan, J.I. Artigas, I. Urriza and O. Jimenez, "Synchronous FPGA-based high-resolution implementations of digital pulse-width modulators," *IEEE Trans. Power Electron.*, vol. 27, no. 5, pp. 2515–2525, May 2012.

[163] Y. Qiu, J. Li, M. Xu, D. S. Ha and F. C. Lee, "Proposed DPWM scheme with improved resolution for switching power converters," in *Proceedings of the 22nd IEEE Applied Power Electronics Conference and Exposition (APEC)*, 2007, pp. 1588–1593.

[164] J. Li, Y. Qiu, Y. Sun, B. Huang, M. Xu, D. S. Ha and F. C. Lee, "High resolution digital duty cycle modulation schemes for voltage regulators," in *Proceedings of the 22nd IEEE Applied Power Electronics Conference and Exposition (APEC)*, 2007, pp. 871–876.

[165] R. Schreirer and G. C. Temes, *Understanding Delta-Sigma Data Converters.* Wiley-IEEE Press, 2004.

[166] M. Norris, L. Platon, E. Alarcon and D. Maksimović, "Quantization noise shaping in digital PWM converters," in *Proceedings of the 38th IEEE Power Electronics Specialists Conference (PESC)*, Jun. 2008, pp. 127–133.

[167] R. J. Van De Plaasche, *CMOS Integrated Analog-To-Digital and Digital-To-Analog Converters*, 2nd ed. Kluwer Academic Publishers, 2003.

[168] C. Kranz, "Complete digital control method for PWM DCDC boost converter," in *Proceedings of the 34th IEEE Power Electronics Specialists Conference (PESC)*, vol. 2, Jun. 2003, pp. 951–956.

[169] B. Mather and D. Maksimović, "Single comparator based A/D converter for output voltage sensing in power factor correction rectifiers," in *Proceedings of the 1st IEEE Energy Conversion Conference and Exposition (ECCE)*, 2009, pp. 1331–1338.

[170] M. Rodriguez, V. M. Lopez, F. J. Azcondo, J. Sebastian and D. Maksimović, "Average inductor current sensor for digitally controlled switched-mode power supplies," *IEEE Trans. Power Electron.*, vol. 27, no. 8, pp. 3795–3806, Aug. 2012.

[171] "IEEE Standard Multivalue Logic System for VHDL Model Interoperability (Std_logic_1164)," *IEEE Std 1164-1993*, Mar. 1993.

[172] "IEEE Standard VHDL Synthesis Packages," *IEEE Std 1076.3-1997*, Aug. 2002.

[173] "IEEE Standard VHDL Language Reference Manual," *IEEE Std 1076-2002*, May 2002.

[174] "IEEE Standard for VHDL Register Transfer Level (RTL) Synthesis," *IEEE Std 1076.6-2004*, Oct. 2004.

[175] "IEEE Standard for Verilog® Hardware Description Language," *IEEE Std 1364-2005*, Apr. 2006.

[176] "IEEE Standard for Floating-Point Arithmetic," *IEEE Std 754-2008*, Aug. 2008.

[177] P. J. Ashenden, *The Designer's Guide to VHDL*, 2nd ed. Morgan Kaufmann Publishers, 2002.

[178] "Verilog® Register Transfer Level Synthesis," *IEC 62142-2005 First edition 2005-06 IEEE Std 1364.1*, 2005.

[179] S. Palnitkar, *Verilog HDL – A Guide to Digital Design and Synthesis*, 2nd ed. Prentice-Hall, 2001.

[180] B. Miao, R. Zane and D. Maksimović, "A modified cross-correlation method for system identification of power converters with digital control," in *Proceedings of the 35th IEEE Power Electronics Specialists Conference (PESC)*, vol. 5, 2004, pp. 3728–3733.

[181] B. Miao, R. Zane and D. Maksimović, "Practical on-line identification of power converter dynamic responses," in *Proceedings of the 20th IEEE Applied Power Electronics Conference and Exposition (APEC)*, vol. 1, 2005, pp. 57–62.

[182] B. Miao, R. Zane and D. Maksimović, "Automated digital controller design for switching converters," in *Proceedings of the 36th IEEE Power Electronics Specialists Conference (PESC)*, 2005, pp. 2729–2735.

[183] J. Morroni, R. Zane and D. Maksimović, "An online stability margin monitor for digitally controlled switched-mode power supplies," *IEEE Trans. Power Electron.*, vol. 24, no. 11, pp. 2639–2648, Nov. 2009.

[184] M. M. Peretz and S. Ben-Yaakov, "Time-domain identification of pulse-width modulated converters," *IET Power Electron.*, vol. 5, no. 2, pp. 166–172, Feb. 2012.

[185] H. K. Khalil, *Nonlinear Systems*, 3rd ed. Prentice-Hall, 2002.

[186] Z. Song and D. V. Sarwate, "The frequency spectrum of pulse width modulated signals," *Signal Processing*, vol. 83, no. 10, pp. 2227–2258, 2003.

[187] R. D. Middlebrook, "Describing function properties of a magnetic pulse-width modulator," in *Proceedings of IEEE Power Processing and Electronics Specialists Conference*, 1972.

INDEX

Digital Control of High-Frequency Switched-Mode Power Converters, First Edition.
Luca Corradini, Dragan Maksimović, Paolo Mattavelli, and Regan Zane.
© 2015 The Institute of Electrical and Electronics Engineers, Inc. Published 2015 by John Wiley & Sons, Inc.

IEEE Press Series on Power Engineering

Series Editor: M. E. El-Hawary, Dalhousie University, Halifax, Nova Scotia, Canada

The mission of IEEE Press Series on Power Engineering is to publish leading-edge books that cover the broad spectrum of current and forward-looking technologies in this fast-moving area. The series attracts highly acclaimed authors from industry/academia to provide accessible coverage of current and emerging topics in power engineering and allied fields. Our target audience includes the power engineering professional who is interested in enhancing their knowledge and perspective in their areas of interest.

45. *Principles of Electrical Safety*
Peter E. Sutherland

46. *Advanced Power Electronics Converters: PWM Converters Processing AC Voltages*
Euzeli Cipriano dos Santos Jr. and Edison Roberto Cabral da Silva

47. *Optimization of Power System Operation, Second Edition*
Jizhong Zhu

48. *Digital Control of High-Frequency Switched-Mode Power Converters*
Luca Corradini, Dragan Maksimovic, Paolo Mattavelli, and Regan Zane

49. *Power System Harmonics and Passive Filter Designs*
J. C. Das

Printed and bound by CPI Group (UK) Ltd, Croydon, CR0 4YY

16/04/2025

14658598-0003